Introduction to Nanoelectronics

Increasing miniaturization of devices, components, and integrated systems requires developments in the capacity to measure, organize, and manipulate matter at the nanoscale. This textbook is a comprehensive, interdisciplinary account of the technology and science that underpin nanoelectronics, covering the underlying physics, nanostructures, nanomaterials, and nanodevices.

Without assuming prior knowledge of quantum physics, this book provides a unifying framework for the basic ideas needed to understand the recent developments in the field. Following an introductory description of recent trends in semiconductor and device nanotechnologies, as well as novel device concepts, materials for nanoelectronics are treated, covering methods of growth, fabrication and characterization. Treatment then moves to an analysis of nanostructures, including recently discovered nanoobjects, and concludes with a discussion of devices that use a "simple" scaling-down approach to copy well-known microelectronic devices, and nanodevices based on new principles that cannot be realized at the macroscale.

Numerous illustrations, homework problems and interactive Java applets help the student to appreciate the basic principles of nanotechnology, and to apply them to real problems. Written in a clear yet rigorous and interdisciplinary manner, this textbook is suitable for advanced undergraduate and graduate students in electrical and electronic engineering, nanoscience, materials, bioengineering, and chemical engineering.

Further resources for this title, including instructor solutions and Java applets, are available online at www.cambridge.org/9780521881722.

Vladimir V. Mitin is a Professor and Chair of the Department of Electrical Engineering at the University of Buffalo, State University of New York. He has co-authored eight books, and over 400 professional publications, including ten patents.

Viatcheslav A. Kochelap is Professor and Head of the Theoretical Physics Department at the Institute of Semiconductor Physics National Academy of Sciences, Ukraine. He has published over 200 journal articles.

Michael A. Stroscio is a Professor in the departments of Electrical and Computer Engineering, and Bioengineering, at the University of Illinois at Chicago. He is a Fellow of the IEEE, the American Physical Society, and the AAAS.

Introduction to Nanoelectronics

Science, Nanotechnology, Engineering, and Applications

VLADIMIR V. MITIN
University of Buffalo, State University of New York

VIATCHESLAV A. KOCHELAP
Institute of Semiconductor Physics National Academy of Sciences, Ukraine

MICHAEL A. STROSCIO
University of Illinois at Chicago

CAMBRIDGE
UNIVERSITY PRESS

CAMBRIDGE UNIVERSITY PRESS
Cambridge, New York, Melbourne, Madrid, Cape Town, Singapore, São Paulo

Cambridge University Press
The Edinburgh Building, Cambridge CB2 8RU, UK

Published in the United States of America by Cambridge University Press, New York

www.cambridge.org
Information on this title: www.cambridge.org/9780521881722

First published 2008

Printed in the United Kingdom at the University Press, Cambridge

A catalog record for this publication is available from the British Library

ISBN 978-0-521-88172-2 hardback

Contents

Preface

Welcome to the amazing *nanoworld*! In this book you will find fundamental principles in nanoscience and basic techniques of measurement, as well as fabrication and manipulation of matter at the nanoscale. The book discusses how these principles, techniques, and technologies are applied to the newest generation of electronics, known as *nanoelectronics*.

The science of atoms and simple molecules, and the science of matter from microstructures to larger scales, are both well established. A remaining, extremely important, size-related challenge is at the nanoscale – roughly the dimensional scales between 10 and 100 molecular diameters – where the fundamental properties of materials are determined and can be engineered. This field of science – nanoscience – is a broad and interdisciplinary field of emerging research and development.

Nanotechnology is concerned with materials, structures, and systems whose components exhibit novel and significantly modified physical, chemical, and biological properties due to their nanoscale sizes. A principal goal of nanotechnology is to control and exploit these properties in structures and devices at atomic, molecular, and supramolecular levels. To realize this goal, it is essential to learn how to fabricate and use these devices efficiently. Nanotechnology has enjoyed explosive growth in the past few years. In particular, nanofabrication techniques have advanced tremendously in recent years. Obviously, revolutionary changes in the ability to measure, organize, and manipulate matter on the nanoscale are highly beneficial for electronics with its persistent trend of downscaling devices, components, and integrated systems. In turn, the miniaturization required by electronics is one of the major driving forces for nanoscience and nanotechnology.

Practical implementations of nanoscience and nanotechnology have great importance, and they depend critically on training people in these fields. Thus, modern education needs to address the rapidly evolving facets of nanoscience and nanotechnology. A new generation of researchers, technologists, and engineers has to be trained in the emerging nanodisciplines. With the purpose of contributing to education in the nanofields, we present this textbook providing a unifying framework for the basic ideas needed to understand recent developments underlying nanoscience and nanotechnology, as applied to nanoelectronics. The book grew out of the authors' research and teaching experience in these subjects. We have found that many of the ideas and achievements in fields underlying nanoscience and nanotechnology can be explained in a relatively simple setting, if the necessary foundational underpinnings are presented properly. We have designed this textbook mainly for *undergraduate students*, who will be trained in diverse fields including nanoscience, physics of material devices, electrical and optical engineering, materials science and engineering, and mechanical engineering. It can be helpful also for training students in bioengineering and chemical engineering. To reach such a broad audience, materials are presented in such a way that an instructor can choose the level of presentation depending on the backgrounds of the students. For example, we have included Chapters 2 and 3 in part for students who have not taken a quantum mechanics course. An analogy with wave fields – elastic waves and optical waves – is exploited widely to introduce wave mechanics of particles and quantum principles, which play key roles in the interpretation of the properties of nanomaterials.

One of us (V.V.M.) has taught the course for students in the second semester of their sophomore year. For students at this level, Chapters 2 and 3 were covered in detail and, consequently, there was not enough time to cover all of the devices that are discussed in Chapter 8. If students using the book have previously taken courses on quantum mechanics and electromagnetics, the instructor may start from Chapter 4. This book may be also used as an introductory graduate or senior undergraduate course. Another of us (M.A.S.) has used Chapters 2 and 3 as the introduction to a graduate course on nanoelectronics for a class with students drawn from electrical engineering, materials engineering, chemical engineering, mechanical engineering, and physics. By covering Chapters 2 and 3 at the beginning of the course, the students can then proceed from this common basis in quantum mechanics and other underlying areas of physics to cover more advanced topics, either in the current text or in other texts such as *Quantum Heterostructures* by V. Mitin, V. Kochelap, and M. Stroscio. The latter approach has been used by M.A.S. in teaching nanoelectronics to graduate students with diverse backgrounds in many disciplines within engineering and the physical sciences. For this purpose, we include details of derivations and mathematical justification of concepts in some sections. These details can be omitted from an undergraduate curriculum.

The book contains homework problems on various subjects. These problems illustrate the basic material and help students to understand and learn the basic principles of the nanoscience and the nanotechnology.

* * * * *

Essentially, the chapters are organized into three main groups.

Chapters 1–3 are of an introductory character. In Chapter 1, we present in concise form the main subject of the book. The recent and diverse trends in semiconductor and device nanotechnologies, as well as novel concepts of nanodevices, are reviewed. These trends make it clear why understanding the fundamentals of nanoscience is of great importance.

Chapters 2 and 3 are written for students who have not taken a quantum mechanics course. An analogy with wave fields (elastic waves and optical waves) is exploited widely to introduce wave mechanics of particles and the quantum principles, which play key roles in the interpretation of the properties of nanomaterials.

In Chapter 2, we explain that the fundamental laws of physics governing particles and material fields in the nanoworld are different from those that apply to familiar macroscopic phenomena. Starting with an analysis of an arbitrary wave field (elastic vibrations in solids, electromagnetic fields, etc.), we find particle-like behavior of this wave field for small wave amplitudes and (or) for spatial scales larger than the wavelength. By analyzing particle motion, we show that at small spatial scales a particle can not be characterized by exact coordinates and momentum and that it behaves rather as an extended wave-like object. This analysis establishes the essence of the wave–particle duality which is an underlying principle of nanophysics.

In Chapter 3, we discuss the basic physical concepts and equations related to the behavior of particles in the nanoworld. We introduce the Schrödinger wave equation for particles and determine the ways in which to calculate observable physical quantities. Keeping in mind the diverse variants of nanostructures, by using wave mechanics we analyze a number of particular examples, which highlight important quantum properties of particles. Many of the examples analyzed can serve as the simplest models of nanostructures and are exploited in later chapters.

Chapters 4 and 5 are devoted to materials used in nanoelectronics, methods of their growth, and fabrication and characterization techniques.

In Chapter 4, we present an overview of the basic materials that are exploited in nanoelectronics. We start with semiconductor materials as the principal candidates for use in nanoelectronics, because they offer great flexibility in the control of the electronic and optical properties, and functions, of nanoelectronic devices. We show how, through proper regimes of growth, doping by impurities, and sequent modifications and processing, one can fabricate nanostructures and nanodevices starting from "bulk-like" materials. Then, we introduce other materials that have properties of great potential in nanoelectronics. Organic semiconductors and carbon nanotubes are discussed.

In Chapter 5, the principal methods of materials growth and nanodevice fabrication are presented. We start with an analysis of fabrication of nanodevices on the basis of perfect materials and continue by considering processing techniques. All stages of fabrication and methods of processing are considered in detail. Then, we discuss special regimes of material growth, when nanostructures (mainly quantum dots) are formed spontaneously due to the growth kinetics. These approaches to the production of nanostructures and nanoelectronic devices actually represent "evolutionary" improvements in the growth and processing methods applied previously in microelectronics. Nanoscale objects like carbon nanotubes and biomolecules require, in general, other techniques for production. These innovative techniques are also highlighted in this chapter.

We pay special attention to the most important characterization techniques, such as atomic-force microscopy, scanning tunneling microscopy, and transmission electron microscopy, among others.

Also in Chapter 5, we review advances in nanotechnology that came from synthetic chemistry and biology. These include chemical and biological methods of surface nanopatterning for preparing nanostructured materials with predefined and synthetically programmable properties. The basic ideas related to these chemical and biological approaches are discussed. Finally, we study the methods of fabrication of a new class of devices commonly known as nanoelectromechanical systems (NEMS).

Chapters 6–8 include analyses of electron properties of nanostructures, traditional low-dimensional systems, and recently discovered nano-objects.

In Chapter 6, transport of charge carriers is analyzed. Important aspects of transport regimes are elucidated by comparing the time and length scales of the carriers with device dimensions and device temporal phenomena related to operating frequencies. Then, we consider the behavior of the electrons in high electric fields, including hot-electron effects. For short devices, we describe dissipative transport and the velocity overshoot effect as well as semiclassical ballistic motion of the electrons. We present ideas on quantum transport in nanoscale devices.

To distinguish the nanostructures already having applications from the newly emerging systems, we refer to the former as traditional low-dimensional structures (quantum wells, quantum wires, and quantum dots). These structures are considered in Chapter 7.

In Chapter 8, we consider newly emerging electronic, optical, and electromechanical devices based on nanostructures. First, we discuss the devices which resemble well-known microelectronic devices using a "simple" scaling-down approach. Examples include such heterostructure devices as the field-effect and bipolar transistors as well as injection bipolar lasers. Then, we study nanodevices based on new physical principles, which can not be realized in microscale devices. Among these are resonant-tunneling devices, hot-electron transistors, single-electron-transfer devices, monopolar injection cascade lasers, nanoelectromechanical devices, and quantum-dot cellular automata. We understand that there are more devices to be reviewed. For example, not enough attention has been paid to progress in silicon device technology. The ideas and results presented provide an understanding of near-future developments in nanoelectronics and optoelectronics that are occurring as a result of advances in nanotechnology. This will encourage students to learn more about nanoelectronics.

The authors have many professional colleagues and friends from numerous countries who must be acknowledged. Without their contributions and sacrifices this work would not have been completed. Special thanks go to the Division of Engineering Education and Centers (EEC) in the Directorate for Engineering and to the Division of Materials Research (DMR) in the Directorate for Mathematical and Physical Sciences (MPS), and especially to the program manager Mary Poats at the National Science Foundation for the partial support of this work through the Nanoscale Science and Engineering (NSE) Nanotechnology in Undergraduate Education (NUE) Program. The help of Dr. Nizami Vagidov in preparing figures and editing the text is especially appreciated.

V.V.M. acknowledges the support and active encouragement of the faculty of the Department of Electrical Engineering and the School of Engineering and Applied Sciences as well as the members of the Center on Hybrid Nanodevices and Systems, especially Dr. Andrei Sergeev, at the University at Buffalo State University of New York. He also extends his thanks to the undergraduate students who took the course EE 240 "Nanotechnology, Engineering, and Science" as well as to Matthew Bell, who was the teaching assistant for the course. Their feedback helped him in the choice and presentation of the material that is presented in this book. Undergraduate students, Garun Vagidov and Jonathan Bell, helped with some editorial work. V.V.M. is also grateful to his family and friends for their strong support and encouragement, as well as for their understanding and forgiveness of the fact that he did not devote enough time to them while working on the book, and especially to his mother, grandson Anthony, and granddaughter Christina whom he missed the most.

V.A.K. thanks his colleagues Professor S. V. Svechnikov, Professor A. E. Belyaev, Professor F. T. Vas'ko, Dr. V. I. Pipa and Dr. B. A. Glavin from the Institute of Semiconductor Physics at Kiev for numerous fruitful discussions of general problems and perspectives of nanoscience and nanoelectronics. The wide spectrum of research directions and creative atmosphere prevailing in the Theoretical Physics Department stimulated his involvement in various nanoscience activities. He acknowledges permanent contacts with graduate and PhD students, which gave him valuable feedback during the work on the textbook. V.A.K. is deeply grateful to all his family members for their understanding and permanent support.

M.A.S. extends his sincere thanks to Deans Prith Banerjee and Larry Kennedy, at the College of Engineering, University of Illinois at Chicago (UIC) for their active encouragement and their longstanding efforts to promote excellence in research at the UIC. M.A.S. gratefully acknowledges the generous support and enlightened encouragement of Richard and Loan Hill. Special thanks go to Dr. Dwight Woolard of the US Army Research Office, Drs. Daniel Johnstone, Todd Steiner, and Kitt Reinhardt of the AFOSR, Dr. Rajinder Khosla, Dr. James W. Mink and Usha Varshney of the National Science Foundation, Dr. Daniel Herr of the Semiconductor Research Corporation, and Dr. John Carrano of the Defense Advanced Research Projects Agency for their encouragement and interest. M.A.S. acknowledges the essential roles that several professional colleagues and friends played in the events leading to his contributions to this book; these people include Professor Richard L. Magin, Head of the Bioengineering Department at the University of Illinois at Chicago (UIC), Professors Robert Trew, Gerald J. Iafrate, M. A. Littlejohn, K. W. Kim, and R. and M. Kolbas, and Dr. Sergiy Komirenko of the North Carolina State University, Professors G. Belenky and S. Luryi and Dr. M. Kisin of the State University of New York at Stony Brook, Professors George I. Haddad, Pallab K. Bhattacharya, and Jasprit Singh, and Dr. J.-P. Sun of the University of Michigan, Professors Karl Hess and J.-P. Leburton at the University of Illinois at Urbana-Champaign, Professor L. F. Register of the University of Texas at Austin, and Professors H. Craig Casey and Steven Teitsworth of Duke University. M.A.S. also thanks family members for their support while this book was being prepared; these include Anthony and Norma Stroscio, Mitra Dutta, and Elizabeth, Gautam, and Marshall Stroscio.

Notation

Symbols

$\mathbf{A}$ – amplitude of wave

$\langle A \rangle$ – average value of A

a – lattice constant

a_0 – length of carbon–carbon bond in carbon nanotubes

$\mathbf{a}_1, \mathbf{a}_2, \mathbf{a}_3$ – basis vectors

$\mathbf{a}_i$ – basis vectors of lattice

$\mathbf{B}$ – magnetic field

C – capacitance

$\mathbf{d}$ – translation vector

D – diffusion coefficient

d_{sp} – spacer thickness

E – energy of a particle

E_F – Fermi level

E_g – bandgap

e – elementary charge

$\mathbf{F}$ – electric field

F_0 – amplitude of electric field

f – frequency

f_{SET} – frequency of Bloch oscillations

$\mathbf{f}$ – vector of force

$\mathcal{F}$ – distribution function

$\mathcal{F}_F$ – Fermi distribution function

G – conductance

G_0 – quantum of conductance

$\hat{\mathcal{H}}$ – Hamiltonian operator

$\mathbf{H}$ – magnetic field

$\mathbf{H}$ – direction of the nearest-neighbor hexagon rows

$\mathcal{H}$ – total energy, Hamiltonian function

h – Planck's constant

h – wave energy density

h_{1D} – wave energy density for a one-dimensional medium

$\hbar$ – Planck's constant divided by 2π

J – current density

I – current

I_T – tunnel current

$\mathcal{I}$ – wave intensity

$\mathbf{i}$ – quantum-mechanical flux of the particles

$\mathbf{k}$ – wavevector

k_B – Boltzmann's constant

L – inductance

L_T – thermal diffusion length

l_e – mean free path between two elastic collisions

l – orbital quantum number

$\mathbf{l}$ – angular momentum

l_ϕ – coherence length

L_E – inelastic scattering length

L_x, L_y, L_z – dimensions of a sample

M – mass of resonator

$\mathbf{M}$ – magnetic dipole moment

m – magnetic quantum number

m^* – effective mass of electron

m_0 – mass of an electron in vacuum

m_{HH} – heavy-hole mass

m_{LH} – light-hole mass

m_{SH} – split-off hole mass

N_s – sheet concentration of donors

N_{depl} – sheet concentration of ionized acceptors

n – principal quantum number

n_s – sheet concentration of electrons

$P(\xi)$ – Hermite polynomial

P_b – Probability of finding electron under the barrier

$\mathbf{q}$ – wavevector

Q – quality factor

Q – amount of deposited material

R – radial function

r – magnitude of radius vector

$\mathbf{r}$ – coordinate vector

r_0 – Bohr's radius

R – radius of quantum dot

R – tube radius of carbon nanotube

R – reflection coefficient

$\mathbf{S}$ – spin – intrinsic angular momentum

S – cross-section

S_z – projection of the spin of electron

s – distance between tip and surface

s – phase velocity of traveling wave

s – spin of electron

t – time

t_{tr} – transit time

T – time period

$\mathbf{T}$ – vector corresponding to tube axis of carbon nanotube

T – ambient temperature

T_e – electron temperature

T_d – translation operator

$u_{\mathbf{k}}(\mathbf{r})$ – Bloch periodic function

u – displacement of atoms from their equilibrium positions

V – potential energy

V – volume

V_b – barrier height

V_0 – volume of primitive cell

v_d – average(drift) velocity

v – velocity

v_h – velocity of hole

W – crystalline potential

U_M – potential energy

z_0 – characteristic length

Z – atomic number

α – dimensionality factor

β – spring constant

$\delta(x)$ – Dirac's delta-function

ϵ – relative mismatch of lattice constants of the substrate and epilayer

ϵ – dielectric constant of the medium

ϵ_0 – permittivity of free space

ε – energy

Φ – potential

Φ_b – built-in Schottky voltage, Schottky barrier

Φ_0 – applied voltage

ϕ – phase

ϕ – polar angle

γ – gyromagnetic ratio

Λ_{1D} – elastic modulus of string

λ – wavelength

μ – electron mobility

μ_{ph} – partial electron mobility, determined by phonon scattering

μ_{im} – partial electron mobility, determined by impurity scattering

ν – set of quantum numbers

ξ – vector of polarization

ξ – dimensionless coordinate

Ω – angular frequency of a particle

Ω – ohm

ω – frequency

ω_q – frequency of harmonic oscillator

$\Psi(\mathbf{r}, t)$ – non-stationary wavefunction

$\Psi^*(\mathbf{r}, t)$ – complex conjugate of wavefunction $\Psi(\mathbf{r}, t)$

$\psi(\mathbf{r})$ – stationary wavefunction

ρ – three-dimensional density

ρ_{1D} – linear density of string

ϱ – density of states

σ – conductivity

Θ – theta-function

θ – polar angle

τ_E – mean free time between two inelastic collisions

τ_d – decay time of flexural vibrations

τ_e – mean free time between two elastic collisions

$\chi(z)$ – wavefunction

χ – electron affinity

Abbreviations

BT – bipolar transistor

CMOS – complementary MOS, i.e., NMOS and PMOS on the same chip

DPN – dip-pen nanolithography

FET – field-effect transistor

JBT – homojunction BT

JFET – junction FET

HBT – heterojuction BT

HEMT – high-electron-mobility transistor

HFET – heterojunction FET

HOMO – highest occupied molecular orbit

LUMO – lowest unoccupied molecular orbit

MES – metal–semiconductor

MESFET – metal–semiconductor FET

MODFET – modulation-doped FET

MOS – metal–oxide–semiconductor

MOSFET – metal–oxide–semiconductor FET

QUIT – quantum interference transistor

RTD – resonant-tunneling diode

SIMOX – separation by implantation of oxygen

SMS – semiconductor–metal–semiconductor

VMT – velocity-modulation transistor

1 Toward the nanoscale

This book provides the foundations and the main ideas emerging from research that underlies the applied field called *nanoelectronics*. Nanoelectronics promises to improve, amplify, and partially substitute for the well-known field of *microelectronics*. The prefix *micro* denotes one *millionth* and, as applied to electronics, it is used to indicate that the characteristic sizes of the smallest features of a conventional electronic device have length scales of approximately a micrometer. The prefix *nano* denotes one *billionth*. Thus, in nanoelectronics the dimensions of the devices should be as many as a thousand times smaller than those of microelectronics.

Such a revolutionary advance toward miniaturization of electronics is based on the recently developed ability to measure, manipulate, and organize matter on the nanoscale – 1 to 100 nanometers, i.e., 1 to 100 billionths of a meter. At the nanoscale, physics, chemistry, biology, materials science, and engineering converge toward the same principles and tools, and form new and broad branches of science and technology that can be called *nanoscience* and *nanotechnology*.

Advancing to the nanoscale is not just a step toward miniaturization, but requires the introduction and consideration of many additional phenomena. At the nanoscale, most phenomena and processes are dominated by quantum physics and they exhibit unique behavior. Fundamental scientific advances are expected to be achieved as knowledge in nanoscience increases. In turn, this will lead to dramatic changes in the ways materials, devices, and systems are understood and created. Innovative nanoscale properties and functions will be achieved through the control of matter at the level of its building blocks: atom-by-atom, molecule-by-molecule, and nanostructure-by-nanostructure. The molecular building blocks of life – proteins, nucleic acids, carbohydrates – are examples of materials that possess impressive properties determined by their size, geometrical folding, and patterns at the nanoscale. Nanotechnology includes the integration of manmade nanostructures into larger material components and systems. Importantly, within these larger-scale systems, the active elements of the system will remain at the nanoscale.

The driving forces underlying developments at the nanoscale have at least two major complementary components – scientific opportunities and technological motivations.

Scientific opportunities

The progress in physics, chemistry, and biology at the nanoscale represents a natural step in advancing knowledge and understanding Nature. Scientific perspectives on this route are conditioned first of all by new quantum phenomena in atomic- and molecular-scale structures and by the interaction of large numbers of these small objects. Indeed, the fundamental laws of physics in the nanoworld differ from those that apply to familiar macroscopic phenomena. Instead of classical physics, that works so well for macroscopic phenomena, the motion of particles and systems in the nanoworld is determined by the so-called wave mechanics or quantum mechanics. A basic principle of nanophysics is the fundamental concept that all matter, including electrons, nuclei, atoms, electromagnetic fields, etc., behaves as both waves and particles. This wave–particle duality of all matter is strikingly apparent at the nanoscale. For dealing with a large number of particles or systems, the statistical laws are important. Statistical physics on the nanoscale is also fundamentally different from that on the macroscale. In general, phenomena that involve very large numbers of small interacting particles or systems follow different rules from those involving only a few of them. Cooperative behavior of many-object systems is revealed clearly at the nanoscale. Besides the phenomena just discussed, there are other classes of phenomena that are important for science at the nanoscale.

It is appropriate here to refer to the famous 1959 lecture of the Nobel Prize laureate Professor Richard Feynman with the title "There is plenty of room at the bottom," where he discussed "the problem of manipulation and controlling things on a small scale." Feynman did not just indicate that there is "room at the bottom," in terms of decreasing the size of things, but also emphasized that there is "*plenty* of room." In his lecture, Feynman justified the inevitable development of concepts and technologies underlying the nanoworld and presented his vision of exciting new discoveries and scientific perspectives at the nanoscale.

Technological motivations

Achievements in nanoscience and nanotechnology will have tremendous multidisciplinary impact. The benefits brought by novel nanotechnologies are expected for many important practical fields of endeavor. These include materials and manufacturing, electronics, computers, telecommunication and information technologies, medicine and health, the environment and energy storage, chemical and biological technologies, and agriculture. Having stated the purpose of this text, we consider now more detailed motivations for the development of electronics at the nanoscale.

In general, progress in electronics is stimulated, in part, by the enormous demands for information and communication technologies as well as by the development of numerous special applications. The continuous demands for steady growth in memory and computational capabilities and for increasing processing and transmission speeds of signals appear to be insatiable. These determine the dominant trends of contemporary microelectronics and optoelectronics. One of the main trends of the progress in electronics was formulated by Intel co-founder Dr. Gordon Moore as the following empirical observation: *the complexity of integrated circuits, with respect to minimum component cost, doubles every 24 months*. This statement formulated forty years ago is known as *Moore's law* and provides an estimate of the rate of progress in the electronics industry. Specifically, Moore's law predicts that the number of the basic devices – transistors – on a microchip doubles every one to two years. This is possible only if progressive scaling down of all electronic components is realized.

Electronics exploits the electrical properties of solid-state materials. A simple and intuitive classification of solids makes a distinction between dielectrics and metals, i.e., dielectrics are non-conducting materials whereas metals are good conducting materials. Semiconductors occupy the place in between these two classes: semiconductor materials are conducting and optically active materials with electrical and optical properties varying over a wide range. Semiconductors are the basic materials for microelectronics and remain the principal candidates for use in nanoelectronic structures because they exhibit great flexibility in terms of allowing the control of the electronic and optical properties and functions of nanoelectronic devices. Accordingly, to a large extent, we will analyze the trends of electronics in the context of semiconductor technology.

It is instructive to illustrate these trends and achievements through the example of Si-based electronics. Indeed, contemporary microelectronics is based almost entirely on silicon technology, because of the unique properties of silicon. This semiconductor material has high mechanical stability as well as good electrical isolation and thermal conductivity. Furthermore, the thin and stable high-resistance oxide, SiO_2, is capable of withstanding high voltages and can be patterned and processed by numerous methods. Silicon technology also enjoys the advantage of a mature growth technology that makes it possible to grow Si substrates (wafers) of larger areas than for other semiconductor materials. The high level of device integration realizable with Si-based electronics technology may be illustrated by the important integrated circuit element of any computer, controller, etc. – the dynamic random access memory (DRAM). The main elements of DRAM based on complementary metal–oxide–semiconductor technology (Si-CMOS) are metal–oxide–semiconductor field-effect transistors (MOSFETs). For Si MOSFETs, channels for flow of electric current are created in the Si substrate between the source and drain contacts, and the currents are controlled by electrodes – metal gates – which are isolated electrically by very thin SiO_2 layers, which have become thinner than 10 nm.

Figure 1.1 illustrates the evolution of the DRAM size and transistor gate size as functions of time. Besides transistors and capacitors, the chip contains metallic line connections: local, intermediate, and global wiring. Figure 1.1 illustrates the steady scaling down of all characteristic sizes and increasing levels of integration. For example, the 64-Mbit DRAM chip contains approximately 10^8 transistors per cm^{-2}, each with feature sizes of the order of 0.3 μm. The transistors in this DRAM as well as those of the more highly integrated 256-Mbit chip operate as conventional devices and obey the laws of classical physics. The next generation of devices is entering the nanoscale regime where quantum mechanics is important; indeed, as we will discuss in this book, quantum mechanics becomes dominant on the scale of approximately one to ten nanometers for devices that operate at room temperature. According to Fig. 1.1, today's technology has already reached the nanoscale and newer device concepts should be implemented before 2010.

One of the factors driving the huge production and wide use of microelectronic systems is the relatively low cost of their fabrication. Moreover, despite their increasing

Table 1.1 A roadmap for Si-based microelectronics (predictions of the Semiconductor Industry Association)

	1995	1998	2001	2004	2007	2010
Memories, DRAM						
Bits per chip	64 M	256 M	1 G	4 G	16 G	64 G
Cost per bit (milli-cent)	0.017	0.007	0.003	0.001	0.0005	0.0002
Cost per chip (US$)	11	18	30	40	80	130
Logic, microprocessors						
Transistors per cm^2	4 M	7 M	13 M	25 M	50 M	90 M
Cost per transistor (milli-cent)	1	0.5	0.2	0.1	0.05	0.02
Power supply (V)	3.3	2.5	1.8	1.5	1.2	0.9
Parameters						
Minimum feature size (μm)	0.35	0.25	0.18	0.13	0.10	0.07
Wafer size (in.)	8	8	12	12	16	16
Electrical defect density per m^2	240	160	140	120	100	25

The data are from U. König, *Physica Scripta*, **T68**, 90, 1996.

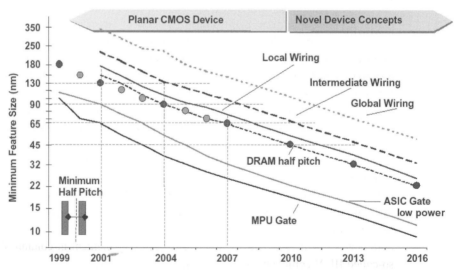

Figure 1.1 Technology nodes and minimum feature sizes from application ITRS Roadmap: MPU, Micro Processing Unit; ASIC, Application-Specific Integrated Circuit. Used with permission, from W. Klingenstein (2002). *Technology Roadmap for Semiconductors.* http://broadband02.ici.ro/program/klingenstein_3d.pdf, page 15. © InfineonTechnologies AG, 2002.

complexity, microelectronic systems continue to be produced at lower costs. In Table 1.1 the costs per bit and costs per chip as well as the associated performance levels are given as functions of the integration level. One can see that every three years the number of bits per chip has increased by a factor of four (even faster than predicted by Moore's law) and the cost per bit has decreased by a factor of two or more.

In the same table, the integration levels of logic circuits and microprocessors are forecast. We see that, for this case, device integration is also large but will increase slightly slower than for DRAMs. The cost of the principal elements of logic circuits – transistors – is significantly greater, but it also tends to decrease. The forecast for the necessary power supply presented in the table portrays a slow, but persistent, decrease. Thus, one can expect favorable trends for the power consumption of microelectronic systems.

The bottom of the table presents the necessary technological parameters for ultra-high integration: minimum feature sizes, diameters of wafers, and electrical defect densities. The large wafer size allows a greater number of devices to be fabricated on a chip. The density of electrical defects (i.e., crystal imperfections which affect electrical properties) is characteristic of the quality of the wafers. Table 1.1 forecasts that wafer diameters will be continuously increased, while the defect density decreases by a factor of six per decade; currently, they must be limited to several tens per m^2.

After this overview of the dominant driving forces in nanoscale development, we will mention briefly other general issues important for this field. These include improving materials, fabrication and measurement techniques on the nanoscale, and novelties in the operation principles of nanodevices.

Improving materials on the nanoscale

In the processes of achieving minimum device sizes and ultra-high levels of integration it is necessary to identify the limiting and critical parameters for improved performance. In reality, these parameters depend on the integrated elements of each individual material system. For example, for transistors two parameters of the host material are of special importance: the ultimate electron velocity and the limiting electric field which does not induce electric breakdown. Further improvements in the parameters can be achieved through materials engineering.

Silicon plays the central role in electronics. However, semiconductors other than silicon can be used. In particular, compound semiconductors constitute a general class of semiconductors that has been used increasingly in recent decades. As examples of forming compound semiconductors, every particular element in column III of the periodic table of elements may be combined with every element in column V to form a so-called III–V compound, which is semiconducting. Then, two or more discrete compounds may be used to form alloys. A common example is aluminum–gallium arsenide, $Al_x Ga_{1-x} As$, where x is the fraction of column III sites in the crystal occupied by Al atoms, and the fraction $1 - x$ is occupied by Ga atoms. As a result, it becomes possible not only to make discrete compounds, but also to realize a *continuous range* of materials for tailoring necessary electronic properties. As for Si technology, the growth of silicon–germanium ($Si_x Ge_{1-x}$) alloys facilitates the control of the properties of materials over a considerable range of the electrical parameters. These techniques are exploited widely in microelectronics.

Further revolutionary modification and engineering of materials can be accomplished by using *heterostructures* with nanoscale features. Heterostructures are structures with two or more abrupt interfaces at the boundaries between the different semiconductor materials. With modern material-growth techniques, it is possible to grow structures with transition regions between adjacent materials that have thicknesses of only one or two atomic monolayers. This allows one to fabricate multilayered semiconductor structures with nanoscale thicknesses.

The simplest multilayered structure has a single heterojunction, i.e., a single-heterojunction structure is made of two different materials. At the interface of such a heterojunction, the electronic properties are changed to improve selected physical characteristics. In particular, electrons can be confined in a thin layer near the interface. In fact, the layers with confined electrons can be made so thin that wave-like behavior – that is, quantum-mechanical behavior – of the electrons becomes apparent. The same phenomena occur for diverse multilayered nanoscale structures that can be grown with high quality.

By using nanostructures, it is possible to modify the electronic properties of a great variety of a nanoscale devices. Indeed, we live in a three-dimensional world, where a particle can, in principle, move in all three directions. Quantum effects on the nanoscale determine the properties of electrons in nanostructures: the nanostructures can be made in such a way that the electron motion becomes *two-dimensional, one-dimensional, or even zero-dimensional*. These nanostructures are known as *low-dimensional* quantum heterostructures and are called quantum wells, quantum wires, and quantum dots, for the cases where the electrons are confined in one, two, and three dimensions, respectively. Such an impressive example of manipulation of the properties of the current-carrying particles clearly illustrates fundamentally new possibilities for electronics that become viable at the nanoscale.

Fabrication techniques on the nanoscale

The progress in heterostructure technology has been made possible largely as a result of new advances in fabrication techniques. In Table 1.2, we provide a very brief summary of some important steps now used in the growth, characterization, and processing of heterostructures. In the period of the 1970s and 1980s, molecular-beam epitaxy was invented, developed, and employed to fabricate high-quality and ultra-thin layers and superlattices. Qualitative electron-beam and X-ray microscope technologies were used to characterize the perfectness of structures, including interface disorder. During this period, lithographic and etching methods suitable for microscale and nanoscale devices were proposed and realized. In the 1980s and later, new epitaxial techniques were developed; these included metal–organic vapor-phase epitaxy and metal–organic molecular-beam epitaxy, among others. These innovations made possible the fabrication of layers with atomic-level accuracy. Lithography and etching methods were improved to the point that they can be used for nanoscale structuring. Desirable spatial-modulation doping by impurities has become possible, including δ-doping – that is, doping of one or a few atomic monolayers.

These approaches to the production of nanostructures and nanoelectronic devices represent "evolutionary" improvements in the growth and processing methods applied previously in microelectronics. Advances in nanotechnology allow, in principle, the utilization of methods and concepts from other areas of science and engineering. Synthetic

Table 1.2 Advances in growth, characterization, and processing of quantum heterostructures

1970s–1980s	Growth and fabrication methods
	Molecular-beam epitaxy
	Ultra-thin-layer fabrication
	Superlattice fabrication
	Characterization methods
	Lithographic microstructuring
	Qualitative electron-beam and X-ray microscopies
1990s–2000s	Growth and fabrication methods
	Metal–organic vapor-phase epitaxy
	Metal–organic molecular-beam epitaxy
	Fabrication to atomic-layer accuracy
	δ-Doping
	Controlled strained layers
	Fabrication methods based on chemistry and biology
	Assembling inorganic nanoblocks with biomolecules
	Characterization methods
	Lithography and etching for nanostructuring
	Dip-pen nanolithography
	Quantitative electron-beam and X-ray microscopies
	Scanning tunneling microscopy (STM)
	Atomic force microscopy (AFM)
	Picosecond and femtosecond spectroscopy
	Terahertz time-domain spectroscopy

chemistry and even biology have much to offer for emerging nanotechnologies. Some fundamental concepts coming from these fields can successfully be exploited for the synthesis of nanomaterials and nanodevices. These include chemical and biological methods of growth of nanoscale objects – such as carbon nanotubes and biomolecules – surface nanopatterning, and preparing nanostructured materials with predefined, synthetically programmable properties from common inorganic building blocks with the help of DNA interconnect molecules, etc.

Improvement in characterization methods for the nanoscale

Progress in the refinement of fabrication techniques for making nanostructures depends on the great improvements made in characterization methods. Some of these methods are indicated in Table 1.2. In particular, composition and dopant distribution, lattice strain, and other parameters within nanostructures must be known with atomic-scale precision. Currently, the manipulation of a single atom (ion) in a solid is possible. New tools – scanning tunneling microscopy and atomic-force microscopy – which portend numerous applications in high-precision fabrication have emerged. Picosecond and femtosecond spectroscopy have progressed substantially and they have been applied to characterize the electronic and lattice properties of heterostructures. Finally, terahertz time-domain spectroscopy was developed, which makes it possible to measure electric signals with time resolution at the level of 10^{-12} seconds.

New principles of device operation at the nanoscale

Fundamental questions arise when conventional principles of device operation fail as a result of entering the nanoscale domain. One of the effects is almost *collisionless* motion – frequently called ballistic motion – of electrons flying through short devices. As mentioned previously, the nanoscale domain is the "realm" of quantum physics. Indeed, scaling down of devices and their integration above the level corresponding to 250 Mbits on a single chip makes it necessary to take into account new regimes and even to modify the principles underlying device operation. Further device downscaling and higher integration densities for information capacities exceeding 1 Gbit per chip imply the need to investigate using *quantum regimes* of operation in future years. Quantum-mechanical effects are not only important for operation of future integration circuits, but also are already used for generation of ultra-high-frequency electromagnetic emission. A relevant example is that of resonant-tunneling phenomena in nanoscale multilayered structures, which creates a foundation for microwave emission up to 1000 GHz.

A number of such device-related quantum effects has been discovered. New physics and new quantum effects always pass ahead of new devices exploiting these effects that have made a substantial impact on device technology. We mention here just a few quantum effects: 1970, the proposal of multilayered structures; 1974, the resonant-tunneling effect; 1978, the modulation doping effect; 1980, conduction of polymers; 1985, the discovery of the buckyball, C_{60}; 1993, the discovery of single-walled carbon nanotubes; and 1996, nanoelectromechanical systems (NEMSs). Some of these effects will be analyzed in this book. Here is a short list of some of the novel quantum devices: 1979, the injection quantum-well laser; 1983; the Microwave DBRTD Oscillator; 1984, the hot-electron transistor; 1998, the quantum-wire carbon nanotube field-effect transistor; 1998, polymer-based transistors and light-emitting devices; 2001, sensors based on NEMSs; 2001, sub-terahertz III–V compound nanoscale field-effect transistor; and 2006, sub-terahertz InP and SiGe bipolar transistors. There is a "delay time" between the discovery of the effect and the device concept, but the delay time is decreasing steadily. The following examples show this tendency. The effect of resonant tunneling was discovered in 1974; the device – the microwave double-barrier resonant-tunneling diode (DBRTD) oscillator – was realized in 1983. The first quantum wires with one-dimensional electron motion were studied in 1986; their first application in lasers occurred in 1995. In both cases the "delay time" was 9 years. The discovery of single-walled carbon nanotubes was made in 1993 and in only 5 years the carbon nanotube transistor was fabricated. The same is valid for the development of nanoelectromechanical systems and their applications for a number of sensors, etc. Thus, for contemporary electronics there is an evident acceleration of the implementation of fundamental physical effects.

Besides quantum effects, reducing device dimensions results in a decrease in the number of electrons participating in the transfer of an electric signal. As a result, nanoscale devices may operate on the basis of single-electron transfer. Various novel single-electron devices have been proposed and demonstrated. By reducing the sizes of quantum dots to 100 Å or less, it is possible to operate with single electrons at temperatures near or close to room temperature.

The great technological advances brought about in mainstream microelectronics and nanoelectronics can be used for the fabrication of other classes of nanodevices. One such approach is based on quantum dots arranged in locally interconnected cellular-automata-like arrays. The fundamental idea of operation of cellular automata is to encode information using the charge configuration of a set of quantum dots. Importantly, in the quantum-dot cellular-automata approach, the information is contained in the arrangement of charges of the dots, rather than in the flow of the charges, i.e., electric current. It can be said that the devices interact by direct Coulomb coupling rather than via the current through the wires.

Another approach employs both electrical and mechanical properties of nanostructures. The new generation of devices and systems based on this approach is commonly referred to as *nanoelectromechanical systems* (NEMSs). Indeed, on the nanoscale a strong enhancement of coupling between electronic and mechanical degrees of freedom occurs. This electromechanical concept may be used for the development of a new class of devices that includes nanomachines, novel sensors, and a variety of other new devices functioning on the nanoscale. Thus, NEMSs may supplement the traditional electronics that works solely with electrical signals.

Nanotechnology for optoelectronics

Thus far, we have reviewed nanoscience and nanotechnology as applied to *electronic* devices, i.e., devices in which electrical properties are exploited and which operate with electrical input and output signals. Another class of devices is comprised of *optoelectronic devices*, which are based on both electrical and optical properties of materials and operate with both optical and electrical signals. An important and growing trend is that optoelectronics complements microelectronics in many applications and systems. First of all, optoelectronics provides means to make electronic systems compatible with lightwave communication technologies. Furthermore, optoelectronics can be used to accomplish the tasks of acquisition, storage, and processing of information. Advances in optoelectronics make significant contributions to the transmission of information via optical fibers (including communication between processing machines as well as within them), to the high-capacity mass storage of information on laser disks, and to a number of other specific applications. Clearly, optoelectronic devices have a huge number of diverse applications.

The principal components of optoelectronic systems are light sources, sensitive optical detectors, and properly designed light waveguides, for example, optical fibers. These devices and passive optical elements are fabricated with optically active semiconductor materials. Semiconductor nanostructures and, in particular, quantum heterostructures provide new means to enhance many optical and electro-optical effects. For example, both of the most widely used light sources – light-emitting diodes and laser diodes –

may be improved greatly when nanostructures such as quantum wells, quantum wires, and quantum dots are exploited as active optical elements.

As for the previously studied case of microelectronics, the trends in optoelectronics involve scaling down the sizes of these devices as well as achieving high levels of integration in systems such as arrays of light diodes, laser arrays, and integrated systems with other electronic elements on the same chip. Optoelectronics benefits substantially through the use of nanotechnology and becomes competitive with its microelectronic counterpart.

In conclusion, the current and projected trends in electronics lead to the use of nanostructures and to the reliance on novel quantum effects as an avenue for realizing further progress. These recent and diverse trends in semiconductor and device technologies as well as in novel device concepts are driving the establishment of a new subdiscipline of electronics based on nanostructures, i.e., nanoelectronics. This subdiscipline and its foundations are studied in this book.

More general information on nanoscience, nanotechnology, and nanostructures, and their potential, may be found in the following reviews:

R. Feynman, "There's plenty of room at the bottom," American Physical Society Meeting, Pasadena, CA, 29 December 1959; originally published in Caltech's *Engineering and Science Magazine*, February 1960; reprinted as R. P. Feynman, "Infinitesimal machinery," *Microelectromechanical Systems*, **2**, 1 (1993); (see, for example, www.zyvex.com/nanotech/feynman.html).

National Nanotechnology Initiative: The Initiative and Its Implementation Plan, National Science and Technology Council, Committee on Technology, Washington DC, 2000 (see, for example, www.nano.gov).

H. Kroemer, "Quasielectric fields and band offsets: teaching electrons new tricks," *Rev. Mod. Phys.*, **73**, 783 (2001).

The International Technology Roadmap for Semiconductors (Semiconductor Industry Association, San Jose, CA, 2002 – update).

2   Particles and waves

2.1   Introduction

The evolution of microelectronics toward reduced device sizes has proceeded to a degree that renders conventional models, approaches, and theories inapplicable. Indeed, for objects with sizes of 100 nanometers or less it is frequently the case that the length scales associated with fundamental physical processes are comparable to the geometrical size of the device; also, fundamental time scales associated with physical processes are of the order of the time parameters for nanodevice operation. Therefore, on the nanoscale the theories and models underlying modern nanoelectronics become more complicated, and rely more and more on basic science.

Generally, in the nanoworld the fundamental laws of physics that govern particles and material fields differ from those that apply to familiar macroscopic phenomena such as the motion of a baseball or a train. Instead of classical mechanics, that works so well for macroscopic phenomena, the motion of particles in the nanoworld is determined by the so-called wave mechanics or quantum mechanics. An underlying principle of central importance for nanophysics is the fundamental concept that *all matter*, including electrons, nuclei, and electromagnetic fields, *behaves as both waves and particles*, that is, wave–particle duality is a basic characteristic of all matter.

At first glance, wave properties and particle properties for the same physical object are hardly compatible. To understand wave–particle duality, we will briefly review, in the following two subsections, the basic properties of particles and waves known from classical physics.

2.2   Classical particles

A particle can be characterized by the momentum vector $\mathbf{p}$ and the kinetic energy E that depends on the momentum. Here and throughout this book we will use the bold fonts for vectors, i.e., $\mathbf{p}$ is the vector and $p \equiv |\mathbf{p}|$ is its absolute value. The change of momentum with time is defined by Newton's second law:

$$\frac{\mathrm{d}\mathbf{p}}{\mathrm{d}t} = \mathbf{f}, \tag{2.1}$$

where t is time and $\mathbf{f}$ is the vector of an external force. From Eq. (2.1) it follows that, if the force is absent, then $\mathrm{d}\mathbf{p}/\mathrm{d}t = 0$, i.e., $\mathbf{p} = $ constant. This is the so-called *momentum conservation law* valid for a mechanical system in the absence of external forces. For simplicity in classical mechanics, we assume that any particle we consider has a very small size in comparison with the space where the particle is located. We refer to such a particle as a *point particle*.

The coordinate vector, $\mathbf{r}$, of a point particle and the particle velocity, $\mathbf{v}$, are related by the well-understood relation

$$\frac{d\mathbf{r}}{dt} = \mathbf{v}. \tag{2.2}$$

To obtain the relationship among the velocity, $\mathbf{v}$, the momentum, $\mathbf{p}$, and the energy of a particle, E, one has to calculate the power associated with the force $\mathbf{f}$ (the work of the force $\mathbf{f}$ on the particle per unit time). So, we have to multiply the left- and right-hand sides of Eq. (2.1) by $\mathbf{v}$, resulting in $\mathbf{v}\, d\mathbf{p}/dt = \mathbf{fv}$. (Remember that the scalar product of two vectors $\mathbf{a}$ and $\mathbf{b}$ is defined as $\mathbf{ab} = a_x b_x + a_y b_y + a_z b_z$.) The right-hand side of the transformed version of Eq. (2.1), $\mathbf{fv}$, is equal to the rate of the energy change dE/dt ($dE/dt = \mathbf{fv}$), and we obtain the relation

$$\frac{dE}{dt} = \mathbf{v}\frac{d\mathbf{p}}{dt}. \tag{2.3}$$

Using the chain rule of function differentiation ($df(x)/dt = (df/dx)(dx/dt)$), we determine how the velocity of the particle is related to the momentum and energy:

$$\mathbf{v} = \frac{dE}{d\mathbf{p}}. \tag{2.4}$$

Here, the derivative with respect to the vector $\mathbf{p}$ also gives the vector $\mathbf{v}$ with components

$$v_x = \frac{dE}{dp_x}, \qquad v_y = \frac{dE}{dp_y}, \qquad \text{and} \qquad v_z = \frac{dE}{dp_z}.$$

Let us consider an important case of a particle moving in a potential field. The force is defined as the derivative of a potential $V(\mathbf{r})$ with respect to the particle coordinate: $\mathbf{f} = -dV/d\mathbf{r}$. Note that, for the vector operator $d/d\mathbf{r} \equiv \{d/dx,\, d/dy,\, d/dz\}$, one often uses another notation: $d/d\mathbf{r} \equiv \nabla$, so that $dV/d\mathbf{r}$ is the so-called *gradient* of function $V(\mathbf{r})$, $\nabla V(\mathbf{r})$. On multiplying the left- and right-hand sides of Eq. (2.1) by $\mathbf{v}$, using the definition of Eq. (2.2) and the chain rule, we find

$$\mathbf{v}\frac{d\mathbf{p}}{dt} + \frac{dV}{d\mathbf{r}}\frac{d\mathbf{r}}{dt} = \frac{d}{dt}(E + V(\mathbf{r})) = 0.$$

The value of the kinetic energy plus the potential energy,

$$\mathcal{H} \equiv E + V(\mathbf{r}), \tag{2.5}$$

represents the *total energy* of the particle, $\mathcal{H}$. The above calculations tell us that the total energy of a particle in a potential field does not change during its motion. So, we have demonstrated the *law of energy conservation*, $d\mathcal{H}/dt = 0$. When $\mathcal{H}$ is considered as a function of two variables $\mathbf{p}$ and $\mathbf{r}$, it is called the Hamiltonian function, or simply the *Hamiltonian*. Remarkably, the partial derivatives of $\mathcal{H}$ give us both fundamental equations (2.1) and (2.4): $d\mathbf{p}/dt = \mathbf{f} = -\partial\mathcal{H}/\partial\mathbf{r}$ and $d\mathbf{r}/dt = \mathbf{v} = \partial\mathcal{H}/\partial\mathbf{p}$. As we shall see, the Hamiltonian of classical physics also plays an important role in the formulation of quantum mechanics.

A point particle moving in free space may be characterized by a mass m and by the kinetic energy:

$$E = \frac{\mathbf{p}^2}{2m}.$$ (2.6)

The latter dependence is frequently referred to as the *energy dispersion*. Here E is an isotropic function of $\mathbf{p}$. From the definition of kinetic energy given by Eq. (2.6) we find that

$$\mathbf{v} = \frac{\mathbf{p}}{m}.$$ (2.7)

So, the velocity and the momentum are collinear vectors. Then, Newton's second law, Eq. (2.1), can be rewritten in its usual form:

$$m \frac{d^2 \mathbf{r}}{dt^2} = \mathbf{f}.$$ (2.8)

Now, Eq. (2.5) takes the form that we will use often in this book:

$$\mathcal{H} \equiv \frac{\mathbf{p}^2}{2m} + V(\mathbf{r}).$$ (2.9)

One of the important results following from classical mechanics is that, if we know the particle position $\mathbf{r}_0$ and its momentum $\mathbf{p}_0$ (or velocity $\mathbf{v}_0$) at an initial moment t_0, from Eqs. (2.1)–(2.4), we can find the position and the momentum (velocity) of the particle at any given moment of time t for any given $\mathbf{f}$ or $V(\mathbf{r})$.

Equations (2.1)–(2.8) are the equations of classical mechanics. All of the variables, such as $\mathbf{r}$, $\mathbf{p}$, E, and $\mathbf{v}$, are continuous variables. Importantly, $|\mathbf{p}|$ can have any value, including zero, i.e., $\mathbf{p} = 0$ and $E = 0$ are allowed.

For a particle, say an electron, moving inside of a crystal (a metal, a dielectric, a semiconductor, etc.), the interaction of this particle with the crystal generally makes the relationship between E and $\mathbf{p}$ – the dispersion relation – more complicated. In particular, E may be an anisotropic function, and the velocity and momentum may be noncollinear vectors. Examples of such energy dependences will be given in the problems for this chapter.

2.3   Classical waves

We are all familiar with a lot of examples of waves and wave processes. These include sound waves in air, sea waves, and elastic waves in solids, electromagnetic waves, and gravitational waves. Generally, in classical physics wave motion arises in extended continuous media with an interaction between the nearest elements of the medium. Such an interaction gives rise to the transfer of a distortion (an excitation) from one element to another and to a propagation of this distortion through the medium. Despite the

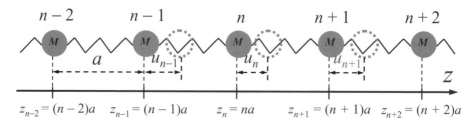

Figure 2.1 A linear chain of identical atoms of mass M: u_n are displacements of atoms from their equilibrium positions and the restoring force acting on the nth atom is $f_n = -\beta(u_n - u_{n+1}) - \beta(u_n - u_{n-1})$. Note: displacements u_n are not shown to scale.

differences in the particular nature of waves, wave motion has much in common for different media. We introduce wave properties by analyzing the following simple model.

We will construct a model of a one-dimensional medium, elements of which are represented by "atoms" connected by massless springs. Vibrations in such a linear atomic chain are governed by the laws of classical mechanics. The chain is supposed to be infinitely long. Let the equilibrium distance between atoms be a. Thus, the equilibrium position of the nth atom is $z_n = na$, and the displacement of this atom from its position is denoted by u_n. Figure 2.1 depicts such a linear chain of identical atoms of mass M. The springs represent interatomic forces, i.e., interaction between nearest elements of the medium. If the displacements of atoms from their equilibrium positions are not too large, the restoring forces in the chain obey Hooke's law,

$$f = -\beta u, \tag{2.10}$$

where u is a change of the spring length, β is the spring constant, and f is the force exerted by the spring. Now, we can apply Eq. (2.10) for the total force, f_n, acting on the nth atom coupled with its two nearest neighbors by two springs as

$$f_n = -\beta(u_n - u_{n+1}) - \beta(u_n - u_{n-1}). \tag{2.11}$$

Hence, the Newton equation of motion, Eq. (2.8), for the nth atom is

$$M \frac{d^2 u_n}{dt^2} = -\beta(2u_n - u_{n-1} - u_{n+1}). \tag{2.12}$$

This set of linear differential equations, Eq. (2.12), in principle, describes wave-like processes. However, we will make a further simplification and modify this discrete set of equations to obtain one equation describing a continuous medium. Such a continuous medium with elastic forces between its elements is, obviously, a string. To make this transformation to the continuous case, we shall consider the discrete coordinate to be continuous, $z_n \to z$, and replace the finite difference in Eq. (2.12) with a derivative:

$$\frac{u_n - u_{n-1}}{a} \to \frac{\partial u}{\partial z} \quad \text{and} \quad \frac{(2u_n - u_{n-1} - u_{n+1})}{a^2} \to -\frac{\partial^2 u}{\partial z^2}. \tag{2.13}$$

Then, we obtain the equation describing the displacement, u, of a string:

$$\rho_{1D} \frac{\partial^2 u}{\partial t^2} - \Lambda_{1D} \frac{\partial^2 u}{\partial z^2} = 0, \tag{2.14}$$

where $\rho_{1D} = M/a$ is the linear density of the string, and $\Lambda_{1D} = \beta a$ is the so-called elastic modulus of the string. The infinite set of *ordinary* differential equations, Eq. (2.12), is replaced by a single *partial* differential equation. Since we started our derivation for a mechanical system, for which the energy (per atom) can be defined, it is instructive to find a similar characteristic for the continuous medium described by Eq. (2.14). Assume, for a moment, that the spring under consideration is of finite large length, L. Then, let us multiply Eq. (2.14) by $\partial u/\partial t$ and integrate it over the length L:

$$\int_0^L dz \left(\rho_{1D} \frac{\partial^2 u}{\partial t^2} \frac{\partial u}{\partial t} - \Lambda_{1D} \frac{\partial^2 u}{\partial z^2} \frac{\partial u}{\partial t} \right) = 0.$$

By integrating the second term by parts by using the standard relationship $\int w \, dv = wv - \int v \, dw$, we find the following identity for a unit length of the string:

$$\frac{\partial}{\partial t} \frac{1}{L} \int_0^L dz \left[\frac{\rho_{1D}}{2} \left(\frac{\partial u}{\partial t} \right)^2 + \frac{\Lambda_{1D}}{2} \left(\frac{\partial u}{\partial z} \right)^2 \right]$$
$$- \frac{\Lambda_{1D}}{L} \left[\left(\frac{\partial u}{\partial z} \frac{\partial u}{\partial t} \right)_{z=L} - \left(\frac{\partial u}{\partial z} \frac{\partial u}{\partial t} \right)_{z=0} \right] = 0.$$

To draw further conclusions, we define an average of a quantity, A, over a piece of the string, Z, as $\overline{A} = (1/Z) \int_{(Z)} A \, dz$. For a long string, the average should not depend on the length, Z, of the piece of string. Now, as $L \to \infty$, the latter identity will be satisfied if the value

$$h_{1D} = \frac{\rho_{1D}}{2} \overline{\left(\frac{\partial u}{\partial t} \right)^2} + \frac{\Lambda_{1D}}{2} \overline{\left(\frac{\partial u}{\partial z} \right)^2} \tag{2.15}$$

is independent of time, i.e., $\partial h_{1D}/\partial t = 0$, and the value h_{1D} is conserved. In fact, the first term on the right-hand side of Eq. (2.15) is, obviously, the density of kinetic energy over the length of the string, while the second term is the elastic (potential) energy. Thus, h_{1D} has the meaning of the energy density for our one-dimensional continuous medium. As expected, the energy density is conserved, if external forces are absent.

Our "one-dimensional" analysis, which assumes that the atoms can move only along a single direction z, can be generalized to a three-dimensional elastic medium; see Problem 4 of this chapter. Now, the displacement becomes a three-dimensional vector, **u**, and, instead of Eq. (2.14), we obtain

$$\rho \frac{\partial^2 \mathbf{u}}{\partial t^2} - \Lambda \left(\frac{\partial^2 \mathbf{u}}{\partial x^2} + \frac{\partial^2 \mathbf{u}}{\partial y^2} + \frac{\partial^2 \mathbf{u}}{\partial z^2} \right) = 0 \tag{2.16}$$

with $\mathbf{r} = \{x, y, z\}$ being the vector coordinate and ρ being the "three-dimensional" density, i.e., the mass of a unit volume of the medium; Λ is the elastic modulus of the medium. Since it was introduced as a single elastic modulus, independent of direction, Eq. (2.16) is valid for an isotropic medium. The energy density of a three-dimensional elastic medium is

$$h = \frac{\rho}{2} \overline{\left(\frac{\partial \mathbf{u}}{\partial t} \right)^2} + \frac{\Lambda}{2} \overline{\left(\frac{\partial \mathbf{u}}{\partial \mathbf{r}} \right)^2}. \tag{2.17}$$

Here, the bar denotes an average over a small volume $\Delta V (\overline{A} = (1/\Delta V) \int_{(\Delta V)} A \, \mathrm{d}V)$, and

$$\frac{\partial \mathbf{u}}{\partial \mathbf{r}} \equiv \frac{\partial u_x}{\partial x} + \frac{\partial u_y}{\partial y} + \frac{\partial u_z}{\partial z} = \mathrm{div}\, \mathbf{u}.$$

Finally, we can rewrite Eq. (2.16) in the standard form of the *wave equation*:

$$\frac{\partial^2 \mathbf{u}}{\partial t^2} - s^2 \left(\frac{\partial^2 \mathbf{u}}{\partial x^2} + \frac{\partial^2 \mathbf{u}}{\partial y^2} + \frac{\partial^2 \mathbf{u}}{\partial z^2} \right) = 0, \tag{2.18}$$

where we introduce a new parameter s, the meaning of which will be clarified below. For our model

$$s = \sqrt{\Lambda/\rho}. \tag{2.19}$$

Although we derived Eq. (2.18) for a particular model of the elastic medium, the equation can be applied to describe a wide class of physical *vector* fields as exemplified by $\mathbf{u}(x, y, z)$ – the displacement field associated with a wave in an elastic medium. If the characteristic of a wave field is a scalar value, say w, in Eq. (2.18), we should simply substitute $\mathbf{u} \rightarrow w$ to describe the case of a *scalar* physical field.

Now we will analyze solutions to Eq. (2.18) for some cases in which the solutions are particularly simple. In many cases, such solutions are associated with wave-like processes. We may look for solutions of the form

$$\mathbf{u}(t, \mathbf{r}) = \mathbf{A} \cos(\mathbf{qr} - \omega t) + \mathbf{B} \sin(\mathbf{qr} - \omega t), \tag{2.20}$$

where $\mathbf{A}$ and $\mathbf{B}$ are arbitrary vectors, ω and $\mathbf{q}$ are unknown parameters; ω is known as the *angular frequency* of the wave and $\mathbf{q}$ is called the *wavevector*. By substituting such a form for $\mathbf{u}(t, \mathbf{r})$ into Eq. (2.18), we easily find that Eq. (2.20) is a solution of Eq. (2.18), if $\omega^2 = s\mathbf{q}^2$. The relationship between ω and $q = |\mathbf{q}|$ is called the *dispersion relation*:

$$\omega = s|\mathbf{q}|. \tag{2.21}$$

Importantly, there is no limitation to the wavevector $\mathbf{q}$: a solution can be found for any $\mathbf{q}$. *This is valid only for infinitely extended media, for which the wavevector can be a "continuous" vector.*

Because the two terms in Eq. (2.20) behave similarly, we can discuss basic properties of these solutions based on the example of "sinusoidal" waves:

$$\mathbf{u}(t, \mathbf{r}) = \mathbf{B} \sin(\mathbf{qr} - \omega t). \tag{2.22}$$

The argument of the sine function is the phase of the wave, $\phi = \mathbf{qr} - \omega t$, and $\mathbf{B}$ is the amplitude of the wave. Let the coordinate $\mathbf{r}$ be given, then we obtain a function that oscillates in time with an angular velocity ω. The frequency defines the rate of change of the phase with time t (radians per unit time). The period of time associated with a single oscillation is

$$T = 2\pi/\omega.$$

Accordingly, T is known as the *period*. If the time t is fixed, Eq. (2.22) represents a function that oscillates as the coordinate changes. These oscillations are characterized by the wavevector, $\mathbf{q}$ (or wavenumber, q). The wavevector defines the rate at which the

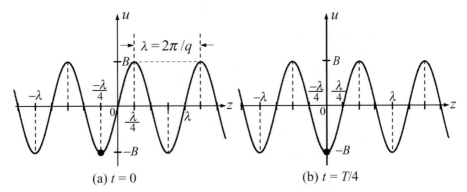

Figure 2.2 A propagating wave $u = B\sin(qz - \omega t)$. At time $t = 0$ the wave $u = B\sin(qz - \omega t)$ at the point $z = -\lambda/4$ has displacement $u = B\sin(-\pi/2) = -B$. At time $t = T/4$ we have $\omega t = (2\pi/T)(T/4) = \pi/2$; the same displacement will occur at the point $z = 0$: $u = B\sin(qz - \omega t) = B\sin(-\pi/2) = -B$.

phase changes with the coordinate (radians per unit length). One can introduce the spatial period to represent the distance for which **u** undergoes an oscillation of one cycle. It is called the *wavelength*,

$$\lambda = 2\pi/q. \tag{2.23}$$

The relationship between the time period and the spatial period is defined by Eqs. (2.21) and (2.23):

$$\lambda = \frac{2\pi}{\omega}s = Ts. \tag{2.24}$$

The waveforms of Eqs. (2.20) and (2.22) are often referred to as *traveling plane waves*. Indeed, their phase changes in a single direction along the wavevector **q** and surfaces of constant phase are planes perpendicular to **q**. Thus, in the case of a plane wave in a three-dimensional medium, the wave parameters do not depend on the two coordinates perpendicular to **q**. A traveling wave is illustrated in Fig. 2.2, where this wave is shown at two different moments of time. Now, we can clarify the meaning of the parameter s. Let the wavevector be directed along the z direction. Then, the wave phase is $\phi = qz - \omega t = q[z - (\omega/q)t]$. So we can conclude that a given magnitude of the phase moves with the velocity $\omega/q = s$, i.e., the parameter s in the wave equation (2.18) is the *phase velocity* of the traveling wave. In the waveform of Eq. (2.22), the constant **B** is the wave amplitude. The average of $\mathbf{u}^2(t, \mathbf{r})$ can be expressed in terms of **B** as $\overline{u^2} = B^2/2$, where we use the formulae $\overline{[\sin(\mathbf{qr} - \omega t)]^2} = \overline{[\cos(\mathbf{qr} - \omega t)]^2} = \frac{1}{2}$. According to Eq. (2.15), the amplitude relates to the wave energy density. Indeed, we note that

$$\frac{\rho}{2}\overline{\left(\frac{\partial \mathbf{u}}{\partial t}\right)^2} + \frac{\Lambda}{2}\overline{\left(\frac{\partial \mathbf{u}}{\partial \mathbf{r}}\right)^2} = \frac{B^2}{2}\left[\rho\omega^2\overline{[\cos(\mathbf{qr} - \omega t)]^2} + q^2\Lambda\overline{[\cos(\mathbf{qr} - \omega t)]^2}\right]$$
$$= \frac{B^2}{2}\left(\frac{\rho\omega^2}{2} + \frac{q^2\Lambda}{2}\right) = \frac{B^2}{2}\left(\frac{\rho\omega^2}{2} + \frac{\rho\omega^2}{2}\right) = \frac{B^2}{2}\rho\omega^2.$$

In obtaining this result, we have taken into account that $q^2 = \omega^2/s^2$ and $s = \sqrt{\Lambda/\rho}$. Thus, we obtain $h = \rho\omega^2 B^2/2$. Importantly, the energy density for traveling plane waves is *independent of the coordinate* and it is proportional to ω^2. If $\omega = 0$, there is no wave and there is no energy associated with it. Note that, for the waveforms of Eqs. (2.20) and (2.22), averaging over a small volume of the medium is equivalent to averaging over the period of oscillations.

For traveling waves, one often introduces also the *wave intensity*, which is the density of the energy flux. This energy flux represents the energy transferred by the wave per unit time through a unit cross-section perpendicular to $\mathbf{q}$. The intensity is a vector directed along $\mathbf{q}$ with the absolute value

$$\mathcal{I} = s \times h = \frac{s\rho\omega^2 B^2}{2}. \tag{2.25}$$

Besides the waveform of Eq. (2.20), a traveling wave can be presented in a complex form,

$$\mathbf{u} = \mathbf{A}e^{i(\mathbf{qr}-\omega t)}, \tag{2.26}$$

where the amplitude $\mathbf{A}$ is generally a complex quantity. Indeed, Eq. (2.26) is a solution to the wave equation (2.18). For some cases, it is convenient to operate with such a complex waveform. However, one should remember that true physical quantities always have real values.

Now we consider the important wave phenomenon known as *wave interference*. Suppose that two sinusoidal waves of the same frequency propagate from different sources through the medium. The sources of the waves are generally at different locations, so the waves reach a point of observation $\mathbf{r}$, in general, with different phase shifts $\varphi_1(\mathbf{r})$ and $\varphi_2(\mathbf{r})$:

$$\mathbf{u}_1(t, \mathbf{r}) = \mathbf{B}_1 \sin(\omega t + \varphi_1(\mathbf{r})), \qquad \mathbf{u}_2(t, \mathbf{r}) = \mathbf{B}_2 \sin(\omega t + \varphi_2(\mathbf{r})). \tag{2.27}$$

The resulting wave field is $\mathbf{u}(t, \mathbf{r}) = \mathbf{u}_1(t, \mathbf{r}) + \mathbf{u}_2(t, \mathbf{r})$. In experiments, instead of the wave amplitude, it is the intensity of the wave that is measured in many cases. From Eq. (2.25) the intensity is proportional to $\overline{\mathbf{u}^2}$. Straightforward calculation of $\overline{\mathbf{u}^2}$ gives us

$$\overline{\mathbf{u}^2} = \frac{1}{2}\left(B_1^2 + B_2^2 + 2\mathbf{B}_1\mathbf{B}_2\cos[\varphi_1(\mathbf{r}) - \varphi_2(\mathbf{r})]\right). \tag{2.28}$$

In deriving Eq. (2.28), we have used the identity

$$\sin x \sin y = \frac{1}{2}\cos(x - y) - \frac{1}{2}\cos(x + y),$$

i.e.,

$$\sin(\omega t + \varphi_1(\mathbf{r}))\sin(\omega t + \varphi_2(\mathbf{r})) = \frac{1}{2}\cos(\varphi_1(\mathbf{r}) - \varphi_2(\mathbf{r})) - \frac{1}{2}\cos(2\omega t + \varphi_1(\mathbf{r}) + \varphi_2(\mathbf{r})),$$

and the fact that the average of $\cos(2\omega t + \varphi_1(\mathbf{r}) + \varphi_2(\mathbf{r}))$ is zero. Thus, the intensity of the resulting wave consists of three contributions: the term related to the wave

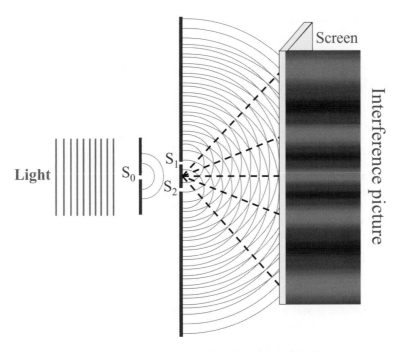

Figure 2.3 A double-slit experiment: S_1 and S_2 form the double slit.

coming from the first source, the term related to the wave from the second source, and an additional term. This third contribution describes the effect of wave interference. It depends critically on the wave phase difference. The interference contribution can be positive (*constructive* interference at $-\pi/2 < (\varphi_1 - \varphi_2) < \pi/2$ when $\cos(\varphi_1 - \varphi_2) > 0$), or negative (*destructive* interference at $\pi/2 < (\varphi_1 - \varphi_2) < 3\pi/2$ when $\cos(\varphi_1 - \varphi_2) < 0$). Importantly, the interference effect can be observed only for waves with the same frequency (otherwise the averaging leads to zero interference contribution). The waves of Eq. (2.27) with time-independent phase shifts are also known as *coherent* waves. If waves are characterized by a phase shift $(\varphi_1 - \varphi_2)$, which jitters randomly in time, the waves are *incoherent* and no interference effect occurs.

The simplest example illustrating the interference effect is the double-slit optical experiment. Let two slits be illuminated by a light wave from a single source. These two slits become two sources of coherent waves. The superposition of these waves generates an interference pattern of fringes, as shown in Fig. 2.3.

The wave analyzed above travels along the vector $\mathbf{q}$. Using the wavevector $-\mathbf{q}$ in Eq. (2.22) at the same frequency, we obtain another wave traveling in the opposite direction. According to Eq. (2.20), a combination of these waves is also a solution to Eq. (2.18):

$$\mathbf{u}(t, \mathbf{r}) = \mathbf{B}_+ \sin(\mathbf{q}\mathbf{r} - \omega t) + \mathbf{B}_- \sin(\mathbf{q}\mathbf{r} + \omega t),$$

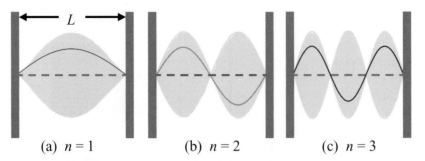

(a) $n = 1$ (b) $n = 2$ (c) $n = 3$

Figure 2.4 Quantization of oscillations in the form of standing waves. The gray area is a standing-wave pattern. Solid lines show amplitudes of oscillations at an instantaneous moment of time. Only waves with integer half-wavelengths exist: $L = n\lambda/2$. Note: for demonstration purposes, the amplitudes are shown not to scale.

where again $\mathbf{B}_+$ and $\mathbf{B}_-$ are arbitrary constant vectors. The two waves can also be interpreted as *incident* and *reflected* waves.

An important case is that of a wave propagating between two reflecting walls placed at $z = 0$ and $z = L$. The waves depend on a single coordinate z: $\mathbf{u}(t, z) = \mathbf{B}_+ \sin(qz - \omega t) + \mathbf{B}_- \sin(qz + \omega t)$. If the walls are rigid there are no displacements at the walls, so we should use the boundary conditions for the waves at $z = 0$ and $z = L$: $\mathbf{u}(z = 0) = \mathbf{u}(z = L) = 0$. One of the many physical embodiments of this situation are represented by waves in a continuum elastic medium bounded by rigid boundaries at $z = 0$ and $z = L$; another is given by a string attached to two fixed boundaries at $z = 0$ and $z = L$. From the boundary condition at $z = 0$, we find $\mathbf{B}_+ - \mathbf{B}_- = 0$, thus $\mathbf{u}(t, z) = \mathbf{B}_+[\sin(qz - \omega t) + \sin(qz + \omega t)] = 2\mathbf{B}_+ \sin(qz) \cos(\omega t)$. (Note: $\sin x + \sin y = 2 \sin[\frac{1}{2}(x + y)]\cos[\frac{1}{2}(x - y)]$). The boundary condition at $z = L$ can be satisfied if, and only if, $\sin(qL) = 0$. The latter requires the so-called "quantization" of the wavevector q, $qL = \pi n$, or

$$q = q_n = \frac{\pi}{L}n, \quad n = 1, 2, 3, \ldots, \tag{2.29}$$

where q_n is called the *wavenumber*. That is, there exist only waves of a special form,

$$\mathbf{u} = \mathbf{B} \sin(q_n z) \cos(\omega_n t), \tag{2.30}$$

at discrete values of the wavevector q_n and frequency $\omega_n = sq_n$. This important class of waves is known as *standing waves*. From Eqs. (2.23) and (2.30), we have the relation $L = (\lambda/2)n$. For standing waves, a strictly integer number of half-waves may be put between the reflecting walls. Figure 2.4 illustrates standing waves of various wavelengths for an arbitrary instant of time. The longest possible wave ($n = 1$) corresponds to $\lambda = 2L$ when a half-wave fits between the reflecting walls; see Fig. 2.4(a). To calculate the energy density of the standing waves, we can use Eq. (2.15). Then, the energy density is $h_n = \rho B^2 \omega_n^2/4$. It is independent of coordinate, as found for the traveling plane wave. Obviously, the energy flux is now zero. It is important to stress again the difference between traveling waves (with arbitrary and continuous values of q including $q \to 0$, hence $\lambda \to \infty$, in infinite media) and standing waves with quantized (or discrete) values of q_n with the minimum wavenumber $q_1 = \pi/L$.

The waveform given by Eq. (2.20) is actually a simple specific example of a more general class of wave fields. Indeed, due to the linear character of the wave equation, a sum of an arbitrary number of partial solutions is also a solution of the equation. This property of waves satisfying linear wave equations is known as the *superposition principle*. According to this principle, we may write the solution to Eq. (2.18) in the general (complex) form as

$$\mathbf{u}(t, \mathbf{r}) = \sum_{\mathbf{q}} \mathbf{A_q} e^{i(\mathbf{qr} - \omega_q t)}. \tag{2.31}$$

Here, the summation is taken over the wavevectors q. In addition, ω_q relates to $\mathbf{q}$ through the dispersion relation (2.21). The amplitudes of waves contributing to $\mathbf{u}$ depend, in general, on the wavevectors. The superposition principle is the basis for many important phenomena, including interference, formation of standing waves, and diffraction.

Electromagnetic waves in free space

One of the most important examples of waves is that of electromagnetic waves, i.e., oscillating electromagnetic fields. These fields are responsible for the most basic properties of matter from the nanoworld to the scale of the universe, and they are exploited for a number of technologies critically important for modern society. The fundamentals and applications of electromagnetic fields constitute a separate and extremely important field of science. In this book, we will deal with electromagnetic fields only briefly in order to illustrate the general character of the wave equation (2.18).

Electromagnetic waves are joint electric and magnetic fields that oscillate in both space and time. In the simplest homogeneous case, both the electric field $\mathbf{F}$ and the magnetic field $\mathbf{H}$ are governed by the wave equation in the form (2.18), where one should replace $\mathbf{u} \to \mathbf{F}$ or $\mathbf{H}$ and $s \to c$, where c is the velocity of light in free space and $\omega = qc$. The wave equation, for example, for the electric field reads as

$$\frac{\partial^2 \mathbf{F}}{\partial t^2} - c^2 \left(\frac{\partial^2 \mathbf{F}}{\partial x^2} + \frac{\partial^2 \mathbf{F}}{\partial y^2} + \frac{\partial^2 \mathbf{F}}{\partial z^2} \right) = 0. \tag{2.32}$$

Now, one can write the electric field in the form of a plane wave similar to Eq. (2.22):

$$\mathbf{F}(\mathbf{r}, t) = \mathbf{b} F_0 \sin(\mathbf{qr} - \omega t),$$

where $\mathbf{F}_0 = \mathbf{b} F_0$, F_0 is the amplitude of the electric field, and $\mathbf{b}$ is the vector of the polarization of the wave which denotes the direction of $\mathbf{F}(\mathbf{r}, t)$. The parameters $\mathbf{q}$ and ω have the same meaning as above: the wavevector and the angular frequency of the wave. Alternatively, it is possible to use a complex form of the plane wave,

$$\mathbf{F}(\mathbf{r}, t) = -i\mathbf{b} F_0 e^{i(\mathbf{qr} - \omega t)}, \tag{2.33}$$

but only the real part of this formula has physical meaning. The same kind of equation may be written for the magnetic field $\mathbf{H}$.

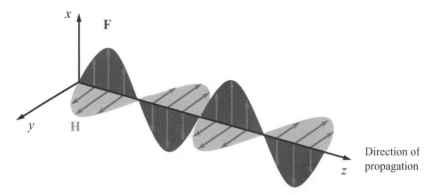

Figure 2.5 A plot of electric, **F**, and magnetic, **H**, fields as functions of z at time $t = 0$.

In free space, the vector **F** is always perpendicular to **q**, so that, if **q** is fixed, the electric field has, in general, *two possible polarizations of the electromagnetic wave* that correspond to the two orthogonal directions perpendicular to **q**. The magnetic field vector **H** is, in turn, perpendicular to both **q** and **F**. Propagation of the electric and magnetic fields is illustrated by Fig. 2.5.

Equations of the form of Eq. (2.32) are valid for a homogeneous or nearly homogeneous dielectric medium, with only the replacement $c \to c/\sqrt{\epsilon}$, where ϵ is the dielectric constant of the medium:

$$\frac{\partial^2 \mathbf{F}}{\partial t^2} - \frac{c^2}{\epsilon} \left(\frac{\partial^2 \mathbf{F}}{\partial x^2} + \frac{\partial^2 \mathbf{F}}{\partial y^2} + \frac{\partial^2 \mathbf{F}}{\partial z^2} \right) = 0. \tag{2.34}$$

Since $\epsilon > 1$, the velocity of light, $c/\sqrt{\epsilon}$, in a dielectric medium is less than that in free space. The frequency, ω, and wavevector, **q**, are related through the dispersion relation

$$\omega = \frac{c}{\sqrt{\epsilon}} |\mathbf{q}|.$$

The energy of the wave can be characterized by the density of the electromagnetic energy, which is defined as

$$W = \epsilon_0 \epsilon \overline{\mathbf{F}^2}(t) = \frac{1}{2} \epsilon_0 \epsilon F_0^2, \tag{2.35}$$

where $\overline{\mathbf{F}^2}(t)$ represents the time average of $F^2(t)$ and ϵ_0 is the permittivity of free space: $\epsilon_0 = 8.851 \times 10^{-12}\ \mathrm{F\,m^{-1}}\,(1\ \mathrm{F} = 1\ \mathrm{C\,V^{-1}})$. We can define the *intensity* of the wave as the energy flux through unit area perpendicular to the wavevector **q**:

$$\mathcal{I} = \frac{c}{2} \epsilon_0 \epsilon F_0^2. \tag{2.36}$$

The above description is associated with the *classical description* of the electromagnetic fields.

In closing this section on classical waves, we summarize that wave motion occurs in extended media and waves themselves are always delocalized physical entities that generally extend over large distances.

2.4 Wave–particle duality

In this section we will study how any physical object may behave simultaneously as a particle and as a wave. To understand this so-called wave–particle duality, we use a "two-way road" from waves to particles and from particles to waves.

From waves to particles

We start with a well-known optical example, which will help us to overcome the gap between classical waves extending over space and discrete particles having certain coordinates. Visible light behaves as an electromagnetic wave exhibiting interference, diffraction, etc. On the other hand, we often see light as a light ray, which can propagate, be reflected, and be deflected. The trajectory of such a light ray is very similar to a particle trajectory. The class of optical effects for which light can be described in terms of rays constitutes the discipline of geometrical optics. The transition from wave optics to geometrical optics is justified if the wavelength of light λ is much smaller than the characteristic scales of the problem being considered: $\lambda \ll L$, where L can be a distance of light propagation, the size of an illuminated object, a scale of inhomogeneity of the medium, etc. Consider the last case with a smooth inhomogeneity, which appears as a "smooth" coordinate dependence of the dielectric constant $\epsilon(\mathbf{r})$ in Eq. (2.34). We will use a waveform similar to Eq. (2.26): $\mathbf{F} = \mathbf{A}e^{i\phi}$. For monochromatic light (light of a single frequency) we can write $\phi = \phi_0 - \omega t$. Here, ϕ_0 depends only on $\mathbf{r}$. We can introduce the wavevector again as $\mathbf{q} = \partial\phi/\partial\mathbf{r} = \partial\phi_0/\partial\mathbf{r}$. In an inhomogeneous medium the wavevector and the wave amplitude depend on the coordinate vector $\mathbf{r}$: $\mathbf{q} = \mathbf{q}(\mathbf{r})$; $\mathbf{A} = \mathbf{A}(\mathbf{r})$. On substituting the chosen waveform, $\mathbf{F} = \mathbf{A}e^{i\phi}$, into Eq. (2.34) we obtain

$$-\omega^2 \mathbf{A}e^{i\phi} - \frac{c^2}{\epsilon(\mathbf{r})}\left(\frac{\partial^2 \mathbf{A}}{\partial x^2}e^{i\phi} + 2i\frac{\partial \mathbf{A}}{\partial x}\frac{\partial \phi_0}{\partial x}e^{i\phi} - \mathbf{A}\left(\frac{\partial \phi_0}{\partial x}\right)^2 e^{i\phi} + \cdots\right) = 0, \quad (2.37)$$

where, for simplicity, we have written only the derivatives with respect to the x-coordinate. Within the geometrical-optics approximation, the phase $\phi(t, \mathbf{r})$ is large: $|\phi| \gg 1$. Moreover, both contributions to the phase are large too: $\omega t, |\phi_0| \gg 1$. Since the phase is large, the leading terms in our calculations, namely the very first term (proportional to $(\partial\phi/\partial t)^2 = \omega^2$) and the third term in the brackets (proportional to $(\partial\phi_0/\partial x)^2$), are quadratic with respect to the phase derivatives. Keeping these leading terms and cancelling out common multipliers, we obtain an equation for ϕ_0:

$$\left(\frac{\partial \phi_0}{\partial \mathbf{r}}\right)^2 = \omega^2 \frac{\epsilon(\mathbf{r})}{c^2}. \quad (2.38)$$

If we find a solution of Eq. (2.38), we can define the so-called *wave surfaces* on which ϕ_0 is constant: $\phi_0(\mathbf{r}) = $ constant. For a given point $\mathbf{r}$, the direction of the wave is determined by the wavevector $\mathbf{q} = d\phi_0/d\mathbf{r}$; the wave direction is perpendicular to the wave surface. After we have found the wave surfaces and calculated $\mathbf{q}(\mathbf{r})$, we may construct the ray trajectories, as presented in Fig. 2.6. In the simplest case of a homogeneous medium,

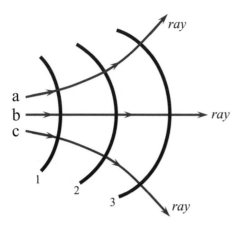

Figure 2.6 A sketch of the construction of ray trajectories by using the wavefronts. Rays a, b, and c are perpendicular to the wavefronts 1, 2, and 3.

where ϵ does not depend on the coordinate, we obtain a trivial result: $\mathbf{q} = $ constant, i.e., the ray propagates along a straight path.

Thus, if the wavelength is small in comparison with the characteristic scales of the system being considered, wave motion can be described in terms of ray trajectories. Up to this point, our discussion has been based totally on classical physics.

Thus far, we have not discussed the wave amplitude (or the wave intensity). The wave amplitudes enter the equations only as parameters that may be defined if the parameters characterizing the light sources are known. However, the wave amplitude is directly related to the kind of interpretation we are using: a classical interpretation or a quantum interpretation. Indeed, it has been established that the classical-physics approach is valid only for large intensities of waves. If the intensity of a wave is small, other laws of quantum physics define the wave field.

Let us consider briefly basic quantum concepts related to wave fields. According to Max Planck and Albert Einstein, electromagnetic waves interacting with matter can be emitted and absorbed in *discrete energy portions (quanta) – photons*. The energy of a photon, E, is proportional to the frequency of the wave:

$$E = \hbar\omega, \tag{2.39}$$

where $\hbar = 1.06 \times 10^{-34}$ J s (Joule-second) is the so-called *reduced Planck constant*; $h = 2\pi\hbar$ is called simply *Planck's constant*. Each photon, like a particle, has a momentum

$$\mathbf{p} = \hbar\mathbf{q}, \tag{2.40}$$

where $\mathbf{q}$ is the wavevector. (Equation (2.40) is formulated for photons in free space.) Apparently, the direction of the momentum vector $\mathbf{p}$ coincides with the direction of propagation of the wave. For each of the two possible polarizations of the radiation, one introduces appropriate characteristics of photons: each polarization of light, $\mathbf{b}$, is related to a certain photon. Thus, a photon can be labeled by two indices, $\{\mathbf{q}, \mathbf{b}\}$.

Table 2.1 A comparison between classical and quantum quantities

Classical quantity	Corresponding quantum quantity
Density of optical energy, W	Number of photons, $N_{q,b} = WV/(\hbar\omega)$
Optical intensity, $\mathcal{I}(\mathbf{r})$	Photon-flux density, $\mathcal{I}(\mathbf{r})/(\hbar\omega)$
Total optical power, P	Photon flux, $P/(\hbar\omega) = N_{q,b}c$

Equation (2.39), which represents the energy of a quantum of light, may be readily generalized for electromagnetic waves in a dielectric medium with dielectric constant ϵ:

$$\hbar\omega = \frac{\hbar c}{\sqrt{\epsilon}}q. \tag{2.41}$$

It is very important that different quanta of an electromagnetic field do not interact with each other, as is reflected by the linear character of the field in free space. An interaction between these modes is possible only in special media. Such media are called *nonlinear optical media*.

By introducing quanta, i.e., particles of electromagnetic field, we may understand Eq. (2.38) and Fig. 2.6 in a new way. Let us multiply this equation by the factor $\hbar^2/(2m)$, where m is an inessential parameter. Then, we may write

$$\frac{\mathbf{p}^2}{2m} - \frac{\hbar^2\omega^2\epsilon(\mathbf{r})}{2mc^2} = 0$$

upon taking into account the relations $\mathbf{q} = d\phi_0/d\mathbf{r}$ and $\mathbf{p} = \hbar\mathbf{q}$. On defining a "potential" energy as $V(\mathbf{r}) = -\hbar^2\omega^2\epsilon(\mathbf{r})/(2mc^2)$, we come to the equation $\mathcal{H} \equiv \mathbf{p}^2/(2m) + V(\mathbf{r}) = 0$, which is a direct analogue of the energy conservation law for a particle (compare this with Eq. (2.9)). That is, Eq. (2.38) describes trajectories of quanta of the wave field.

In quantum theory, instead of the wave intensity one introduces the *number of quanta* (or *photons*), $N_{q,b}$. If each of the quanta transfers the energy $\hbar\omega$, then the intensity, defined as the energy flux density, is $\mathcal{I} = \hbar\omega c N_{q,b}/V$, where V is volume and $N_{q,b}/V$ is the density of quanta with wavevector $\mathbf{q}$. Because the quantum picture has to coincide with the classical picture for a large number of photons, one can match the latter relationship and Eq. (2.36) for the classical intensity. From this comparison, it is possible to find the relationship between the classical amplitude of the electric field, F_0, and the number of quanta:

$$F_0 = \sqrt{\frac{2\hbar\omega N_{q,b}}{\epsilon_0\epsilon V}}. \tag{2.42}$$

Our discussion of the relationship between electromagnetic waves and photons provides an example of the wave–particle duality that is ubiquitous in quantum physics. A comparison of some characteristics of electromagnetic fields in the classical and quantum interpretations is given in Table 2.1. For a wave of small intensity (small number of photons, $N_{q,b}$) the quantum description is more suitable, whereas for a substantial intensity (large $N_{q,b}$) a classical wave interpretation may be used.

From particles to waves

Consider a particle, say an electron. On the basis of classical mechanics one can attempt to characterize it by a mass, m, and a vector representing the momentum, $\mathbf{p}$. In classical physics we can know with certainty that a particle is at a certain position in space, $\mathbf{r}$. From a quantum-mechanical point of view a particle is characterized by the *wavefunction*, $\psi(\mathbf{r})$. Frequently, quantum mechanics is referred to as "*wave mechanics*." The wavefunction is, in general, distributed in space. The main property of the wavefunction is that the value $|\psi(\mathbf{r})|^2\, d\mathbf{r}$ gives the probability of finding a particle inside of a small volume $d\mathbf{r}$ around point $\mathbf{r}$. Thus, the wavefunction, ψ, may be interpreted as the probability amplitude corresponding to the probability density for finding a particle at a particular point of space, $\mathbf{r}$. That is, there is a direct analogue between $|\psi|^2$ and the square of the electromagnetic wave magnitude $|F_0|^2$, which determines the photon density, as discussed previously.

Probabilistic behavior is one of the key features of quantum mechanics; thus, a word of explanation is necessary to define what is meant by "probability" in this context. To understand probability in a quantum-mechanical context, it is convenient to have in mind the following situation. Imagine an ensemble of similarly prepared systems. By "similarly prepared" we mean identical systems as far as any physical measurement is concerned. Now, if a measurement is made on one of the systems to determine whether a particle is in a particular volume element, the result will be definite: either a particle is there, or it is not. When the same measurements are made on a large number of similarly prepared systems, the number of times a particle is found in the fixed volume is taken as a measure of the probability of finding a particle in the elementary volume.

For the simplest case of a particle in free space, the wavefunction has the form of a plane wave, as introduced previously:

$$\Psi(\mathbf{r}, t) = A\mathrm{e}^{\mathrm{i}\phi} = A\mathrm{e}^{\mathrm{i}(\mathbf{kr}-\Omega t)}, \tag{2.43}$$

where $\phi = \mathbf{kr} - \Omega t$ is the phase, $\mathbf{k}$ is the wavevector of the particle, A is the amplitude of the wave, and Ω is the angular frequency associated with the energy of the particle. The wavevector, or more precisely its magnitude $k = |\mathbf{k}|$, is related to the wavelength of a particle, λ:

$$\lambda = \frac{2\pi}{k}. \tag{2.44}$$

According to the de Broglie relationship, the momentum of a particle is related to a wavelength associated with the particle through the equation

$$\lambda = \frac{2\pi\hbar}{p}. \tag{2.45}$$

From Eqs. (2.44) and (2.45) we obtain a relation between the wavenumber and the momentum of a particle: $p = \hbar k$, or, in the vector form,

$$\mathbf{p} = \hbar\mathbf{k}. \tag{2.46}$$

This relationship coincides exactly with Eq. (2.40) for quantization of electromagnetic waves. This simplest case demonstrates the way in which one may attribute to a particle both particle-like and wave-like properties.

Within the plane-wave description of a free particle, it follows from Eq. (2.43) that there is an equal probability of finding a particle in any point of space:

$$|\Psi(\mathbf{r}, t)|^2 = \Psi(\mathbf{r}, t)\Psi^*(\mathbf{r}, t) = |A|^2 = \text{constant},$$

where $\Psi^*(\mathbf{r}, t)$ is the complex conjugate of $\Psi(\mathbf{r}, t)$.

This result appears to be in contradiction with the classical description of a particle. Just as for the previously discussed case of electromagnetic fields, this contradiction can be resolved by introducing the *uncertainty principle*. This principle has the form of conditions restricting the *range* of coordinates and the *range* of momenta that can be measured simultaneously for a particle. Uncertainties in the quantities $\Delta\mathbf{p}$ and $\Delta\mathbf{r}$ have to satisfy the following inequalities:

$$\Delta p_x\,\Delta x \geq h, \qquad \Delta p_y\,\Delta y \geq h, \qquad \Delta p_z\,\Delta z \geq h. \qquad (2.47)$$

Thus, if a particle is localized in a region of space of width Δx, the uncertainty in the x-component of its momentum will be greater than or equal to $h/\Delta x$. The particle described by the wavefunction of Eq. (2.43) has a certain momentum $\mathbf{p} = \hbar\mathbf{k}$, but is completely delocalized in space.

Note that the phase of the wave ϕ in Eq. (2.43) depends on time. The angular frequency of the oscillations of this phase is related to the energy of the particle E through $\Omega \equiv E/\hbar$. The latter again coincides *exactly* with the energy of the wave quanta introduced by Eq. (2.39). This kind of temporal phase dependence $\propto e^{iEt/\hbar}$ and remains valid for any complex system under stationary conditions, including the condition of a constant external field.

Another important fact is that the superposition principle discussed previously for classical waves is valid for particle waves. Thus, typical wave phenomena such as interference and diffraction should be observed for particles. One of the requirements necessary to observe these effects is coherence of the waves participating in the superposition. In "particle language" this means, first of all, that particles should be monoenergetic. Indeed, direct experiments with monoenergetic electrons have proved the occurrence of interference and diffraction of the electron waves. Since the famous 1927 Davisson–Germer experiment on diffraction of electrons by metal crystals, numerous experiments confirming the wave nature of particles have been done. Recent (1989) experiments by Akira Tonomura with diffraction of electrons involved repeating a double-slit experiment that had been performed with light; see Fig. 2.7. In these experiments, direct confirmation of wave-like properties of electrons was obtained. The experimental setup of Tonomura's experiment is presented in Fig. 2.7(a). It consists of (i) an electron gun that emits, one by one, electrons with high velocity; (ii) an electron biprism (electron splitter); (iii) a detector of diffracted electrons; and (iv) a CCD (charge-coupled device) camera that records and displays the positions of the registered electrons. Ten electrons per second were emitted by the source. For the first several minutes the picture on the CCD screen reflected a chaotic distribution of electrons. Gradually the build-up of registered electrons

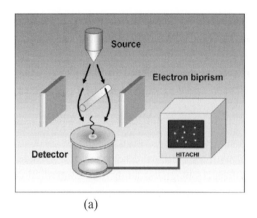 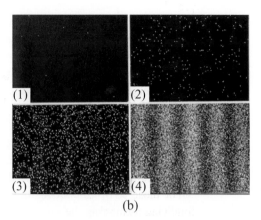

(a)   (b)

Figure 2.7 Tonomura's experiment. (a) A schematic diagram of the double-slit experiment. (b) The build-up of interference fringes at various times: pictures on the monitor after (1) 10 electrons, (2) 200 electrons, (3) 6000 electrons, and (4) 140 000 electrons had been detected. Electrons were emitted at a rate of 10 per second. Used with permission, from A. Tonomura. (2006). *Double-slit Experiment*. (http://hqrd.hitachi.co.jp/globaldoubleslit.cfm), figures 1 and 2. © Hitachi, Ltd. 1994, 2006. All rights reserved.

produced an interference pattern characteristic of experiments involving diffraction of light in a two-slit experiment.

To estimate the particle wavelength and to understand the consequences of the uncertainty principle, let us assume that a free electron moves with a velocity of about 10^7 cm s^{-1}. The mass of a free electron is $m_0 = 9.11 \times 10^{-28}$ g; thus, $p_0 = m_0 v = 9 \times 10^{-21}$ g cm s^{-1}, $k_0 = p_0/\hbar = 8.7 \times 10^6$ cm^{-1}, and the de Broglie wavelength of a free electron is $\lambda_0 = 2\pi/k_0 = 7.2 \times 10^{-7}$ cm $= 72$ Å. If we need to measure both the position and the momentum of the electron, and we impose the limit of 10% accuracy on the value of its momentum, i.e., $\Delta p_0 = 9 \times 10^{-22}$ g cm s^{-1} and $\Delta k_0 = 8.7 \times 10^5$ cm^{-1}, we cannot predict the position of this electron with an accuracy greater than $\Delta x = h/\Delta p_0 = 2\pi/\Delta k_0 = 7.2 \times 10^{-6}$ cm $= 720$ Å. This value is about ten times greater than the wavelength of the electron!

According to our estimates, we see that electron wavelengths have very small values. For a material particle with a larger mass, the wavelength is even smaller. That is why in most cases of ordinary life we do not observe wave-like behavior of particles. As we will learn later, the wavelength of the electron in solids is ten times larger than that in vacuum, i.e., up to 1000 Å $= 100$ nm (or more). This is why, for an electron in the situations encountered in nanoelectronics, wave-like physical properties are its key attributes and they determine the basic properties of this nanoworld.

2.5   Closing remarks

This chapter emphasizes a fundamental property of all existing matter, which may be called *wave–particle duality*. Starting with an analysis of an arbitrary wave field (elastic vibrations in solids, electromagnetic fields, etc.), we found particle-like behavior of this wave field at small wave amplitudes and (or) spatial scales larger than the wavelength. By analyzing particle motion, we have seen that at small spatial scales a particle can not be characterized by exact coordinates and momentum (velocity) and behaves rather as an extended wave-like object. Thus, wave properties and particle properties for the same physical object are compatible. This is the essence of wave–particle duality; indeed, this is an underlying principle of central importance for nanophysics.

Contemporary microelectronics and optoelectronics exploit electric and light phenomena, which are determined entirely by the properties of electrons in solids. Estimates of the de Broglie wavelength of electrons in solids give us a value of about 10 nm and larger, i.e., emerging nanoelectronics inevitably should be based on the wave mechanics.

For those who want to look deeper into wave–particle duality we recommend the following textbooks:

> R. P. Feynman, R. B. Leighton, and M. Sands, *Lectures on Physics*, vol. 3 (New York, Addison-Wesley, 1964).
> W. E. Lamb Jr. *The Interpretation of Quantum Mechanics*, edited and annotated by J. Mehra (Princeton, MA, Rinton Press, 2001).

The problems presented below have two aims: to help the reader to attain a better understanding of the definitions and principles stated in this chapter, and to illustrate some unusual behavior of the electrons in solids.

In the following chapters, we will introduce basic definitions and equations of quantum (wave) mechanics and analyze simple instructive examples that illustrate the main quantitative and qualitative features of nanophysics.

2.6   Problems

1.   For many semiconductor materials used in contemporary electronics, the relationship between the momentum and energy of an electron is given by the implicit formula

$$\frac{\mathbf{p}^2}{2m^*} = E\left(1 + \frac{E}{E_g}\right),$$

where m^* is the so-called effective mass of the electron and $1/E_g$ is the so-called non-parabolicity parameter. The formula has two solutions for unknown E: for electrons ($E > 0$) and another for the so-called holes ($E < 0$).

(a) Find both solutions for E.
(b) Derive the expression for $\mathbf{v}$ for electrons only.
(c) Determine the electron velocity in free space and compare it with the expression derived in (b).

As an example, consider GaAs, for which $m^* = 0.067m_0$ and $E_g = 1.42$ eV.
Note: the expression for the kinetic energy of an electron in free space is

$$E = \frac{\mathbf{p}^2}{2m_0}.$$

2. Some metals and semiconductor materials instead of having an isotropic parabolic energy dispersion, $E = \mathbf{p}^2/(2m^*)$, have anisotropic parabolic energy dispersion,

$$E = \left(\frac{1}{m^*}\right)_{ij} p_i p_j, \tag{2.48}$$

i.e., for such cases "the electron mass" is no longer a scalar, but is instead a tensor. Let the reciprocal effective-mass tensor $(1/m^*)_{ij}$ be

$$\left(\frac{1}{m^*}\right)_{ij} = \begin{pmatrix} m_t^{-1} & 0 \\ 0 & m_l^{-1} \end{pmatrix}, \tag{2.49}$$

where $i, j = x, y$. Here, the parameters m_t and m_l are the so-called transverse and longitudinal effective masses of the conduction electron. For this case, the dispersion relation (2.6) is simplified to

$$E = \frac{p_x^2}{2m_t} + \frac{p_y^2}{2m_l}. \tag{2.50}$$

For the given momentum vector $\mathbf{p} = \{|\mathbf{p}|\sin\theta, |\mathbf{p}|\cos\theta\}$

(a) calculate the velocity $\mathbf{v} = dE/d\mathbf{p}$, and
(b) plot momentum vectors and velocity vectors corresponding to momentum vectors with three values of θ, $\theta = 30°, 45°$, and $60°$, and take $|\mathbf{p}| = 1$. Consider n-Ge, for which $m_t = 0.019m_0$ and $m_l = 0.95m_0$.

Notice that the directions of the velocity vectors and momentum vectors do not coincide (i.e., they are not collinear).

3. Consider a particle of mass m, which moves along a single coordinate z. Assume that the potential force is $f = -\beta z$. Find the corresponding potential energy. Show that the general solution to Eq. (2.8) is $A\sin(\sqrt{\beta/m}\, t + \phi)$, with A and ϕ being arbitrary parameters. That is, the particle oscillates around the point $z = 0$ with the frequency $\omega = \sqrt{\beta/m}$. Particular values of parameters A and ϕ determine the magnitude and the phase of these oscillations.

4. To illustrate the method of generalization of the one-dimensional model, one can consider a square (two-dimensional) lattice with identical atoms placed at the corners of the squares. Each atom interacts with its nearest neighbors, i.e., with four atoms, as shown in Fig. 2.8. The equilibrium position of each atom is determined now by two integer numbers, say n and m. The coordinates of the atoms can be written as $\mathbf{r}_{n,m} = \{x_n, y_m\} = \{na, ma\}$. The atom with a given n and m has neighbors with coordinates $\mathbf{r}_{n-1,m}$, $\mathbf{r}_{n+1,m}$, $\mathbf{r}_{n,m-1}$, and $\mathbf{r}_{n,m+1}$. If $\mathbf{u}_{n,m}$ are displacements of the atoms from their equilibrium positions, their new vector-coordinates are $\mathbf{r}'_{n,m} = \mathbf{r}_{n,m} + \mathbf{u}_{n,m}$. Now the displacements $\mathbf{u}_{n,m}$ and the forces are vectors. According to Hooke's law, the force acting on the $\{n, m\}$th atom from its nearest neighbors is

$$\mathbf{f}_{n,m} = -\beta\big[(\mathbf{u}_{n,m} - \mathbf{u}_{n-1,m}) + (\mathbf{u}_{n,m} - \mathbf{u}_{n+1,m}) + (\mathbf{u}_{n,m} - \mathbf{u}_{n,m-1}) + (\mathbf{u}_{n,m} - \mathbf{u}_{n,m+1})\big].$$

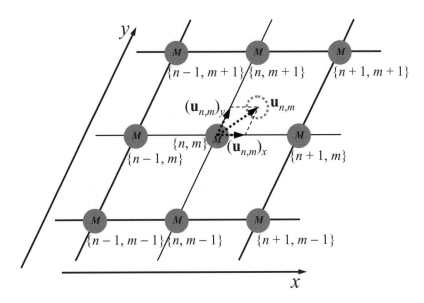

Figure 2.8 A square two-dimensional lattice.

Calculate $\mathbf{f}_{n,m}$, keeping only terms that are linear with respect to displacements $\mathbf{u}_{n,m}$. Show that the force $\mathbf{f}_{n,m}$ has a form similar to Eq. (2.11), i.e.,

$$(f_{n,m})_x = -\beta\big[(u_{n,m})_x - (u_{n+1,m})_x + (u_{n,m})_x - (u_{n-1,m})_x\big],$$

etc. Write down the Newton equations for the displacements $\mathbf{u}_{n,m}$. Using the previously discussed replacement of the discrete system by the continuous medium, obtain a two-dimensional wave equation. Find expressions for the density and the elastic modulus of the two-dimensional elastic medium considered.

5.   The simplest example of the interference effect is given by the superposition of two sinusoidal waves ($\mathbf{u}_1 = \mathbf{B}\sin\varphi_1(\mathbf{r}, t)$ and $\mathbf{u}_2 = \mathbf{B}\sin\varphi_2(\mathbf{r}, t)$). The resulting wave is a sinusoidal wave of the form $\mathbf{u}_{\text{int}} = \mathbf{B}_{\text{int}}\sin\varphi_3(\mathbf{r}, t)$.

(a) Find $\varphi_3(\mathbf{r}, t)$.
(b) Find $\mathbf{B}_{\text{int}}$.
(c) For the specific case of $\varphi_1(\mathbf{r}, t) = \tilde{\varphi}_1(\mathbf{r}) + \omega t$ and $\varphi_2(\mathbf{r}, t) = \tilde{\varphi}_2(\mathbf{r}) + \omega t$, find $\varphi_3(\mathbf{r}, t)$ and $\mathbf{B}_{\text{int}}$, and discuss the differences between your answers for this specific case and answers (a) and (b) obtained before.

6.   In addition to the solutions analyzed in Section 2.2, the wave equation (2.14) has an infinite number of solutions of other forms. Let $\mathbf{u}(z)$ be an arbitrary function that satisfies Eq. (2.14), for which first and second derivatives $d\mathbf{u}/dz$, and $d^2\mathbf{u}/dz^2$ can be calculated. Show that functions $\mathbf{u}(z \pm st)$ are the solutions to the wave equation.

7.   Using the energy density equation

$$h = \frac{\rho}{2}\overline{\left(\frac{\partial \mathbf{u}}{\partial t}\right)^2} + \frac{\Lambda}{2}\overline{\left(\frac{\partial \mathbf{u}}{\partial \mathbf{r}}\right)^2}$$

and the equation for a standing wave

$$\mathbf{u} = \mathbf{B}\sin(q_n z)\cos(\omega_n t),$$

where $|\mathbf{B}| = 1$ cm, $s = 340$ m s^{-1}, $\rho = 1.29$ kg m^{-3}, $V = L_x L_y L_z$, and $L_x = L_y = L_z = 2$ m.

(a) Calculate the energy density of the wave for $n = 1, 2,$ and 3.
(b) Calculate the total energy of the wave for $n = 1, 2,$ and 3.

8.   Consider a wave field inside of a box of dimensions L_x, L_y, and L_z. The walls of the box are supposed to reflect waves (a "mirror" box). Find three-dimensional solutions of Eq. (2.18). Calculate wavenumbers and frequencies of the standing waves. Calculate the total energy of the standing waves.

9.   The superposition principle of Eq. (2.31) allows one to present any wave field as a combination of plane waves. The function $u(t, z) = Be^{-[(z-st)/\Delta z]^2}$ is a wave pulse propagating through the medium with the velocity s. By using the Fourier transform of u, present this wave field as a sum of plane waves.

10.   Consider visible (yellow) light with the wavelength $\lambda = 600$ nm (600×10^{-9} m). Assume a light wave propagating in free space with the intensity density $\mathcal{I} = 1$ mW cm^{-2} (10^{-3} J s^{-1} cm^{-2}). Calculate the electric field amplitude F_0 (in units of V cm^{-1}). Find the wavevector, photon momentum, and energy. Estimate the density of quanta $N_{q,b}$.

3 Wave mechanics

3.1 Introduction

In this chapter we discuss the basic physical concepts and equations related to the behavior of particles in the nanoworld. We introduce the Schrödinger wave equation for particles and determine the ways to calculate observable physical quantities. We find that, in wave mechanics, the motion of a particle confined to a finite volume is always characterized by discrete values of the energy and standing-wave-like wavefunctions, i.e., such a motion is quantized. While motion in an infinite space (i.e., free motion) is not quantized and is described by propagating waves, the energy of the particle is characterized by a continuous range of values.

Keeping in mind the diverse variants of nanostructures, by using wave mechanics we analyze some particular examples, which highlight important quantum properties of particles. Many of the examples analyzed can serve as the simplest models of nanostructures and will be exploited in following chapters to understand the fundamentals of processes in nanoelectronics.

3.2 The Schrödinger wave equation

From the previous chapter, we conclude that nanosize physical systems are quantum-mechanical systems, inasmuch as their sizes are comparable to typical de Broglie wavelengths of the particles composing these systems. In dealing with quantum-mechanical systems, one aims at determining the wavefunction of a single particle or of the whole system. As we will demonstrate in the subsequent discussion, knowledge of the wavefunction in quantum mechanics is sufficient to describe completely a particle or even a system of particles. This means that, if we know the wavefunction of a system, we can, in principle, calculate all macroscopic parameters that define the properties of the system.

The wavefunction of a particle Ψ satisfies the principal equation of quantum mechanics, the *time-dependent Schrödinger wave equation*,

$$i\hbar \frac{\partial \Psi}{\partial t} - \hat{\mathcal{H}}\Psi = 0, \tag{3.1}$$

where, $\hbar$, the reduced Planck constant, $\hbar = h/(2\pi)$, was introduced earlier, and the operator $\hat{\mathcal{H}}$ is the Hamiltonian of the system:

$$\hat{\mathcal{H}} = -\frac{\hbar^2 \nabla^2}{2m} + V(\mathbf{r}). \tag{3.2}$$

In quantum mechanics, the Hamiltonian is an operator, in contrast to the case in classical mechanics, where it is a function. Let us consider the consequences of the basic equation of quantum mechanics, Eq. (3.1). The Hamiltonian operator in quantum mechanics is constructed by using the classical form of the Hamiltonian of Eq. (2.9), where the particle momentum $\mathbf{p}$ is replaced by the momentum operator $\hat{\mathbf{p}} = -i\hbar \, \partial/\partial\mathbf{r}$. Thus, the first term in Eq. (3.2) is the operator of the kinetic energy, with

$$\nabla^2 = \frac{\partial^2}{\partial x^2} + \frac{\partial^2}{\partial y^2} + \frac{\partial^2}{\partial z^2} \tag{3.3}$$

being the Laplacian operator. On comparing Eqs. (3.1) and (3.2) for particles with Eq. (2.18) for wave fields, we may notice that both equations have second derivatives with respect to the spatial variable, $\mathbf{r}$, and, correspondingly, first and second derivatives with respect to time, t. Despite the latter difference, the solutions of Eq. (3.1) are expected to be in wave-like form.

If the potential $V(\mathbf{r})$ is time-independent, one can separate the dependences on the time and spatial coordinates:

$$\Psi(\mathbf{r}, t) = e^{-iEt/\hbar} \psi(\mathbf{r}), \tag{3.4}$$

where $\psi(\mathbf{r})$ is a complex function of the spatial coordinates only. The time-dependent wavefunction, $\Psi(\mathbf{r}, t)$, is often called the *non-stationary wavefunction*, while $\psi(\mathbf{r})$ is referred to as a *stationary wavefunction*. By substituting Eq. (3.4) into Eq. (3.1), one gets the time-independent Schrödinger equation:

$$\left(-\frac{\hbar^2 \nabla^2}{2m} + V(\mathbf{r}) \right) \psi(\mathbf{r}) = E\psi(\mathbf{r}). \tag{3.5}$$

In Eqs. (3.4) and (3.5), E is the total energy of a particle. One of the important properties of solutions to Eq. (3.5) is the orthonormality of solutions $\psi_i(\mathbf{r})$ and $\psi_j(\mathbf{r})$ corresponding to different values of the energy, E_i and E_j:

$$\int \psi_i^*(\mathbf{r})\psi_j(\mathbf{r})d\mathbf{r} \propto \delta_{ij}, \tag{3.6}$$

where

$$\delta_{ij} = \begin{cases} 1, & \text{if } i = j, \\ 0, & \text{if } i \neq j. \end{cases}$$

The major task of quantum mechanics is to solve the Schrödinger wave equation, Eq. (3.1).

As we have mentioned already, the wavefunction of a particle in free space ($V(\mathbf{r}) = 0$) has a plane-wave form $\Psi(\mathbf{r}, t) = Ae^{i(\mathbf{kr}-\Omega t)}$ (Eq. (2.43)). On substituting this

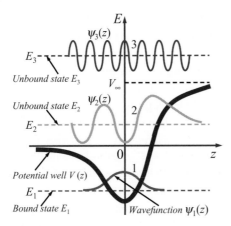

Figure 3.1 Three types of solutions of the Schrödinger equation for a one-dimensional well of arbitrary form.

wavefunction into the Schrödinger equation (3.1), one gets the relationship between the electron wavevector and its energy,

$$E = \hbar\Omega = \frac{\hbar^2}{2m}\left(k_x^2 + k_y^2 + k_z^2\right) = \frac{p^2}{2m}, \tag{3.7}$$

which coincides with the classical relationship between the particle's momentum, p, and its energy, E.

In general, the exact value of the energy E characterizes the system only for the so-called *stationary-state case*, when the potential energy and, therefore, the Hamiltonian do not depend on time. In addition, for a stationary situation, a determination of energy requires an infinite time of observation (measurement). If this time, Δt, is finite, the inaccuracy of a measurement of the energy, ΔE, should satisfy the following inequality:

$$\Delta E\, \Delta t \geq h. \tag{3.8}$$

This is the *uncertainty relation* between the energy and the time. Inequalities (2.47) and (3.8) constitute fundamental uncertainty relations in quantum physics.

Equation (3.5) has the form known as an eigenvalue equation. The energy, E, is its eigenvalue and the wavefunction, $\psi(\mathbf{r})$, is its eigenfunction. The eigenvalue E may run over discrete or continuum values, depending on the shape of the potential function, $V(\mathbf{r})$, and the boundary conditions.

In order to illustrate both possible types of solutions of the Schrödinger equation and energy states, as well as to clarify the tasks which arise, let us consider the one-dimensional problem for a system with potential energy $V(\mathbf{r}) = V(z)$ shown in Fig. 3.1. Here the vertical axis depicts the energy E, and z represents only one space coordinate. The potential has a negative minimum at $z = 0$, tends to zero as $z \to -\infty$, and saturates to a finite value V_∞ at $z \to \infty$. This potential is the most general form of a potential well.

At this point, a short qualitative discussion serves to emphasize that the boundary conditions define the type of solution. These solutions will be obtained and discussed in more detail in due course. Among the possible solutions of the Schrödinger equation (3.5) with the chosen potential, there can exist solutions with negative energy $E = E_i < 0$. The wavefunction corresponding to energy $E = E_1 < 0$ is represented in Fig. 3.1 by curve 1. One of the peculiarities of solutions with negative energy is that the spatial region with classically allowed motion, where the kinetic energy $p^2/(2m) = E - V(z) > 0$, is certainly restricted. In the classically forbidden regions where $V(z) > E$, the ψ-function decays as $|z| \to \infty$. The states of particles, like those just described, are the so-called *bound states* and they are characterized by a *discrete energy spectrum*. Consider next the possible solutions for the energy range $0 \le E \le V_\infty$ as shown by the line labeled E_2 in Fig. 3.1. These solutions exist for any values of E; they are finite as $z \to -\infty$ and penetrate slightly into the barrier region $V(z) > E$ as shown by curve 2 in Fig. 3.1. In the limit $z \to -\infty$, these solutions may be represented as a sum of two waves traveling in opposite directions: one wave is incident on, and the other is reflected from, the barrier. For each energy in the range $0 \le E \le V_\infty$ there is only one solution satisfying the physical requirements. For any energy $E > V_\infty$ there exist two independent solutions; see curve 3 corresponding to the energy E_3. One solution may be chosen in the form of the wave propagating from left to right. At $z \to \infty$ it has only one component, namely the wave overcoming the barrier, and as $z \to -\infty$ there is a superposition of incident and reflected waves. It must be emphasized that the wave reflected from the barrier, when its energy exceeds the height of the barrier, arises only in quantum physics. The other wavefunction can be chosen in the form of waves propagating from right to left. The latter is an example of continuum energy spectra with energies $E > V_\infty$ when wavefunctions are finite far away from the potential relief. These considerations demonstrate the importance of boundary conditions for Eq. (3.5): if the decay of wavefunctions (i.e., $\psi \to 0$ at $z \to \pm\infty$) away from the potential well is required, the discrete energies and the bound states can be found and they are of principal interest; if the boundary conditions correspond to the incident wave, continuous energy spectra will be obtained. To conclude the discussion of boundary conditions, let us mention that we often use potential energies with discontinuities. In this case, at the point of discontinuity of the potential, we should require *both continuity of the wavefunction and continuity of the derivative of the wavefunction with respect to coordinates*.

Since the Schrödinger equation is linear, it is clear that, if a function Ψ is a solution of the equation, then any function of the form of constant $\times \Psi$ is also a solution of the same equation. To eliminate this ambiguity, we have to take into account the probabilistic character of the wavefunction. Indeed, if a physical system is enclosed in a finite volume, the actual probability of finding a particle in this volume must equal 1, i.e.,

$$\int |\Psi(\mathbf{r}, t)|^2 \, d\mathbf{r} = 1. \tag{3.9}$$

Equation (3.9) is called the normalization condition. It provides the condition needed to determine the constant multiplicative factor of the wavefunction for the case of a system of finite size.

If an infinite volume is under consideration and the integral of Eq. (3.9) does not exist, there are other normalization conditions instead of Eq. (3.9). Consider, for example, the so-called scattering problem, in which electrons come from infinity and are scattered by a local potential. This case corresponds to energies E_2 and E_3 for particle motion in the potential sketched in Fig. 3.1. For this problem one can assume the incident wave to be a plane wave with a given amplitude A: $\Psi(z, t) = A\mathrm{e}^{\mathrm{i}kz}\mathrm{e}^{-\mathrm{i}Et/\hbar}$. Then, the amount of scattered waves will be proportional to the amplitude A, due to the linearity of the Schrödinger equation. Very often the last condition is referred to as an initial condition instead of as a boundary condition because we are dealing with the state before the scattering and states after the scattering.

In an overview of quantum mechanics it is very useful to introduce the density of particle flow, **i**. In classical physics, the density of flow is a vector, which specifies the direction of particle flow and has a modulus equal to the number of particles crossing a unit area perpendicular to the area per unit time. In quantum mechanics, this quantity is given by the following formula

$$\mathbf{i} = -\frac{\mathrm{i}\hbar}{2m}(\psi^* \nabla \psi - \psi \nabla \psi^*). \tag{3.10}$$

For example, the density of flow of particles described by the plane wave of Eq. (2.43) is straightforwardly found to be

$$\mathbf{i} = \frac{\hbar \mathbf{k}}{m}|A|^2 = \frac{\mathbf{p}}{m}|A|^2 = \mathbf{v}|A|^2.$$

Average values of physical quantities

In light of the probabilistic character of the description of quantum-mechanical systems, we have to clarify how to determine the average values of quantities that characterize such systems. The simplest case is the calculation of the average coordinate of a particle. Indeed, the absolute square of the normalized wavefunction gives the actual probability per unit volume of finding a particle at a particular point in space, as studied in Section 2.4. Hence, the average value of a particular coordinate, say z, is given by

$$\langle z \rangle = \int \psi^* z \psi \, \mathrm{d}\mathbf{r} = \int z|\psi|^2 \, \mathrm{d}\mathbf{r}. \tag{3.11}$$

Thus, the integral over space gives the mean, or *expectation* value, of coordinate z. It must be stressed again that the meaning of the expectation value is the average of a number of measurements of the coordinate z carried out over an ensemble of identical particles.

Equation (3.11) can be generalized to a more general form for the calculation of the expectation value of any observable a:

$$\langle a \rangle = \langle \hat{A} \rangle = \int \psi^* \hat{A} \psi \, \mathrm{d}\mathbf{r}, \tag{3.12}$$

where the operator $\hat{A}$ is associated with the observable a. From the definition of the expectation value, Eq. (3.12), one can see that, if ψ_a is an eigenfunction of the operator $\hat{A}$ and corresponds to a certain eigenvalue a,

$$\hat{A}\psi_a = a\psi_a, \tag{3.13}$$

then the expectation value of the physical observable a coincides with the eigenvalue: $\langle a \rangle = a$. For example, if the wavefunction is the solution of the Schrödinger equation (3.5), one may calculate the mean energy, $\langle E \rangle$:

$$\langle E \rangle = \langle \hat{\mathcal{H}} \rangle = \int \psi_E^* \hat{\mathcal{H}} \psi_E \, \mathrm{d}\mathbf{r} = E. \tag{3.14}$$

However, if a particle were in a state with no well-defined energy, say a state that is characterized by the superposition of the solutions ψ_{E_i},

$$\psi = \sum_i C_i \psi_{E_i}, \tag{3.15}$$

one would obtain the average energy in the form

$$\langle E \rangle = \sum_i |C_i|^2 E_i \quad \text{with} \quad \sum_i |C_i|^2 = 1. \tag{3.16}$$

Here, we take into account the orthogonalization and normalization conditions of Eqs. (3.6) and (3.9).

It is necessary to explain the differences between the two cases given by Eqs. (3.14) and (3.16). The first case is related to a system characterized by a wavefunction, which is an eigenfunction of the operator $\hat{A}$; in this particular case $\hat{A} = \hat{\mathcal{H}}$. The second case corresponds to the situation described by a superposition of the eigenfunctions of the same operator. Measurements of the value of the energy for the first case would reproducibly give the same result, E. In the second case, the measurements would give us different probabilistic results: energies E_i will be measured with their probabilities $|C_i|^2$, and only their average, $\langle E \rangle$, remains the same.

Importantly, by using the Schrödinger equation, straightforward calculations of derivatives of the average vector-coordinate $\langle \mathbf{r} \rangle$ with respect to time show that

$$m \frac{\mathrm{d}^2}{\mathrm{d}t^2} \langle \mathbf{r} \rangle = -\left\langle \frac{\mathrm{d}V(\mathbf{r})}{\mathrm{d}\mathbf{r}} \right\rangle,$$

i.e., the classical Newton equation (2.8) is recovered in terms of expectation values obtained using wave mechanics.

Thus, the Schrödinger equation describes the evolution of the wavefunction of the quantum-mechanical system of particles. Its solution with proper boundary or/and initial conditions gives us all the information necessary to calculate macroscopic parameters of the physical system and device operation being analyzed.

3.3   Wave mechanics of particles: selected examples

The main principles of quantum mechanics have been discussed in Section 3.2. To understand new peculiarities of particles arising due to the wave–particle duality, these basic principles of wave mechanics are applied here in several instructive examples. The simplest of these are related to the so-called one-dimensional case in which only one-dimensional solutions are considered.

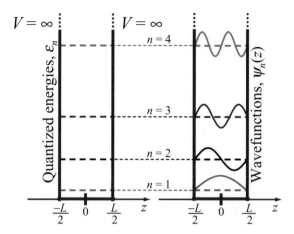

Figure 3.2 Solutions of the Schrödinger equation for a quantum well with infinite barriers. Eigenenergies, ε_n, are defined as $\varepsilon_n = n^2\varepsilon_1$, where $\varepsilon_1 = \hbar^2\pi^2/(2mL^2)$.

Of course, we live in a "three-dimensional world." Thus, wavefunctions of real particles are dependent on the three-dimensional coordinate vector **r**. However, often, the potential energy is dependent on a single coordinate, or two coordinates. For such cases, there exists a significant simplification of the wave-mechanical description. Indeed, let V be a function of the z-coordinate only. Then, we can introduce the two-dimensional wavevector $\mathbf{k}_{\|} = \{k_x,\ k_y\}$ and write the wavefunction as

$$\Psi(t, \mathbf{r}) = e^{i(k_xx+k_yy-\Omega t)}\psi(z), \tag{3.17}$$

i.e., the wavefunction is a plane wave propagating in the $\{x,\ y\}$-plane with the amplitude modulated in the z direction. By substituting Eq. (3.17) into Eq. (3.1), we find the one-dimensional equation

$$\left(\frac{\hbar^2}{2m}\frac{d^2}{dz^2} + V(z) - \varepsilon\right)\psi(z) = 0, \tag{3.18}$$

where

$$\varepsilon = E - \frac{\hbar^2k_{\|}^2}{2m}. \tag{3.19}$$

Here E, $\hbar^2k_{\|}^2/(2m)$, and ε can be interpreted as the total energy, the kinetic energy of "free motion" along the x- and y-coordinates, and the energy associated with motion in the potential $V(z)$, respectively. Equations (3.17) and (3.18) demonstrate for the case being considered that the motion along the z-coordinate is independent of motion in the two other directions and can be analyzed in terms of one-dimensional motion.

A particle between two impenetrable walls

Initially, let us consider the one-dimensional case, when a particle is placed between two impenetrable (rigid) walls at $z = \pm L/2$, as depicted in Fig. 3.2. Thus, L is the width of a one-dimensional well with infinitely high potential barriers, namely $V(z) = 0$ inside of the well and $V(z) = \infty$ outside of the well:

$$V(z) = \begin{cases} 0, & \text{for } |z| < L/2, \\ +\infty, & \text{for } |z| > L/2. \end{cases} \tag{3.20}$$

The spatially dependent part of the wavefunction $\psi(z)$ satisfies the time-independent Schrödinger equation (3.18):

$$\left(-\frac{\hbar^2}{2m} \frac{d^2}{dz^2} - \varepsilon \right) \psi(z) = 0. \tag{3.21}$$

In the general case, the solution to Eq. (3.21) is

$$\psi(z) = Ae^{iKz} + Be^{-iKz}, \tag{3.22}$$

where we have introduced

$$K = \sqrt{\frac{2m\varepsilon}{\hbar^2}}. \tag{3.23}$$

Impenetrable walls are viewed as boundaries that exclude the particle. Specifically, for the case of impenetrable walls, the probability of finding the particle at $z \le -L/2$ or $z \ge L/2$ is zero, i.e.,

$$\psi(-L/2) = \psi(L/2) = 0. \tag{3.24}$$

These are, thus, the boundary conditions for the wavefunction of a particle in a quantum well with impenetrable walls. By applying these boundary conditions to the wavefunction (3.22), we obtain

$$Ae^{-iKL/2} + Be^{iKL/2} = 0 \quad \text{for } z = -L/2,$$
$$Ae^{iKL/2} + Be^{-iKL/2} = 0 \quad \text{for } z = L/2. \tag{3.25}$$

A nontrivial solution of this algebraic system of equations exists if, and only if the following determinant is equal to zero:

$$\begin{vmatrix} e^{-iKL/2} & e^{iKL/2} \\ e^{iKL/2} & e^{-iKL/2} \end{vmatrix} = 0,$$

which results in

$$\sin(KL) = 0 \quad \text{or} \quad KL = \pi n. \tag{3.26}$$

The latter equation determines the "eigenvalues," K_n:

$$K_n = \frac{\pi}{L} n, \quad n = \pm 1, \pm 2, \pm 3, \dots; \tag{3.27}$$

$n = 0$ is excluded because for $K_0 = 0$ Eqs. (3.25) and (3.22) give $\psi(z) \equiv 0$. From Eq. (3.23) and (3.27), we find possible energies of this particle:

$$\varepsilon_n = \frac{\hbar^2 K_n^2}{2m} = \frac{\hbar^2 \pi^2}{2mL^2} n^2. \tag{3.28}$$

By substituting Eq. (3.27) into Eq. (3.25), it is easy to check that at a given n the relationship between coefficients A and B is $B = -e^{i\pi n} A = (-1)^{n+1} A$. Thus, if n is an odd integer number, we obtain symmetric wavefunctions from Eq. (3.22):

$$\psi_n(z) = \sqrt{\frac{2}{L}} \cos\left(\frac{\pi n z}{L}\right), \quad n = \pm 1, \ \pm 3, \ \ldots \tag{3.29}$$

If n is an even integer, we obtain anti-symmetric wavefunctions:

$$\psi_n(z) = \sqrt{\frac{2}{L}} \sin\left(\frac{\pi n z}{L}\right), \quad n = \pm 2, \ \pm 4, \ \ldots \tag{3.30}$$

In these functions, the factor $\sqrt{2/L}$ arises as a result of the normalization condition:

$$\int_{-\infty}^{\infty} |\psi_n(z)|^2 \, dz = 1. \tag{3.31}$$

The integer n is called the *quantum number*. Actually, we see that the energies ε_n do not depend on the sign of the quantum number n. The same is valid for the physically important quantity $|\psi_n(z)|^2$. Thus, we can use only, say, positive quantum numbers, $n > 0$. Let us write down the four lowest states explicitly:

$$n = 1, \qquad \varepsilon_1 = \frac{\hbar^2 \pi^2}{2mL^2}, \qquad \psi_1 = \sqrt{\frac{2}{L}} \cos\left(\frac{\pi z}{L}\right);$$

$$n = 2, \qquad \varepsilon_2 = 4\varepsilon_1, \qquad \psi_2 = \sqrt{\frac{2}{L}} \sin\left(\frac{2\pi z}{L}\right);$$

$$n = 3, \qquad \varepsilon_3 = 9\varepsilon_1, \qquad \psi_3 = \sqrt{\frac{2}{L}} \cos\left(\frac{3\pi z}{L}\right);$$

$$n = 4, \qquad \varepsilon_4 = 16\varepsilon_1, \qquad \psi_4 = \sqrt{\frac{2}{L}} \sin\left(\frac{4\pi z}{L}\right). \tag{3.32}$$

These solutions allow one to draw some important conclusions.

First, the energy spectrum of the particle confined by a potential well is discrete. That is, instead of a continuous change of energy – as is characteristic in classical physics – a quantum particle placed in a well may have only certain discrete energy values. In other words, the energy spectrum becomes *quantized*, as shown in Fig. 3.2. Often, one refers to this type of the spectrum as a *set of discrete energy levels*. Interestingly, for the case being considered, the interlevel distances, $\varepsilon_{n+1} - \varepsilon_n$, increase with the number of levels n.

Second, the lowest energy level (usually called the *ground state*) is not zero; it is finite. That is, the particle can not have zero energy! Actually, this is a direct consequence of the uncertainty principle. Indeed, a particle placed in a well is localized in a space region of size L, i.e., the uncertainty Δz is less than or equal to L. According to Eq. (2.47), such a localization leads to an uncertainty in the momentum $\Delta p \geq \Delta p_z \sim h/L$. Then, the momentum of the particle can be estimated by $p \geq \Delta p$, i.e., the momentum is not zero and there is a non-zero total energy.

Third, the wavefunctions are, in fact, *standing waves*, and strictly integer numbers of half-waves may exist between these impenetrable walls. This result is mathematically

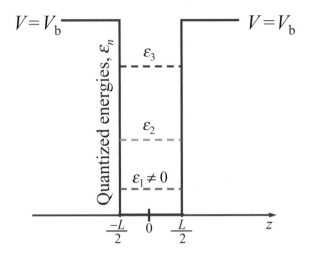

Figure 3.3 Eigenenergies of a particle in a quantum well with finite potential barriers.

similar to the case of the quantization of standing elastic waves between rigid walls depicted in Fig. 2.4.

A particle in a quantum well with finite potential barriers

Now we consider the case of particle confinement in a potential well with potential barriers of finite height. We accept the following idealized form of the potential as illustrated in Fig. 3.3:

$$V(z) = \begin{cases} 0, & \text{for } |z| \leq L/2, \\ V_b, & \text{for } |z| \geq L/2, \end{cases} \tag{3.33}$$

where V_b and L are the height and the width of the well, respectively. In this case, since the energy spectrum is quantized such a potential well is frequently referred to as a *quantum well with finite potential barriers*.

Using our classical-physics intuition, we may expect that a particle will be confined in the well if its energy $\varepsilon < V_b$. Let us focus just on this case. First, we have to solve Eq. (3.18). Inside the well, $V(z) = 0$, and the solution is a simple combination of sine and cosine functions, as in the previous case of Eqs. (3.29) and (3.30). So, it is convenient to rewrite it in a form that combines them,

$$\psi(z) = C \cos(k_w z) + D \sin(k_w z), \quad \text{for } |z| \leq L/2, \tag{3.34}$$

where

$$k_w = \sqrt{2m\varepsilon/\hbar^2}, \tag{3.35}$$

and C and D are arbitrary constants. Outside the well the solution has the form

$$\psi(z) = \begin{cases} Ae^{-\kappa_b(z-L/2)}, & \text{for } z \geq L/2, \\ Be^{\kappa_b(z+L/2)}, & \text{for } z \leq -L/2, \end{cases} \tag{3.36}$$

where $\kappa_b = \sqrt{-2m(\varepsilon - V_b)/\hbar^2}$. As a result of the symmetry of the problem, we can choose either even or odd combinations in Eqs. (3.36) and (3.34). Consequently, continuity of the wavefunction implies that $A = B$ for the even solutions and $A = -B$ for the odd solutions. Now we have two constants for odd solutions and two for even solutions. In particular, for even solutions we find

$$\psi(z) = \begin{cases} C\cos(k_w z), & \text{for } |z| \leq L/2, \\ Ae^{\mp \kappa_b(z \mp L/2)}, & \text{for } |z| \geq L/2, \end{cases} \tag{3.37}$$

where the signs "$-$" and "$+$" correspond to positive and negative values of z, respectively. The next step in finding the solution is to match the wavefunctions and their derivatives with respect to z at the points $z = \pm L/2$. For example, for even solutions we obtain from Eq. (3.37) the system of algebraic equations

$$C\cos(k_w L/2) - A = 0,$$
$$Ck_w \sin(k_w L/2) - A\kappa_b = 0. \tag{3.38}$$

This system of algebraic equations has solutions if the corresponding determinant is equal to zero. So, for even and odd solutions we obtain

$$\tan\left(\frac{k_w L}{2}\right) = \frac{\kappa_b}{k_w}, \tag{3.39}$$

$$\cot\left(\frac{k_w L}{2}\right) = -\frac{\kappa_b}{k_w}. \tag{3.40}$$

These trigonometric equations can be solved numerically, but it is more instructive to analyze them graphically. Let us transform them into the following equations:

$$\cos(k_w L/2) = \pm k_w/k_0, \quad \text{for } \tan(k_w L/2) > 0, \tag{3.41}$$
$$\sin(k_w L/2) = \pm k_w/k_0, \quad \text{for } \cot(k_w L/2) < 0, \tag{3.42}$$

where $k_0 = \sqrt{2mV_b/\hbar^2}$. The signs "$+$" and "$-$" in Eq. (3.41) are to be chosen when values of $\cos(k_w L/2)$ are positive and when they are negative, respectively. The same is valid for "$+$" and "$-$" regarding $\sin(k_w L/2)$ in Eq. (3.42) for the odd solutions.

The left-, $\mathcal{L}$, and right-, $\mathcal{R}$, hand sides of Eqs. (3.41) and (3.42) can be displayed on the same plot as the functions of k_w; see Fig. 3.4. The right-hand sides of Eqs. (3.41) and (3.42) are linear functions with slope equal to k_0^{-1}. The left-hand side is a cosine or sine function. Intersections of these two curves give us values $k_{w,n}$ for which our problem has solutions satisfying all necessary conditions.

To analyze the results, we note that the problem is characterized by two independent parameters: the height of the well, V_b, and the width, L. We can fix one of these parameters and vary the other. Let us vary the height of the well V_b, i.e., the parameter k_0. In this case the left-hand side of Eq. (3.41) and the corresponding curves $\mathcal{L}$ in Fig. 3.4 do not change, but the slope of the line, k_w/k_0, is controlled by k_0. We can see that at small k_0 (small V_b), when the slope is large, as shown by the dashed–dotted line in Fig. 3.4, there is only one solution $k_{w,1}$ corresponding to small k_0. This first solution exists at any k_0 and gives the first energy level $\varepsilon_1 = \hbar^2 k_{w,1}^2/(2m)$. As k_0 increases a new energy level emerges at

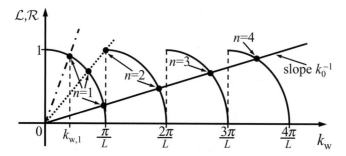

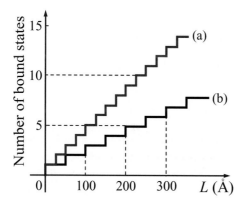

Figure 3.4 A graphical solution for Eqs. (3.41) and (3.42). Solutions correspond to the intersections of the curves. The dash–dotted line corresponds to small κ_0, resulting in only one solution of Eq. (3.41). The dotted line corresponds to a critical value of κ_0 at which a second level appears in the well as a result of the solution of Eq. (3.42). The solid line corresponds to an intermediate value of κ_0 leading to four solutions.

Figure 3.5 The number of bound states of a square well plotted as a function of the well thickness: (a) $V_b = 224$ meV, $m = 0.067m_0$; (b) $V_b = 150$ meV, $m = 0.4m_0$.

$k_w = k_0 = \pi/L$, as is shown by the dotted line in Fig. 3.4, with energy slightly below V_b, ($\varepsilon_2 \approx V_b$). When k_0 increases further, the first and second levels become deeper and a third level occurs in the well, and so on. Indeed, new levels occur when the parameter $\sqrt{(2mV_bL^2)/(\pi^2\hbar^2)}$ becomes an integer.

Figure 3.5 depicts the number of bound states as a function of well thickness for two specific values of V_b and for two particle masses. The parameters chosen are characteristic for artificial quantum wells fabricated on the basis of AlGaAs/GaAs materials (see the next chapter). One can see that for "thin" quantum wells ($L < 100$ Å) only a few energy levels can occur.

The first example we discussed is related to an infinitely deep well, when $k_0 \to \infty$. In this case, the slope of the linear function in Fig. 3.4 tends to zero, solutions correspond to $k_w = \pi n/L$, and the energy levels are given by Eq. (3.28) as discussed earlier.

Thus, as for the first example, for a quantum well with walls of finite height, we have also found discrete energy levels. The lowest energy level (the ground state) is also non-zero. However, the number of discrete levels is finite. Another important phenomenon is the *penetration of a particle* under a barrier. Equation (3.36) describes the wavefunction under a barrier. To explain this effect, consider the energy (2.9) for a classical particle: $\varepsilon = p^2/(2m) + V(z)$. Since p^2 is always positive, particle motion is possible in the spatial region with $\varepsilon - V(z) \geq 0$. For $\varepsilon < V_b$, this "classically allowed" region coincides with the region "within" the quantum well. The motion of such a classical particle is always restricted to a finite interval of the coordinates and we never find the classical particle outside of the region where its total energy is less than the potential energy: i.e., a classical particle treats any barrier as an impenetrable barrier. The quantum-mechanical analysis shows that the wavefunctions are finite at any coordinate and we can find a particle even in classically forbidden regions, or, as one says, *under the barrier*.

The effect of penetration of a particle under a barrier is the so-called *tunneling effect*. It is principally a quantum-mechanical phenomenon. The probability of finding the particle rapidly decreases as we move away from the classically allowed interval.

A confining potential with quadratic coordinate dependence

Now we will briefly analyze the very important case of particle confinement in a potential with quadratic coordinate dependence:

$$V(z) = \frac{1}{2}\frac{\mathrm{d}^2 V}{\mathrm{d}z^2} z^2 \equiv \frac{1}{2}m\omega^2 z^2, \tag{3.43}$$

where we introduce intentionally specific notation for the second derivative of the potential V. Indeed, in classical physics, the force corresponding to this potential is a linear function: $f(z) = -m\omega^2 z$ and the solutions to Newton's equation (2.8) are functions oscillating with frequency ω; see Problem 3 of Chapter 2. This type of particle motion is that of a *harmonic oscillator*. Moreover, the first non-zero term in the Taylor expansion of any potential near its minimum is quadratic with respect to displacement from its minimum as in Eq. (3.43). This is why the example of the harmonic oscillator has wide applicability.

For the harmonic oscillator potential, the time-independent Schrödinger equation (3.18) can be rewritten as

$$\frac{\mathrm{d}^2\psi}{\mathrm{d}z^2} + \frac{2m}{\hbar^2}\left(\varepsilon - \frac{m\omega^2 z^2}{2}\right)\psi = 0.$$

It is convenient to introduce the characteristic length, $z_0 = \sqrt{\hbar/(m\omega)}$. In dimensionless coordinates, $\xi = z/z_0$, the wave equation takes the simple form

$$\frac{\mathrm{d}^2\psi}{\mathrm{d}\xi^2} + \left(\frac{2\varepsilon}{\hbar\omega} - \xi^2\right)\psi = 0. \tag{3.44}$$

Consider first ψ at very large coordinates: $\xi^2 \gg 2\varepsilon/(\hbar\omega)$, when the second term in brackets dominates. Then, Eq. (3.44) reduces to

$$\frac{\mathrm{d}^2\psi}{\mathrm{d}\xi^2} - \xi^2\psi = 0. \tag{3.45}$$

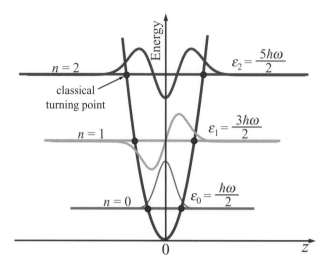

Figure 3.6 Eigenenergies and eigenfunctions of a particle in a parabolic quantum well (harmonic oscillator).

The asymmetric solution of the latter equation is proportional to $e^{\pm\xi^2/2}$. Taking into account that the solutions should be finite at $|\xi| \to \infty$, we can look for a wavefunction that will be a solution of Eq. (3.44) for all values of ξ in the form

$$\psi(\xi) = e^{-\xi^2/2} \times P(\xi).$$

It can be shown that, to satisfy the normalization condition (3.31) for wavefunctions, the unknown function $P(\xi)$ should be a polynomial of a special form, which is called a *Hermite polynomial*. This is possible only at certain values of the energy:

$$\varepsilon_n = \hbar\omega\left(n + \frac{1}{2}\right), \quad n = 0, 1, 2, \ldots \quad (3.46)$$

For each value of n, the function $P(\xi)$ can be calculated by using derivatives:

$$P_n(\xi) = (-1)^n e^{\xi^2} \frac{d^n}{d\xi^n} e^{-\xi^2}.$$

Returning to the coordinate z, we present the three lowest states as

$$n = 0, \qquad \varepsilon_0 = \frac{1}{2}\hbar\omega, \qquad \psi_0(z) = \left(\frac{m\omega}{\pi\hbar}\right)^{1/4} e^{-z^2/(2z_0^2)};$$

$$n = 1, \qquad \varepsilon_1 = \frac{3}{2}\hbar\omega, \qquad \psi_1(z) = \left(\frac{m\omega}{\pi\hbar}\right)^{1/4} \frac{z}{z_0} e^{-z^2/(2z_0^2)}; \qquad (3.47)$$

$$n = 2, \qquad \varepsilon_2 = \frac{5}{2}\hbar\omega, \qquad \psi_2(z) = \left(\frac{m\omega}{4\pi\hbar}\right)^{1/4} \left(-1 + 2\frac{z^2}{z_0^2}\right) e^{-z^2/(2z_0^2)}.$$

The quadratic potential, energy positions, and wavefunctions are shown in Fig. 3.6. Importantly, in contrast to the previous examples, the interlevel distance is strictly constant and equals $\hbar\omega$. This case can be classified as the *case of equidistant energy levels*.

As in the previous example, the particle penetrates, i.e., tunnels under, a barrier of finite value. For comparison, we indicate that the classically allowed region of motion for the particle is restricted by the two crossings of the curve $V(z)$ and the line $E = $ constant, as illustrated by Fig. 3.6.

Concluding these examples of particles placed into confining potentials, we emphasize the fundamental significance of Planck's constant, $\hbar$: all quantized energies and interlevel distances are proportional to $\hbar$. If we formally suppose that $\hbar \to 0$, the effects of energy quantization, tunneling, etc., will disappear. That is, this limit corresponds to classical mechanics.

Quantized electromagnetic waves as harmonic oscillators

The harmonic oscillator model has a number of applications. The most important is quantization of electromagnetic waves. This is one more illustration of the fundamental principle of wave–particle duality.

In Section 2.3, we described briefly the quantization of electromagnetic waves. The crossover from a classical wave characterized by the wavevector $\mathbf{q}$ and polarization ξ to its quantized analogue is based on introducing the energy quantum given by Eq. (2.39) and the photon number $N_{\mathbf{q},\xi}$. The energy of the wave must increase linearly with $N_{\mathbf{q},\xi}$. In other words, energy states of the field are equidistant, just as for the harmonic oscillator studied above. This link between the quantized field and the harmonic oscillators is developed as follows. According to quantum physics, the electromagnetic field consists of an infinite number of *modes* (waves), each of which is characterized by a wavevector, $\mathbf{q}$, and a specific polarization, ξ. Each mode $\{\mathbf{q}, \xi\}$ may be described in terms of a harmonic oscillator of frequency $\omega = cq/\sqrt{\epsilon}$. Here, c and ϵ are the speed of light and the dielectric constant, respectively. Correspondingly, the energy separation between levels of this quantum-mechanical oscillator is $\hbar\omega = (\hbar c/\sqrt{\epsilon})q$ (see Eq. (2.41)). This oscillator may be in the non-excited state, which manifests the ground-state or *zero-point vibrations* of the electromagnetic field. The existence of this zero-point energy is a purely quantum-mechanical phenomenon, as manifested by the finite lowest energy for particles in a confined potential. The oscillator may be excited to some higher energy level. Let the integer $N_{\mathbf{q},\xi}$ be a quantum number of this level; then, the energy of the electromagnetic field associated with the oscillator in mode $\{\mathbf{q}, \xi\}$ is

$$W_{\mathbf{q},\xi} V = \left(N_{\mathbf{q},\xi} + \frac{1}{2} \right) \hbar\omega, \qquad (3.48)$$

where $W_{\mathbf{q},\xi}$ is the energy density of the mode and V is the volume of the system. Equation (3.48) is the quantum analogue of the energy density given by the classical equation (2.35). The excited-level number, $N_{\mathbf{q},\xi}$, is interpreted as the number of quanta, or the number of photons in the mode under consideration. At large photon numbers, $N_{\mathbf{q},\xi} \gg \frac{1}{2}$, we immediately obtain the classical relationships given in Table 2.1.

The use of such an interpretation of wave-field quantization allows one to explain all known effects of the interaction of electromagnetic fields with matter. For example,

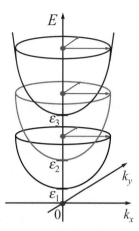

Figure 3.7 An example of two-dimensional subbands. The subbands are represented by the paraboloids in $\{E, k_x, k_y\}$-space; ε_1, ε_2, and ε_3 indicate the bottoms of subbands.

the existence of zero-point vibrations completely explains spontaneous emission from atoms, an effect that does not exist in classical physics.

Low-dimensional subbands and "low-dimensional" systems

Now we return to the analysis of quantization effects related to particles. By generalizing the examples of confined potentials previously considered, we can conclude that the energy spectrum for a potential confining a particle in one dimension is

$$E = E_{n,\mathbf{k}_\parallel} = \varepsilon_n + \frac{\hbar^2}{2m}\left(k_x^2 + k_y^2\right). \qquad (3.49)$$

The energy spectrum depends on three parameters: two continuous parameters, k_x and k_y, and the discrete quantum number, n. Such a spectrum is interpreted as a *set of energy subbands*: a set of paraboloids with minima at quantized energies ε_n, as shown in Fig. 3.7. According to Eq. (3.49), a particle characterized by a given quantum number n can move freely only in the $\{x, y\}$-plane. We can say that the particle possesses only two degrees of freedom. In such a case, the particle may be viewed as a particle that is free to move in two dimensions. The two-dimensional character of particles becomes apparent when energy intervals between subbands are large. This, in turn, is possible when the potential well has a width in the nanometer regime. Such a "quantum well" with low-dimensional particles (electrons) can be fabricated by using thin (tens of nanometers) semiconductor layers. Simple examples are just thin suspended films or free-standing structures (depending on the fabrication method), as illustrated by Fig. 3.8. Another example is a thin layer placed between two other materials. To explain this example, we will define in Section 4.5 the energy required to remove an electron from a semiconductor to a vacuum outside of the semiconductor – the so-called *electron affinity*. If a thin film possesses a large electron affinity and surrounding materials are of smaller affinities, the difference between the electron affinities explains the formation of a potential well for the electrons. Figure 3.9

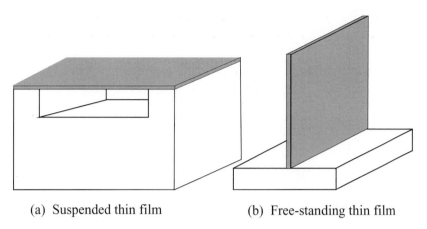

(a) Suspended thin film   (b) Free-standing thin film

Figure 3.8 Examples of suspended and free-standing thin films.

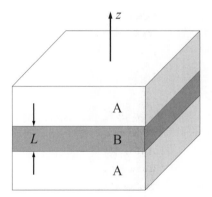

Figure 3.9 A three-layer structure where layers A create potential barriers for electrons from layer B.

illustrates a heterostructure system forming quantum wells, which are important for nanoelectronics.

Quite similarly, confinement of a particle in two directions leads to additional energy quantization and leaves only one degree of freedom for particle-wave propagation. The energy for this case is

$$E = E_{n_1,n_2,k_x} = \varepsilon_{n_1,n_2} + \frac{\hbar^2}{2m}k_x^2, \tag{3.50}$$

where the quantized energy ε_{n_1,n_2} depends on two integer quantum numbers, n_1 and n_2, and motion along the propagation direction (assumed to be the x direction) is characterized by the one-dimensional wavevector k_x. Thus, the spectrum consists of a set of one-dimensional subbands as shown in Fig. 3.10 (compare the imaginary slices of the parabaloids of Fig. 3.7 at fixed values of k_y with the parabolas of Fig. 3.10). Such a quantized particle is free in one dimension and the corresponding artificial structure is called a *quantum wire*. Currently, there exist several technological methods for the fabrication

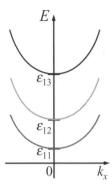

Figure 3.10 An example of one-dimensional subbands. The subbands are presented by the parabolas in $\{E, k_x\}$-space; ε_{11}, ε_{12}, and ε_{13} indicate the bottoms of subbands.

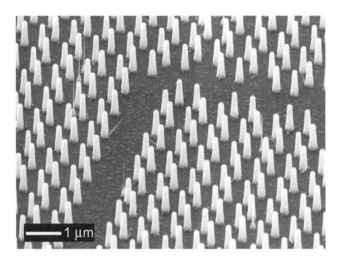

Figure 3.11 A scanning electron microscope image of free-standing InP quantum wires. Reprinted with permission from Thomas Mårtensson, Patrick Carlberg, *et al*., "Nanowire arrays defined by nanoimprint lithograph," *Nano Lett.*, **4** (4), 699–702 (2004). © American Chemical Society.

of quantum wires. For illustration, in Fig. 3.11, an array of free-standing quantum wires made from InP is presented.

In the following sections we will show that transformation of spectra from three-dimensional to low-dimensional energy subbands drastically changes the behavior of the particles.

Quantum-box, dot, and "zero-dimensional" systems

So far, we have considered examples where a particle is confined in one or two directions. This leads to quantization of the electron spectrum, resulting in two-dimensional or one-dimensional energy subbands. But there still is at least one direction for free propagation of the particle along the barriers of confining potentials. The advances in semiconductor

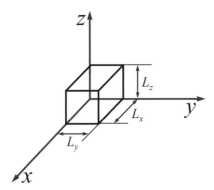

Figure 3.12 A sketch of a quantum box, $L_x \times L_y \times L_z$, embedded in a matrix.

technology allow one to go further and fabricate heterostructure-based structures, in which all existing degrees of freedom of electron propagation are quantized. These so-called *quantum dots*, or *quantum boxes*, or *zero-dimensional systems*, are much like artificial atoms, and they exhibit extremely interesting behavior.

To consider the energy spectrum of a zero-dimensional system, we have to study the Schrödinger equation (3.5) with a confining potential that is a function of all three coordinates and that confines the electron in all three directions. The simplest case is the quantum box in the form of a parallelepiped with impenetrable walls. The corresponding potential, $V(x, y, z)$, is

$$V(x, y, z) = \begin{cases} 0, & \text{inside of the box,} \\ +\infty, & \text{outside of the box,} \end{cases} \tag{3.51}$$

where the box is restricted by the conditions $0 \le x \le L_x$, $0 \le y \le L_y$, and $0 \le z \le L_z$; see Fig. 3.12. Using the results of the one-dimensional analysis discussed previously, one can write down the solutions of the Schrödinger equation for a box:

$$E_{n_1,n_2,n_3} = \frac{\hbar^2 \pi^2}{2m_0} \left(\frac{n_1^2}{L_x^2} + \frac{n_2^2}{L_y^2} + \frac{n_3^2}{L_z^2} \right), \quad n_1, n_2, n_3 = 1, 2, 3, \ldots,$$

$$\psi_{n_1,n_2,n_3}(x, y, z) = \sqrt{\frac{8}{L_x L_y L_z}} \sin\left(\frac{\pi x n_1}{L_x} \right) \sin\left(\frac{\pi y n_2}{L_y} \right) \sin\left(\frac{\pi z n_3}{L_z} \right). \tag{3.52}$$

Of fundamental importance is the fact that E_{n_1,n_2,n_3} is the total electron energy, in contrast to the previous cases, where the solution for the bound states in a quantum well and quantum wire gave us only the energy spectrum associated with the transverse confinement. Another unique feature is the presence of three discrete quantum numbers, n_1, n_2, and n_3, resulting straightforwardly from the existence of three directions of quantization. Thus, we obtain three-fold discrete energy levels and wavefunctions localized in all three dimensions of the quantum box. It is important to stress here that the elimination of a degree of freedom brings about a quantum number that substitutes for a continuous variable, i.e., a component of momentum (or wavevector) corresponding to that particular degree of freedom. In a quantum dot of a parallelepipedal shape we have three quantum numbers, n_1, n_2, and n_3, that substitute for the three components of the wavevector, $\mathbf{k}$: k_x, k_y, and k_z.

The discrete spectrum in a quantum box and the lack of free propagation of a particle in any direction are the main features distinguishing quantum boxes from quantum wells and quantum wires. As is well known, these features are typical for atomic systems as well.

The similarity with atoms is seen from another example of a potential – the spherical dot. In this case, the potential can be modelled as

$$V(\mathbf{r}) = \begin{cases} 0, & \text{for } r \leq R, \\ +\infty, & \text{for } r \geq R. \end{cases} \tag{3.53}$$

It is instructive to compare this result with Eq. (3.20). We assume that the center of the dot coincides with the center of coordinates; r is the magnitude of the radius-vector, $\mathbf{r}$, and R is the radius of the spherical dot.

As we have found above, a quantum number (continuous or discrete) can be attributed to each of the degrees of freedom of the particle. For a problem with spherical symmetry, like that given by Eq. (3.53), it is convenient to introduce the following three degrees of freedom: radial motion (along the vector coordinate) and two rotations, which can be described by two polar angles, θ and ϕ. As in the case of a parallelepiped-shaped quantum dot, the three degrees of freedom will bring about three quantum numbers. One, associated with the motion along the radius, r, substitutes for the wavevector component k_r, and the two other quantum numbers, associated with rotations, l and m, substitute for the wavevector components k_θ and k_ϕ.

The electron wavefunction in a spherical quantum dot can be found in the form

$$\psi_{n,l,m}(r, \theta, \phi) = R_n(r)Y_{lm}(\theta, \phi), \tag{3.54}$$

where $R_n(r)$ is the radial function and $Y_{lm}(\theta, \phi)$ are the so-called *spherical functions*, which are angle-dependent. We will consider the simplest case of a wavefunction with spherical symmetry, which corresponds to $Y_{00} = 1$. For spherical symmetry both quantum numbers associated with rotations are equal to zero: $l = m = 0$; this corresponds to $k_\theta = k_\phi = 0$, i.e., the wavevector, $\mathbf{k}$, is parallel to the radius, $\mathbf{r}$. On rewriting the radial function as $R_n(r) = \chi_n(r)/r$ we find the Schrödinger equation for $\chi_n(r)$ to be similar to that in Cartesian coordinates:

$$\left(-\frac{\hbar^2}{2m}\frac{\partial^2}{\partial r^2} - E \right)\chi_n(r) = 0.$$

Thus, as a result of the spherical symmetry the problem reduces formally to a one-dimensional equation. For the case in which there is an infinitely high barrier surrounding the quantum dot, the wavefunction vanishes for $r > R$ and it follows that $R_n(r) = 0$ for $r > R$. One can obtain the solutions of this equation as

$$R_n(r) = \frac{A \sin(k_w r)}{r}, \tag{3.55}$$

where $k_w = \sqrt{2m_0 E}/\hbar$ at $r < R$, $R_n(r) = 0$ at $r > R$, and A is an arbitrary constant. The condition that $R_n(r) = 0$ for $r > R$ leads to the result that $\sin(k_w R) = 0$. This equation

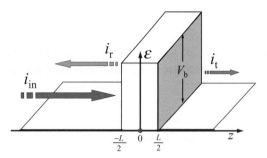

Figure 3.13 Transmission of electrons through the barrier region.

is the same as Eq. (3.26) but with KL replaced by $k_w R$. So, k_w is quantized: $k_w = \pi n / R$ (compare this with Eq. (3.27) and its analogy with Eq. (3.28)) One can obtain

$$E_{n,0,0} = \frac{\hbar^2 \pi^2 n^2}{2m R^2}, \quad n = 1, 2, \ldots, \tag{3.56}$$

where zero indexes show that the values of l and m are equal to zero. This series of levels is the same as for the one-dimensional quantum well with infinitely high barriers.

Quantum reflection, transmission and tunneling effects

Now we consider quantum effects arising for continuous energy spectra of particles. Consider the simplest potential in the form of a rectangular barrier, as shown in Fig. 3.13:

$$V(z) = \begin{cases} V_b, & \text{for } |z| \leq L/2, \\ 0, & \text{for } |z| > L/2, \end{cases} \tag{3.57}$$

where V_b, the barrier height, is greater than zero, and L is the barrier width. In such a potential, the particle is not confined. The problem should be formulated as a scattering problem: an incident particle moves, for example, from the left (from $z \to -\infty$), then it can be reflected back from the barrier or transmitted through the barrier into the region $z > L/2$. The main goal of the analysis is to find the probabilities of reflection and transmission. It is convenient to define these quantities as the ratios of reflected, i_r, and transmitted, i_t, particle fluxes to the incident flux i_{in}:

$$R(\varepsilon) = \frac{i_r}{i_{in}} \quad \text{and} \quad T(\varepsilon) = \frac{i_t}{i_{in}}. \tag{3.58}$$

The fluxes should be calculated according to the definition of Eq. (3.10).

There exist two distinct energy intervals, such that the behavior of the particle is expected to exhibit different characteristics depending on which energy interval the particle's energy is in: $0 < \varepsilon < V_b$ and $\varepsilon > V_b$. In classical physics, for the first energy interval, obviously the reflection coefficient is $R = 1$, while the transmission is $T = 0$; for the second energy interval $R = 0$ and $T = 1$.

In wave mechanics, we have to solve Eq. (3.18) with the potential (3.57). The solutions are

$$\psi(z) = \begin{cases} e^{ikz} + re^{-ikz}, & z \leq -L/2, \\ ae^{-\kappa z} + be^{\kappa z}, & -L/2 \leq z \leq L/2, \\ te^{ikz}, & z \geq L/2. \end{cases} \tag{3.59}$$

To the left of the barrier we introduce the incident wave with a unit magnitude, e^{ikz} and a reflected wave, re^{-ikz}. To the right of the barrier there is only a transmitted wave, te^{ikz}. Then, the exponential factors are

$$k = \frac{\sqrt{2m_0\varepsilon}}{\hbar} \quad \text{and} \quad \kappa = \frac{\sqrt{2m_0(V_b - \varepsilon)}}{\hbar}.$$

For $\varepsilon < V_b$, κ is a real number, whereas for $\varepsilon > V_b$, κ is an imaginary number. The parameters r, a, b, and t are still arbitrary functions of ε, which we find by matching these solutions and their derivatives at the walls of the barrier, $z = \pm L/2$. The fluxes at $z \to \pm\infty$ can be calculated in terms of r and t: $i_{in} = 1$, $i_r = |r|^2$, and $i_t = |t|^2$. Omitting the procedure for calculation of r and t, we write down the obvious relationship

$$R + T = 1,$$

and the two different results for the two energy intervals are as follows:

$$T = \frac{1}{1 + \left(\dfrac{k^2 + \kappa^2}{2k\kappa}\right)^2 \sinh^2(2\kappa L)}, \quad \varepsilon < V_b, \tag{3.60}$$

$$T = \frac{1}{1 + \left(\dfrac{k^2 - K^2}{2kK}\right)^2 \sin^2(2KL)}, \quad \varepsilon > V_b. \tag{3.61}$$

In the above formulas we used $K^2 = -\kappa^2$.

Let us consider first the case of an energy less than the barrier height, i.e., Eq. (3.60). Classical results can be obtained formally at the limit of zero Planck's constant, $\hbar$. Indeed, when $\hbar \to 0$, we obtain $\kappa \to \infty$, $\sinh(2\kappa L) \to \infty$ and $T \to 0$. That is, no transmission through the barrier occurs, as expected. In reality, this classical result is realized for high and wide barriers (V_b, $L \to \infty$). However, at finite values of these parameters, we always obtain a finite probability of particle transmission through the barrier, which is, as we discussed previously, the tunneling effect. For $\kappa L \gg 1$, $\sinh(2\kappa L) = \frac{1}{2}e^{2\kappa L}$, the second term in the denominator of Eq. (3.60) predominates and we can approximate the formula as

$$T \approx \frac{16k^2\kappa^2}{(k^2 + \kappa^2)^2} e^{-4\kappa L}. \tag{3.62}$$

Thus, the tunneling effect is determined primarily by the exponential factor. According to Eq. (3.59) for the wavefunction, the probability of finding the particle under the barrier is also exponentially small.

For the second case, corresponding to an energy greater than the barrier height, when classical physics gives strictly $T = 1$, from the quantum-mechanical equation (3.61), we

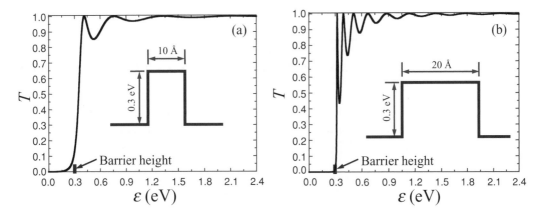

Figure 3.14 Dependences of the transmission coefficient, T, on electron incident energy, ε, for various thicknesses of the barrier, L: (a) $L = 10$ Å and (b) $L = 20$ Å. The height of the barrier $V_b = 0.3$ eV.

find that T reaches 1 only at $\sin(2KL) = 0$, i.e., at certain "resonance" energies, when $2KL = \pi n$, with n being an integer. Otherwise, $T < 1$ with minima at $2KL = (n + \frac{1}{2})\pi$. All of this means that there exists a reflection of the particle even at large energy. Figures 3.14(a) and 3.14(b) illustrate the transmission coefficient for the barrier shown in the inset. Both energy intervals, $0 < \varepsilon < V_b$ and $\varepsilon > V_b$, are presented. The transmission coefficient for $\varepsilon < V_b$ is finite due to tunneling, but it is small; whereas for $\varepsilon > V_b$ the coefficient is close to 1. Oscillations in T with energy are explained by another quantum effect – "overbarrier reflection" of the particles.

Concluding this analysis, we emphasize the great significance of quantum effects in continuous energy spectra, and that these effects determine the basic electrical and optical properties of nanostructures.

3.4 Atoms and atomic orbitals

It is instructive to compare the simple model of a spherically symmetric potential of Eq. (3.53) and the simplest hydrogen atom. This atom consists of the positive proton (nucleus) and the negatively charged electron, which interact according to the Coulomb law:

$$V(\mathbf{r}) = -\frac{1}{4\pi\epsilon_0}\frac{e^2}{r}, \tag{3.63}$$

where ϵ_0 is the permittivity of free space, e is the elementary electrical charge, and r is the distance between proton and electron. The negative sign in Eq. (3.63) indicates that the electron and the proton are attracted to each other.

The potential of Eq. (3.63), shown in Fig. 3.15, has some similarities to that of a quantum dot with impenetrable walls. This similarity allows us to suppose that the energy of the electron in the hydrogen atom would be quantized. There are two major differences

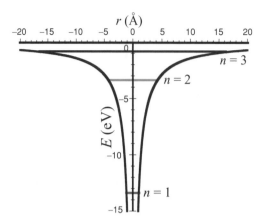

Figure 3.15 Eigenenergies of an electron in a hydrogen-atom potential well.

between these potentials: (i) the profiles of the potentials (compare Eqs. (3.53) and (3.63)) and (ii) the reference points for zero energy. Namely, the zero energy was located in the quantum dot and the electron should have infinite energy to escape into the potential walls, whereas in the hydrogen atom the energy of a free electron is taken to be zero at $r = \pm\infty$ and the energy at $r = 0$ is equal to $-\infty$. For further analysis of the hydrogen-atom model, the free-electron mass, m_0, is used in this section because the motion of nuclei can be neglected due to the fact that the mass of the nucleus is almost 2000 times greater than the free-electron mass.

The ground state of the hydrogen atom may be described by the so-called *radially symmetric function*, $\psi(r)$:

$$\psi(r) = \frac{1}{\sqrt{\pi r_0}} e^{-r/r_0}, \tag{3.64}$$

where r_0 is the characteristic radius of the ground state. It can be estimated as follows. Let us evaluate the kinetic energy of an electron confined in the sphere of radius r_0 by using the uncertainty conditions of Eq. (2.47): $p \approx \hbar/r_0$; then, the kinetic energy can be evaluated as

$$\frac{p^2}{2m_0} \approx \frac{\hbar^2}{2m_0 r_0^2}. \tag{3.65}$$

Thus, the total electron energy, $\mathcal{H}$, at $r = r_0$ is

$$\mathcal{H}(r_0) = V(r_0) + \frac{p^2}{2m_0} = -\frac{1}{4\pi\epsilon_0}\frac{e^2}{r_0} + \frac{\hbar^2}{2m_0 r_0^2}. \tag{3.66}$$

To find the r_0 that corresponds to the minimum of the total energy, $\mathcal{H}$, we find the first derivative of $\mathcal{H}(r_0)$ with respect to r_0 and equate the result to zero:

$$\frac{d\mathcal{H}(r_0)}{dr_0} = \frac{1}{4\pi\epsilon_0}\frac{e^2}{r_0^2} - \frac{2\hbar^2}{2m_0 r_0^3} = 0. \tag{3.67}$$

The solution of Eq. (3.67) gives the formula for r_0:

$$r_0 = \frac{4\pi\epsilon_0\hbar^2}{m_0 e^2},$$ (3.68)

which is called the *Bohr radius*. Now we can find the energy of the ground state E_1:

$$E_1 = -\frac{1}{4\pi\epsilon_0}\frac{e^2}{2r_0} = -\frac{m_0 e^4}{(4\pi\epsilon_0)^2 2\hbar^2}.$$ (3.69)

Numerical estimates give the following values for the energy and the radius of the ground state of the hydrogen atom: $E_1 = -13.5\,\text{eV}$ and $r_0 = 0.529\,\text{Å}$ ($1\,\text{Å} = 10^{-10}\,\text{m} = 0.1\,\text{nm}$).

An arbitrary atomic state is characterized by the following three quantum numbers, n, l, and m:

principal number $n = 1,\ 2,\ 3,\ \ldots;$
orbital number $l = 0,\ 1,\ 2,\ \ldots,\ n-1;$ or $l = \text{s, p, d,}\ \ldots;$
magnetic number $m = 0,\ \pm 1,\ \pm 2,\ \ldots,\ \pm l.$

The series of energy levels for the hydrogen atom is the following: 1s, 2s, 2p, 3s, 3p, 3d, ... The ground-state wavefunction of Eq. (3.64) corresponds to the case of $n = 1$ and $l = m = 0$. This wavefunction has the eigenenergy E_1 of Eq. (3.69). The general expression for the nth energy level of the hydrogen atom is given by

$$E_n = \frac{E_1}{n^2}.$$ (3.70)

Importantly, the energy spectrum of the hydrogen atom depends on only one discrete number n and does not depend on the orbital and magnetic numbers l and m. This means that some states of the atom can have the same energy. This situation is called *degeneracy* of the states.

Results from a number of detailed optical experiments have supported this classification of energy spectra of hydrogen atoms and confirmed the energy values with high accuracy. However, it turns out that nature is more complicated. Additional non-optical experiments revealed a new property of quantum systems, which has no analogue in classical physics.

Indeed, in classical physics a particle with angular momentum $\mathbf{l}$ can have a magnetic dipole moment $\mathbf{M} = \gamma\mathbf{l}$ with the coefficient γ called the *gyromagnetic ratio*. The latter parameter is associated with the fundamental parameters of the electrons:

$$\gamma = -\frac{e}{2m_0 c},$$ (3.71)

where e, m_0, and c are the elementary charge, free-electron mass, and speed of light, respectively. If an external magnetic field, $\mathbf{B}$, is applied, an additional potential energy of the particle in the magnetic field arises:

$$U_\text{M} = -\textbf{BM}. \tag{3.72}$$

Let the magnetic field be nonuniform, say it depends on the z-coordinate. Then the force acting on the particle will be

$$f_z = -\frac{\partial U_\text{M}}{\partial z} = \gamma M_z \frac{\partial B}{\partial z}. \tag{3.73}$$

As a result, a beam of atoms with magnetic moments will split into components corresponding to different values of M_z. In special experiments, known as Stern–Gerlach experiments, with hydrogen atoms in the ground s state, when $l = m = 0$ and thus $M_z = 0$, a splitting of the beam into *two* symmetrically deflected components has been detected. It provided evidence of the existence of a dipole magnetic moment, which we have not taken into account yet. It has been assumed that the electron has an intrinsic angular momentum which is responsible for the results of the experiment. This intrinsic angular momentum **S** is called *spin*. The projection of the spin, S_z, of an electron in the magnetic field, **B**, takes on the half-integer values $+\frac{1}{2}$ and $-\frac{1}{2}$. The spin effect explains also the results of a number of other experiments with electrons. Moreover, it has been found similarly that some spin characteristic can be attributed to any particle. For example, photons – light quanta – have a spin equal to 1.

Thus, the complete description of an energy state of the hydrogen atom should include additionally the spin number S_z. Now the series of states including the spin degeneracy can be written down as 1s(2), 2s(2), 2p(6), 3s(2), 3p(6), 3d(10), . . . The numbers in brackets indicate the degeneracies of the states. For example, the nomenclature 2p(6) describes six states of 2p type ($l = 1$) with the same energy, but with different spins ($\pm\frac{1}{2}$) and different magnetic numbers ($m = -1, 0, +1$).

Spin is a very important characteristic. It is responsible for a number of quantum effects. Moreover, it determines the character of the population of different energy levels by particles: in a system consisting of identical electrons *only two (with different spins) or fewer electrons can be found in the same state at the same time*. Since spin is included among the quantum numbers, two electrons with opposite spins and all other quantum numbers being the same should be considered as electrons in different quantum states. The Pauli exclusion principle can be reformulated as follows: **no two electrons can be in the same state**. This is the so-called *Pauli exclusion principle*. This is one more quantum principle. The Pauli exclusion principle is especially important for *many-electron* systems.

The classification of energy levels used for the hydrogen atoms may be applied to many-electron atoms. In the case of many-electron atoms, we should take the nuclear charge as Ze, where Z is the atomic number in the Periodic Table of elements (Fig. 3.16), and take into account the Pauli exclusion principle. The electronic structure of complex atoms can be understood in terms of the filling up of energy levels. One can introduce internal energy shells and external energy shells (or *valence* energy states). The latter are filled up by the outermost valence electrons and play the major role in determining the chemical behavior of complex atoms.

1	2	3	4	5	6	7	8	9	10	11	12	13	14	15	16	17	18
1 **H** 1.008																1 **H** 1.008	2 **He** 4.003
3 **Li** 6.941	4 **Be** 9.012						Transition Metals					5 **B** 10.81	6 **C** 12.01	7 **N** 14.01	8 **O** 16.00	9 **F** 19.00	10 **Ne** 20.18
11 **Na** 22.99	12 **Mg** 24.30											13 **Al** 26.98	14 **Si** 28.09	15 **P** 30.97	16 **S** 32.06	17 **Cl** 35.45	18 **Ar** 39.95
19 **K** 39.10	20 **Ca** 40.08	21 **Sc** 44.96	22 **Ti** 47.87	23 **V** 50.94	24 **Cr** 52.00	25 **Mn** 54.94	26 **Fe** 55.85	27 **Co** 58.93	28 **Ni** 58.69	29 **Cu** 63.55	30 **Zn** 65.39	31 **Ga** 69.72	32 **Ge** 72.61	33 **As** 74.92	34 **Se** 78.96	35 **Br** 79.90	36 **Kr** 83.80
37 **Rb** 85.47	38 **Sr** 87.62	39 **Y** 88.91	40 **Zr** 91.22	41 **Nb** 92.91	42 **Mo** 95.94	43 **Tc** (97.91)	44 **Ru** 101.1	45 **Rh** 102.9	46 **Pd** 106.4	47 **Ag** 107.9	48 **Cd** 112.4	49 **In** 114.8	50 **Sn** 118.7	51 **Sb** 121.8	52 **Te** 127.6	53 **I** 126.9	54 **Xe** 131.3
55 **Cs** 132.9	56 **Ba** 137.3	71 **Lu** 175.0	72 **Hf** 178.5	73 **Ta** 180.9	74 **W** 183.8	75 **Re** 186.2	76 **Os** 190.2	77 **Ir** 192.2	78 **Pt** 195.1	79 **Au** 197.0	80 **Hg** 200.6	81 **Tl** 204.4	82 **Pb** 207.2	83 **Pi** 209.0	84 **Po** (209.0)	85 **At** (210.0)	86 **Rn** (222.0)
87 **Fr** (223.0)	88 **Ra** (226.0)	103 **Lr** (262.1)	104 **Rf** (261.1)	105 **Db** (262.1)	106 **Sg** (263.1)	107 **Bh** (264.1)	108 **Hs** (265.1)	109 **Mt** (266.1)	110 (269.1)	111 (272.1)	112 (277.1)						

Lanthanides:

57	58	59	60	61	62	63	64	65	66	67	68	69	70
La	**La**	**Pr**	**Nd**	**Pm**	**Sm**	**Eu**	**Gd**	**Tb**	**Dy**	**Ho**	**Er**	**Tm**	**Yb**
138.9	140.1	140.9	144.2	(144.9)	150.4	152.0	157.2	158.9	162.5	164.0	167.3	168.9	173.0

Actinides:

89	90	91	92	93	94	95	96	97	98	99	100	101	102
Ac	**Th**	**Pa**	**U**	**Np**	**Pu**	**Am**	**Cm**	**Bk**	**Cf**	**Es**	**Fm**	**Md**	**No**
(227.0)	232.0	231.0	238.0	(237.0)	(244.1)	(243.1)	(247.1)	(247.1)	(251.1)	(252.1)	(257.1)	(258.1)	(259.1)

Figure 3.16 The Periodic Table of elements.

While the energy levels of the hydrogen atom exhibit some degree of degeneracy, i.e., they are independent of the angular momentum described by the orbital number l and its projection described by the magnetic number m, in many-electron atoms, the energies depend not only on the principal quantum number n, but also on the orbital number l. Therefore, the electrons can be subdivided into "subshells" corresponding to different orbital numbers l. The spectroscopic notations for these subshells are the same as above: s, p, d, f, ... For a given l there are $2l + 1$ values of m and two values of the spin $S_z = \pm\frac{1}{2}$. Thus, the total number of electrons that can be placed in subshell l is $2(2l + 1)$. Periods of the Periodic Table of elements are constructed according to filling of the shells. Hydrogen, H, is the first element of the table. It has one electron occupying the 1s state, i.e., $l = 0$. The next atom is helium, He. It has two electrons, which occupy the lowest s states in the same shell, but have opposite spins; this is referred to as the $1s^2$ electron configuration. So, the 1s shell is completely filled by two electrons.

The next period of the table begins with lithium, Li. It has three electrons, with the third electron sitting on the 2s-type level. Thus, we may immediately represent the electron configuration of Li as $1s^2 2s^1$. After lithium follows beryllium, Be, with the configuration $1s^2 2s^2$. In the next element boron, B, the fifth electron begins to fill up the next subshell: $1s^2 2s^2 2p^1$. For the p shell, we have $l = 1$ and $m = -1$, 0, 1. Thus, the valence electron in Be can take three values of the magnetic quantum number, m, and two values of the spin, S_z. In general, we may put six electrons in the p shell, and so on. For nanoelectronics the elements of group IV and groups III and V that constitute semiconductor crystals are important: silicon (Si, whose atomic number, and number of electrons, is 14), germanium (Ge, atomic number 32), gallium (Ga, atomic

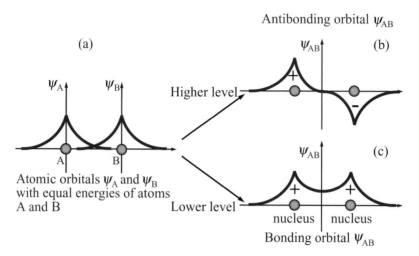

Figure 3.17 Formation of bonding and antibonding atomic orbitals (two atoms).

number 31), arsenic (As, atomic number 33), etc. Their electron configurations are as follows:

group IV: Si : $1s^2 2s^2 2p^6 3s^2 3p^2$,   Ge: $1s^2 2s^2 2p^6 3s^2 3p^6 3d^{10} 4s^2 4p^2$;

group III: Ga: $1s^2 2s^2 2p^6 3s^2 3p^6 3d^{10} 4s^2 4p^1$;   (3.74)

group V: As: $1s^2 2s^2 2p^6 3s^2 3p^6 3d^{10} 4s^2 4p^3$.

One can see that, both for Si and for Ge, the outer valence electron configurations are very similar (the same number of electrons in the same p state). Thus, we can expect that their chemical and physical properties are also similar.

Atoms with completed subshells are stable and chemically inert. Examples are helium, He, with the configuration $1s^2 2s^2$; neon, Ne, with the configuration $1s^2 2s^2 2p^6$; and argon, Ar, with the configuration $1s^2 2s^2 2p^6 3d^{10} 4s^2 4p^6$.

In general, atoms with not completely filled-up subshells, that is, those which have valence electrons, are expected to be chemically active, i.e., they can form chemical bonds. Coupling between atoms in molecules and solids is determined by the type of the electron wavefunctions. Often, these wavefunctions are called *atomic orbitals*. For example, the wavefunction of a two-atom molecule should be a combination (a hybrid) of two orbitals. Such a hybrid can be either a symmetric function, or an antisymmetric one. Let this hybrid be a symmetric function (constructive interference of the orbitals), as shown in Fig. 3.17(c). Then, there is a finite probability of finding the electron between the nuclei. The presence of electrons between two nuclei, A and B, attracts both nuclei and keeps them together. The corresponding state is called a *bonding state* of two atoms and the corresponding wavefunction, ψ_{AB}, is called the *bonding orbital*. The antisymmetric hybrid (destructive interference of the orbitals) does not bond atoms and is called an *antibonding state*, see Fig. 3.17(b). The simplest example of the bonding effect is given by the two-atom hydrogen molecule: the bonding state constructed from s states of the hydrogen atoms corresponds to an energy lower than the energy of two uncoupled

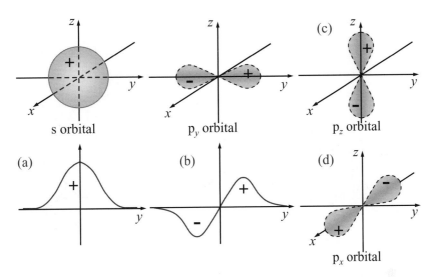

Figure 3.18 Atomic orbitals: (a) s, (b) p_y, (c) p_z, and (d) p_x.

hydrogen atoms. This results in a stable H_2 molecule. The antibonding state corresponds to an energy greater than the energy of two free hydrogen atoms. In the antibonding state the atoms repel each other.

Actually, the formation of bonding and antibonding states of two hydrogen atoms is a particular example of a more general and very important result: when two identical atoms approach each other closely, their energies change and, instead of a single atomic (degenerate) energy level, two levels arise, as illustrated in Figs. 3.17(b) and 3.17(c). This is the so-called *energy splitting*. Thus, bonding and antibonding orbitals correspond to these two levels. Three atoms drawn together are characterized by three close levels, etc.

Now, consider in detail atoms of group IV. As presented above, for these atoms the outer shell has four electrons: two in s states and two in p states, i.e., their configuration can be thought as a *core* plus the s^2p^2 shell. Thus, atoms of group IV can form four bonds with other atoms in molecules or solids (one bond for one electron). The wavefunctions of these outer-shell states have very different spatial configurations. Figure 3.18 illustrates the wavefunctions (orbitals) for an s state and for p states with different angular momentum projections m. Thus, the s orbital is spherically symmetric without any angular dependence. The 3p orbitals are anisotropic and can be considered as "perpendicular" to each other. The hybrid orbitals forming bonds with other atoms have two electrons with opposite spins. In fact, because of interaction between electrons the s and p orbitals overlap significantly and realistic bonds are formed as linear combinations of s and p orbitals, which are called sp^3 *hybridized orbitals*. Importantly, when the three directed p orbitals are hybridized with the s orbital, the negative parts of the p orbitals are almost cancelled out, so that mainly positive parts remain, as shown in Fig. 3.19(a). Thus, the hybridized orbitals present "directed" bonds in space, as shown in Fig. 3.19(b). These directed bonds of atoms are responsible for the particular crystal structures of elements

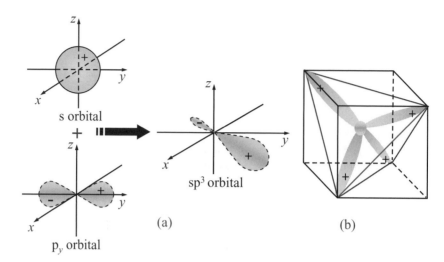

Figure 3.19 Formation of hybridized sp^3 orbitals (a) and four sp^3 hybridized orbitals in Si (b).

in group IV. In the next chapter, we will analyze these crystals using the definitions discussed above.

3.5   Closing remarks

In this chapter we have presented basic definitions and equations of quantum (wave) mechanics. We found that two cases of particle motion should be distinguished: (i) quantized motion always occurs with in a finite volume (for example, in a potential well); (ii) free motion occurs in an infinite space (practically, in a large volume). We discussed the differences in particle behavior between these two cases.

We analyzed simple instructive examples that illustrate the main qualitative features of quantum physics. Then, we briefly discussed the electronic properties of atoms, which we should use for the analysis of nanoelectronic systems. Many of the basic quantum-mechanical concepts introduced above will be applied in the following chapters to study semiconductor nanostructures and the principles of operation of nanodevices.

For those who want to look deeper into the basic principles of quantum physics we recommend the following textbooks:

L. Schiff, *Quantum Mechanics* (New York, McGraw-Hill, 1968).
D. Saxon, *Elementary Quantum Mechanics* (San Francisco, CA, Holden-Day, Inc., 1968).
R. P. Feynman, R. B. Leighton, and M. Sands, *Lectures on Physics*, vol. 3 (New York, Addison-Wesley, 1964).

The problems presented below illustrate the main definitions given in this chapter and allow one to practice with the simplest quantum-mechanical examples.

3.6   Problems

1.   Suppose that a particle is placed into a region of space of volume V. Using the normalization condition of Eq. (3.9), find the amplitude A of the plane wave (2.43).

2.   Assume that the wavefunction is a superposition of two plane waves of the same energy: $\Psi = \psi_1 + \psi_2$ with $\psi_{1,2} = A_{1,2}e^{i(\mathbf{k}_{1,2}\mathbf{r}-\Omega t)}$. Calculate the probability density of finding a particle at the point $\mathbf{r}$, i.e., calculate $|\Psi(\mathbf{r})|^2$. Discuss the interference effect and its dependence on the wavevectors $\mathbf{k}_1$ and $\mathbf{k}_2$.

3.   Assume that an electron with mass $m_0 = 9.8 \times 10^{-31}$ kg is placed in a quantum well with two impenetrable walls and that the distance between the walls is $L = 10^{-6}$ cm. Calculate the three lowest subband bottom energies: ε_1, ε_2, and ε_3. For these stationary states, find the probability density of finding the electron at the middle of the well at $z = 0$.

4.   Consider an electron, which is placed into a quantum well with width $L = 10^{-6}$ cm and barrier height $V_b = 300$ meV. Calculate the lowest energy level and compare your result with results for the previous problem. Since the probability density for finding the particle at a point z is $|\psi(z)|^2$, the probability of finding the particle in the classically forbidden regions is

$$\mathcal{P} = 2 \times \int_{L/2}^{\infty} |\psi(z)|^2 \, dz.$$

Use the solution for the lowest energy level to calculate $\mathcal{P}$. Find the number of discrete energy levels in such a quantum well.

5.   The hydrogen molecule, H_2, has the frequency of oscillations $\omega = 8.2 \times 10^{14}$ s^{-1}. The reduced mass, which determines the relative displacement of hydrogen atoms, is $m = 0.84 \times 10^{-27}$ kg, which is half of the mass of a hydrogen atom. Using the model of a harmonic oscillator, estimate the characteristic displacement, z_0, during vibrations of this molecule.

6.   In classical physics, for a system under equilibrium conditions, the equipartition principle is valid. According to this principle the *average energy* of any oscillator is $k_B T$, where $k_B = 1.38 \times 10^{-23}$ J K^{-1} is Boltzmann's constant and T is the ambient temperature. Calculate the energy of zero-point vibrations of an electromagnetic wave of wavelength 10 μm. Calculate the average number of photons, $N_{\mathbf{q},\xi}$, that will be excited with this wavelength if the ambient temperature is equal to 500 K.

7.   Consider two quantum objects, a cubic box and a spherical dot surrounded by impenetrable walls. Suppose that the volumes of the classically allowed regions are equal. Compare the ground-state levels for the box and the dot.

8. By using approximate formulae for the transmission coefficient, calculate the probability of tunneling of the conduction electron through an AlGaAs layer of thickness 20 Å embedded within a GaAs matrix, for a barrier height equal to 0.3 eV, with the electron mass in GaAs $m = 0.067m_0$, where $m_0 = 9.8 \times 10^{-31}$ kg is the mass of a free electron.

9. One of the experimental facts which constitutes one of the fundamentals of wave mechanics is the observation of discrete optical spectra (spectral lines) of emission of hydrogen atoms. Using the formulae for the energy levels of a hydrogen atom, E_n, calculate a general expression for the emission frequencies. Find the wavelengths for several lines of the so-called Lyman series, for which one of the states participating in emission is the ground state 1: $1 \leftrightarrow 2$, $1 \leftrightarrow 3$, $1 \leftrightarrow 4$, ...

4 Materials for nanoelectronics

4.1 Introduction

After the previous introduction to the general properties of particles and waves on the nanoscale, we shall now overview the basic materials which are exploited in nanoelectronics. As discussed in Chapter 1, electronics and optoelectronics primarily exploit the electrical and optical properties of solid-state materials. The simplest and most intuitive classification of solids distinguishes between dielectrics, i.e., non-conducting materials, and metals, i.e., good conducting materials. Semiconductors occupy a place in between these two classes: semiconductor materials are conducting and optically active materials with electrical and optical properties varying over a wide range. Semiconductors are the principal candidates for use in nanoelectronic structures because they allow great flexibility in the control of the electrical and optical properties and functions of nanoelectronic devices.

The semiconductors exploited in microelectronics are, in general, crystalline materials. Through proper regimes of growth, subsequent modifications and processing, doping by impurities, etc., one can fabricate nanostructures and nanodevices starting from these "bulk-like" materials.

Other physical objects that demonstrate promising properties for nanoelectronics were discovered recently, for example carbon nanotubes. These wire-like and extended objects are of a few nanometers in cross-section. They can be produced with good control of their basic properties; in particular, they can be fabricated as either semiconductors or metals. Various types of processing techniques have been shown to be viable for the fabrication of electronic nanodevices from carbon nanotubes.

In this chapter, we consider various materials that have applications for nanoelectronics. We start with the classification of dielectrics, semiconductors, and metals. Then, we define electron energy spectra, which determine the basic properties of the electrons in crystals. For nanoelectronics, a critical issue is the engineering of electron spectra, which can be realized in heterostructures. Thus, we analyze basic types of semiconductor heterostructures. Finally, we briefly describe organic semiconductors and carbon-based nanomaterials, among them carbon nanotubes and such nano-objects as fullerenes.

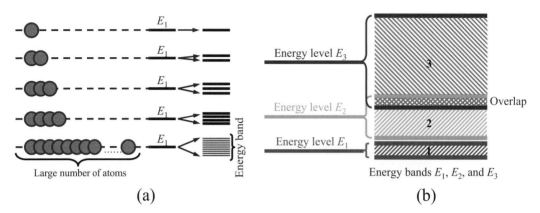

Figure 4.1 (a) Formation of an energy band. The ground energy level E_1 of a single atom evolves into an energy band when many identical atoms are interacting with each other. (b) Energy bands 1, 2, and 3 that correspond to single-atom energy levels E_1, E_2, and E_3, respectively.

4.2   Semiconductors

In every solid, electrons can be characterized in terms of their energy levels. In crystals, allowed electron energies typically have an energy band structure that may be understood as follows. In Section 3.4, we pointed out that, if two atoms, each with the same energy, come close to each other, the composite two-atom system is characterized by two close energy levels. Similarly, for a system of N atoms, every energy level of the isolated atom is split into N closely spaced levels. This assembly of close levels may be considered as an *energy band*. Figure 4.1(a) illustrates schematically the formation of such an energy band from a single atomic level. Since a single atom has a series of energy levels, the electron energies in a crystal constitute a series of energy bands that may be separated by energy gaps, or may overlap as illustrated by Fig. 4.1(b). As soon as the energy bands are formed, the electrons should be thought of as collectivized: they can no longer be attributed to specific atoms, since the energy bands characterize the whole system of N atoms.

A crucial point is the filling of the bands by electrons. We shall use the Pauli exclusion principle, which we considered in the previous chapter. That is, no two electrons can be in the same state. It is possible, as an example, for two electrons to be in the same energy state, but these electrons must be in different spin states; thus the electrons are in fact in different overall states. Under equilibrium conditions and at low ambient temperature, the lowest energy levels should be populated. As we will see later, the most important electrons are those in the upper populated bands. Then, principally, we obtain two possible cases.

First, all bands are completely filled and the filled bands are separated from the upper (empty) bands by an energy gap. This is the case illustrated by Figs. 4.2(a) and 4.2(b) for *dielectrics (insulators)* with bandgaps $E_g > 5$ eV and for semiconductors with bandgaps $E_g < 5$ eV, respectively. Actually, there is no difference between filling up energy bands for a dielectric (insulator) and for a semiconductor. The difference is in the energy gap

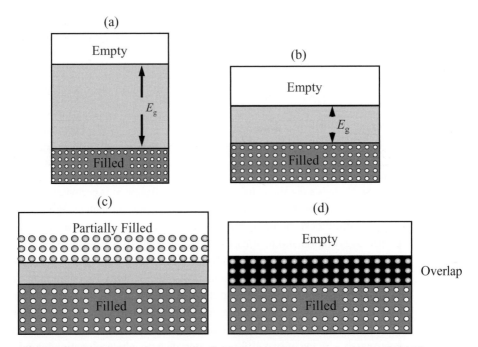

Figure 4.2 Energy bands in the cases of a dielectric, a semiconductor, and a metal. (a) The case of a dielectric: filled valence band and empty conduction band; $E_g > 5$ eV. (b) The case of a semiconductor: filled valence and empty conduction band (at low temperature); $E_g < 5$ eV. (c) Electrons in the partially filled band may gain energy from an electric field and, as a result, transfer to empty ("free") levels of higher energy, thus exhibiting electric conductivity. Accordingly, a structure with a partially filled band corresponds to a metal. (d) For overlapping bands the available electrons fill states in both bands.

between the upper filled band and the next empty band: for semiconductors this energy gap is much smaller than it is for dielectrics, as is illustrated in Figs. 4.2(a) and 4.2(b). This upper band is empty, at least at low temperature. Later, we will study the electronic band structure of semiconductor materials in more detail.

Second, the upper bands that contain electrons are not completely filled as in Figs. 4.2(c) and 4.2(d); this case corresponds to a metal. Indeed, to exhibit conductivity under an applied electric field, an electron should experience acceleration and a gain in energy. That is, an electron should be able to gain a small amount of energy and be transferred to close, but higher, energy levels. If all energy levels (a whole band) are already filled up, the electron can not participate in conduction processes. This is the case of a *dielectric (an insulator)*. In contrast, if the energy band is not completely filled and there are empty energy levels available for the electrons, in an electric field the electron can move to upper levels and gain energy in a process that corresponds to electrical conductivity. This is the case of a *metal*.

Semiconductors include elements from the central portion of the Periodic Table of elements – columns II to VI, as shown in Table 4.1. At the center of Table 4.1 is silicon, Si, the backbone material of modern electronics. Silicon plays the central role in electronics

Table 4.1 The central portion of the Periodic Table of elements

Group II	Group III	Group IV	Group V	Group VI
Be	B	C	N	O
Mg	Al	Si	P	S
Zn	Ga	Ge	As	Se
Cd	In	Sn	Sb	Te
Hg	Tl	Pb	Bi	Po

just as steel plays a dominant role in metallurgy. Below Si is germanium, Ge. Nowadays, Ge is rarely used by itself; however, Ge–Si alloys play an increasingly important role in today's electronics technology. Besides the elemental materials, contemporary electronics also uses combinations of elements from group III and group V, and combinations of elements of group II and group VI, as well as some more complicated combinations. These combinations are called *compound semiconductors*. By combining each element from group III with N, P, As, Sb, and Bi from group V, 25 different III–V compounds can be formed. The most widely used compound semiconductor is GaAs (gallium arsenide) and all III–V semiconductors are used to fabricate so-called *heterostructures*. A heterostructure is made of two different materials with a heterojunction boundary between them. The specific choice of heterostructure depends on the application.

Two or more compounds may be used to form *alloys*. A common example is aluminum gallium arsenide, $Al_xGa_{1-x}As$, where x is the fraction of group III sites in the crystal occupied by Al atoms, and $1 - x$ is the fraction of group III sites occupied by Ga atoms. Hence, now we have, not just 25 discrete compounds, but a continuous range of materials. As with the III–V compounds, every element shown in column II may be used together with every element in column VI to create II–VI compounds, and again, by combining more than two of these elements, it is possible to create a continuous range of materials. As a result, it is possible to make compositionally different IV–IV, III–V, and II–VI compounds.

4.3   Crystal lattices: bonding in crystals

We start with the definition of crystals. A *crystal* is a solid in which the constituent atoms are arranged in a certain *periodic* fashion. That is, one can introduce a basic arrangement of atoms that is repeated throughout the entire solid. In other words, a crystal is characterized by a strictly periodic internal structure. Not all solids are crystals. In Fig. 4.3, for comparison, we present a crystalline solid (a), a solid without any periodicity (a so-called *amorphous* solid) (b), and a solid in which only small regions are of a single-crystal material (a so-called *polycrystalline* solid) (c). As might be expected, crystalline materials can be the most perfect and controllable materials. Before studying periodic arrangements of atoms in crystals, we shall discuss different types of bonding in crystals.

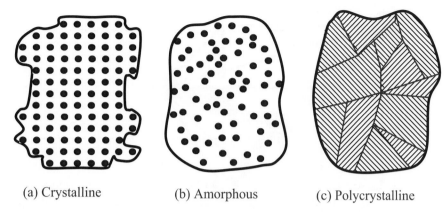

(a) Crystalline   (b) Amorphous   (c) Polycrystalline

Figure 4.3 Three types of solids: (a) ordered crystalline and (b) amorphous materials are illustrated by microscopic views of the atoms, whereas (c) a polycrystalline structure is illustrated by a more macroscopic view of adjacent single-crystalline regions, each of which has a crystalline structure as in (a).

Ionic crystals

Ionic crystals are made up of positive and negative ions. The ionic bond results primarily from attractive mutual electrostatic interaction of neighboring ions with opposite charges. There is also a repulsive interaction with other neighbors of the same charge. Attraction and repulsion together result in a balancing of forces that leads to the atoms being in stable equilibrium positions in such an ionic crystal. As for the electronic configuration in a crystal, it corresponds to a closed (completely filled) outer electronic shell. A good example of an ionic crystal is NaCl (salt). Neutral sodium, Na, and chlorine, Cl, atoms have the configurations Na^{11} ($1s^2 2s^2 2p^6 3s^1$) and Cl^{17} ($1s^2 2s^2 2p^6 3s^2 3p^5$), respectively. That is, the Na atom has only one valence electron, while one electron is necessary to complete the shell in the Cl atom. It turns out that the stable electronic configuration develops when the Na atom gives one valence electron to the Cl atom. Both of them become ions, with opposite charges, and the pair has the closed outer-shell configuration (like inert gases such as helium, He, and neon, Ne). The inner shells are, of course, completely filled both before and after binding of the two atoms. In general, for all elements with almost closed shells, there is a tendency to form ionic bonds and ionic crystals. These crystals are usually dielectrics (insulators).

Covalent crystals

Covalent bonding is typical for atoms with a low level of filling up of the outer shell. An excellent example is provided by a Si crystal. As we discussed in the previous chapter, the electron configuration of Si can be represented as core $+3s^2 3p^2$. To complete the outer $3s^2 3p^2$ shell, a silicon atom in a crystal forms *four* bonds with four neighboring silicon atoms. The symmetry of the hybrid sp^3 orbitals dictates that these neighboring atoms should be situated in the corners of a tetrahedron as shown in Fig. 4.4. Then, the central

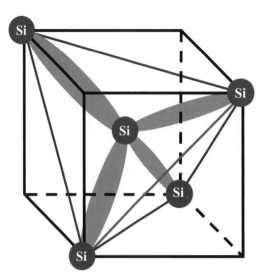

Figure 4.4 Four sp^3-hybrid bonding orbitals in a crystal of Si.

Si atom and each of its nearest-neighbor Si atoms share two electrons. This provides so-called *covalent* (chemical) bonds (symmetric combinations of the sp^3 orbitals) in the Si crystal. The four bonding sp^3 orbitals form an energy band that is completely *filled* by the valence electrons. This band is called the *valence* band. The antisymmetric combination of the sp^3 orbitals with destructive interference, as in Fig. 3.17, leads to antibonding states and to the first empty energy band. The latter band is called the *conduction* band of the Si crystal. Not surprisingly, a covalent bond of this type plays a major role for Ge, which is also from group IV of the Periodic Table of elements.

In fact, both of the types of atomic bonding discussed previously may exist simultaneously in a crystal, for example, in III–V compounds. Indeed, the electron configurations in Ga and As are core $+4s^24p^1$ (Ga) and core $+4s^24p^3$ (As). When a GaAs crystal is formed, the As atom gives one valence electron to the Ga atom, which makes them both ions. The Coulomb interaction of these ions contributes to the ionic bonding in III–V compounds. But now each ion has only two 2s electrons and two 2p electrons, which are not enough to fill the shell completely. Therefore, the rest of the bonding goes through the formation of the sp^3-hybridized orbitals. This is the covalent contribution to the crystal bonding. We can conclude that the III–V compounds are materials with mixed bonding – partly ionic and partly covalent.

Metals

Metals, such as Na, K, and Ca, consist of ions regularly situated in space. Each atom contributes an electron to an electron "sea," in which the ions are embedded. The system as a whole is neutral and stable. The electrons contribute significantly to the binding energy of metals.

Table 4.2 Binding energies for different types of crystals

Type of crystal coupling	Crystal	Energy per atom (eV)
Ionic	NaCl	7.9
	LiF	10.4
Covalent	Diamond, C	7.4
	Si	3.7
	Ge	3.7
Metallic	Na	1.1
	Fe	4.1
	Al	2.4
Molecular and	CH_4	0.1
inert-gas crystals	Ar	0.8

There exist other bonding mechanisms in solids, in molecular crystals and inert-gas crystals. However, these crystals have small binding energies, that is, they are not stable enough to be used widely in electronics. In Table 4.2, for comparison, we present binding energies per atom for different types of crystals. We can see that ionic and covalent crystals typically have binding energies in the range 1–10 eV, whereas molecular and inert-gas crystals are weakly coupled systems. Metals have intermediate coupling. The binding energy of a crystal is an important parameter, since it determines the stability of the crystal, its aging time, the applicability of various treatment processes, etc.

Crystal lattices

Now, we return to the discussion of crystal periodicity. The periodic arrangement of atoms (ions) in a crystal forms the *lattice*. The positions of atoms in the lattice are defined as the *sites*. In principle, atoms always perform small-amplitude oscillations around the sites. However, in many cases we can neglect these small-amplitude oscillations and think of a crystal as a system of regularly distributed atoms (ions). In such a perfect and periodic crystal lattice, we can identify a region called a *unit cell*. Such a unit cell is a representative of the entire lattice, since the crystal can be built by regular repeats in space of this element. The smallest unit cell is called the *primitive cell* of the lattice. The importance of the unit cell lies in the fact that by studying this representative element one can analyze a number of properties of the entire crystal. The primitive cell determines the fundamental characteristics of the crystal, including the basic electronic properties.

One of the most important properties of a perfect crystalline lattice is its *translational symmetry*. Translational symmetry is the property of the crystal's being "carried" into itself under parallel translation in certain directions and for certain distances. For any three-dimensional lattice it is possible to define three fundamental noncoplanar *primitive translation vectors* (*basis vectors*) $\mathbf{a}_1$, $\mathbf{a}_2$, and $\mathbf{a}_3$, such that the position of any lattice site can be defined by the vector $\mathbf{R} = n_1\mathbf{a}_1 + n_2\mathbf{a}_2 + n_3\mathbf{a}_3$, where n_1, n_2, and n_3 are arbitrary integers. If we construct the parallelepiped using the basis vectors, $\mathbf{a}_1$, $\mathbf{a}_2$, and $\mathbf{a}_3$, we obtain just the primitive cell.

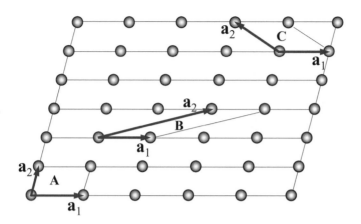

Figure 4.5 A two-dimensional lattice. Three unit cells are illustrated by A, B, and C. Two basis vectors are illustrated by $\mathbf{a}_1$ and $\mathbf{a}_2$.

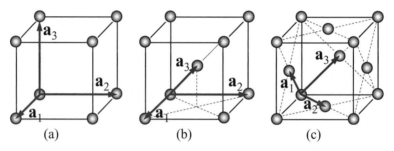

(a)   (b)   (c)

Figure 4.6 Three types of cubic lattice: (a) simple, (b) body-centered, and (c) face-centered. Basis vectors of the three different primitive cells are $\mathbf{a}_1$, $\mathbf{a}_2$, and $\mathbf{a}_3$.

Translational symmetry is illustrated in Fig. 4.5, where for simplicity we present a two-dimensional lattice of which several unit cells (A, B, and C) are shown. The cell A obviously has the smallest magnitudes of vectors $\mathbf{a}_1$ and $\mathbf{a}_2$. Thus, the basis vectors of the primitive cell are the vectors $\mathbf{a}_1$ and $\mathbf{a}_2$ from the unit cell A. An arbitrary lattice site is at the point $\mathbf{R} = n_1\mathbf{a}_1 + n_2\mathbf{a}_2$ with integers n_1 and n_2. To visualize the translational symmetry of this lattice one can start from any point of the lattice and find all other equivalent positions in space by just applying translations that are integer multiples of the basis vectors.

Since many of the crystals used in electronics are of the so-called *cubic symmetry*, here we consider briefly such cubic lattices. For them, the unit cell may be selected in the form of a cube. There are three different types of cubic lattices. The *simple cubic* lattice has atoms located at each corner of the cube, as shown in Fig. 4.6(a). The *body-centered* cubic lattice has an additional atom at the center of the cube, as shown in Fig. 4.6(b). The third type is the *face-centered* cubic lattice, which has atoms at the corners and at the centers of the six faces as depicted in Fig. 4.6(c).

Specifically, the basic lattice structure for diamond, C, silicon, Si, and germanium, Ge, is the so-called *diamond lattice*. The diamond lattice consists of two face-centered cubic structures with the second structure being shifted by a quarter of a diagonal of the

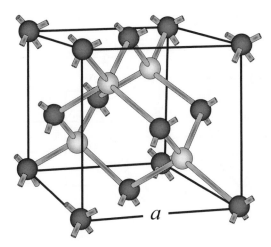

Figure 4.7 The diamond lattice is of the face-centered type of cubic lattices. The tetrahedral bonding arrangement of neighboring atoms is clear.

first cube, or by a distance $\mathbf{a}_1/4 + \mathbf{a}_2/4 + \mathbf{a}_3/4$ from the first face-centered structure; here vectors $\mathbf{a}_1$, $\mathbf{a}_2$, and $\mathbf{a}_3$ are vectors of the first cube as defined in Fig. 4.6(a). In Fig. 4.7 the atoms of the first face-centered structure are shown in black and some atoms of the second one are gray. Thus, a diamond lattice contains twice as many atoms per unit volume as does a face-centered cubic lattice. The four nearest-neighbor atoms to each atom are shown in complementary shading for easier visualization. The parameter a which characterizes the cubic lattice is the so-called *lattice constant*. The lattice of volume $V = a^3$ consists of eight atoms.

Besides the translations, the crystal symmetry contains other symmetry elements, for example, specific rotations around high-symmetry axes. In cubic crystals, axes directed along the basis vectors are equivalent and they are the symmetry axes. It is convenient to use a system of coordinates built on the basis vectors. Three symmetry axes may be denoted as [100], [010], and [001]. (Here we have used notations that are common in crystallography: [100] is a unit vector in the x direction, [010] in the y direction, and [001] in the z direction. All important symmetry directions of a cube are shown in Fig. 4.8(a). The directions of the type [110] and [111] are also important crystal directions. It is evident that these directions are equivalent to the opposite ones, [$\bar{1}\bar{1}0$] and [$\bar{1}\bar{1}\bar{1}$], as well as to analogous ones. In Fig. 4.8(a), we show the symmetry directions of a cubic crystal. If the crystal is carried into itself on rotation through an angle $2\pi/n$ about some axis that passes through the crystal, then this axis is said to be an n-fold axis C_n. For example, in a cubic crystal there are three four-fold axes C_4 and four three-fold axes C_3, as shown in Fig. 4.8(b). The symmetry elements of the lattice make analysis of crystals' properties much simpler.

4.4 Electron energy bands

Summarizing the above analysis, we conclude that a crystal consists of nuclei and electrons. The valence electrons are collectivized by all nuclei and we can expect them to

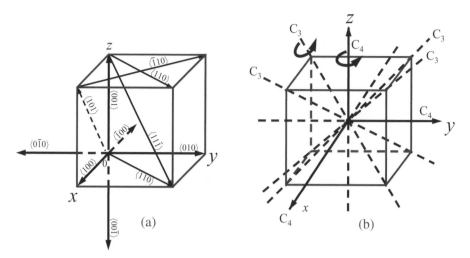

Figure 4.8 (a) Symmetry directions for cubic crystals; (b) rotational symmetry directions in cubic crystals.

be relatively weakly coupled to atoms. This allows one to think about a crystal as a system with two relatively independent subsystems: the atom (ion) subsystem and the electron subsystem. In this section, we will consider the electron subsystem of crystals. Actually, in a crystal an electron moves in the electrostatic potential created by positively charged ions and all other electrons. This potential is frequently referred to as the *crystalline potential*, $W(\mathbf{r})$. We will consider one-particle states for electrons in an ideal crystal and we will present a classification of these states and find a general form for the wavefunctions and energies.

An electron in a crystalline potential

For an ideal crystal the crystalline potential is periodic with the period of the crystalline lattice. Let $\mathbf{a}_j$, with $j = 1, 2$, and 3, be the three basis vectors of the lattice that define the three primitive translations. The periodicity of the crystalline potential, $W(\mathbf{r})$, implies that

$$W\left(\mathbf{r} + \sum_{j=1}^{3} n_j \mathbf{a}_j\right) = W(\mathbf{r}), \tag{4.1}$$

where $\mathbf{r}$ is an arbitrary point of the crystal and n_j are some integers. The one-particle wavefunction should satisfy the time-independent Schrödinger equation

$$\hat{\mathcal{H}}\psi(\mathbf{r}) = \left(-\frac{\hbar^2}{2m_0}\nabla^2 + W(\mathbf{r})\right)\psi(\mathbf{r}) = E\psi(\mathbf{r}), \tag{4.2}$$

where m_0 is the free-electron mass and $\hat{\mathcal{H}}$ is the crystalline Hamiltonian. Equation (4.2) neglects interactions between electrons and that is why the wavefunction, $\psi(\mathbf{r})$, is often called the *one-particle wavefunction*.

Because of the potential periodicity of Eq. (4.1), the wavefunctions $\psi(\mathbf{r})$ may be classified and presented in a special form. To find this form we introduce the translation operator, $\hat{T}_\mathbf{d}$, that acts on the coordinate vector $\mathbf{r}$ as

$$\hat{T}_\mathbf{d}\mathbf{r} = \mathbf{r} + \mathbf{d}, \quad \mathbf{d} = \sum_{j=1}^{3} n_j \mathbf{a}_j. \tag{4.3}$$

On applying this operator to the wavefunction we find that the function $\hat{T}_\mathbf{d}\psi(\mathbf{r}) \equiv \psi(\mathbf{r} + \mathbf{d})$ is also a solution of Eq. (4.2) for the same energy E. Let us assume that the electron state with energy E is not degenerate. Then, we conclude that the two wavefunctions $\psi(\mathbf{r})$ and $\psi(\mathbf{r} + \mathbf{d})$ can differ only by some multiplier, $C_\mathbf{d}$:

$$\psi(\mathbf{r} + \mathbf{d}) = C_\mathbf{d}\psi(\mathbf{r}). \tag{4.4}$$

From the normalization condition

$$\int |\psi(\mathbf{r} + \mathbf{d})|^2 \, d\mathbf{r} = 1 \tag{4.5}$$

we obtain $|C_\mathbf{d}|^2 = 1$. Two different translations $\mathbf{d}_1$ and $\mathbf{d}_2$ should lead to the same result as the single translation $\mathbf{d} = \mathbf{d}_1 + \mathbf{d}_2$, i.e., $C_{\mathbf{d}_1} C_{\mathbf{d}_2} = C_{\mathbf{d}_1+\mathbf{d}_2}$. From this result it follows that $C_\mathbf{d}$ may be represented in an exponential form:

$$C_\mathbf{d} = e^{i\mathbf{k}\mathbf{d}} = \exp\left(i\mathbf{k}\sum_{j=1}^{3} n_j \mathbf{a}_j\right), \tag{4.6}$$

where $\mathbf{k}$ is a constant vector. Thus, from Eq. (4.4) we get the wavefunction in *Bloch form*,

$$\psi(\mathbf{r}) = e^{-i\mathbf{k}\mathbf{d}}\psi(\mathbf{r} + \mathbf{d}) = e^{i\mathbf{k}\mathbf{r}}u_\mathbf{k}(\mathbf{r}), \tag{4.7}$$

where

$$u_\mathbf{k}(\mathbf{r}) = e^{-i\mathbf{k}(\mathbf{r}+\mathbf{d})}\psi(\mathbf{r} + \mathbf{d}). \tag{4.8}$$

One can check that the so-called *Bloch function* $u_\mathbf{k}(\mathbf{r})$ is a periodic function:

$$u_\mathbf{k}(\mathbf{r} + \mathbf{d}') = u_\mathbf{k}(\mathbf{r}), \quad \mathbf{d}' = \sum_{j=1}^{3} n_j \mathbf{a}_j.$$

Therefore, the stationary one-particle wavefunction in a crystalline potential has the form of a plane wave modulated by the Bloch function with the lattice periodicity. The vector $\mathbf{k}$ is called the *wavevector of the electron* in the crystal. This wavevector is one of the quantum numbers of electron states in crystals.

By applying the so-called *cyclic boundary conditions* to the crystal with a number of periods N_j along the direction $\mathbf{a}_j$,

$$\psi(\mathbf{r} + N_j\mathbf{a}_j) = \psi(\mathbf{r}), \quad N_j \to \infty, \tag{4.9}$$

we find for $\mathbf{k}$

$$\mathbf{k}\mathbf{a}_j N_j = 2\pi n_j, \quad n_j = 1, 2, 3, \ldots, N_j. \tag{4.10}$$

These allowed quasi-continuum values of $\mathbf{k}$ form the so-called *first Brillouin zone* of the crystal. They are just those energy bands which we discussed above by using simple qualitative considerations. It is important that the symmetry of the Brillouin zone in **k**-space is determined by the crystal symmetry.

Let the one-particle energy corresponding to the wavevector **k** be $E = E(\mathbf{k})$. If the wavevector changes within the Brillouin zone, one gets a continuum energy band; i.e., *an electron energy band*. At fixed **k**, the Schrödinger equation (4.2) has a number of solutions in the Bloch form:

$$\psi_{\alpha,\mathbf{k}}(\mathbf{r}) = \frac{1}{\sqrt{V}} e^{i\mathbf{k}\mathbf{r}} u_{\alpha,\mathbf{k}}, \tag{4.11}$$

where α enumerates these solutions and, thus, the energy bands. Owing to the crystal's periodicity, the Bloch function can be calculated within a single primitive cell. In Eq. (4.11) we normalize the wavefunction $\psi_{\alpha,\mathbf{k}}$ for the crystal volume $V = NV_0$; $N = N_1 \times N_2 \times N_3$ and V_0 are the number and volume of the primitive crystal cell, respectively. From the normalization of the wavefunction $\psi_{\alpha,\mathbf{k}}(\mathbf{r})$ one obtains

$$\frac{1}{V_0} \int_{V_0} |u_{\alpha,\mathbf{k}}|^2 \, d\mathbf{r} = 1, \tag{4.12}$$

where the integral is calculated over the primitive cell. This formula allows one to estimate the order of the value of $u_{\alpha,\mathbf{k}}$: $|u_{\alpha,\mathbf{k}}| \approx 1$.

Thus, through this analysis we have established an extremely important property of the electrons in crystalline solids: despite the interaction of an electron with atoms and other electrons, in a perfect lattice the electron behaves much like a free electron. The electron can be characterized by a wavevector **k** and, thus, it possesses the *momentum $\hbar\mathbf{k}$*. By considering phenomena that have spatial scales much greater than distances between atoms (ions) in the primitive cell, we may omit the Bloch function $u_{\alpha,\mathbf{k}}$ and describe the electrons by a wavefunction in the form of the plane wave $\psi_{\mathbf{k}}(\mathbf{r}) = A \exp(i\mathbf{k}\mathbf{r})$, just as for a free particle. However, the wavevector changes inside the Brillouin zone in a manner that is specific for a given crystal and, in general, the energy dispersions $E = E_{\alpha}(\mathbf{k})$ can differ from that of the free electron considerably.

The holes

According to the discussion given in Section 4.2, some of the energy bands are completely filled, while the others are almost empty. For our purposes, two of the bands are of great importance: the upper filled band and the lowest empty band. They are called the *valence*, $E_v(\mathbf{k})$, and the *conduction*, $E_c(\mathbf{k})$, band, respectively.

One of the ways to get an electron into the conduction band is to transfer an electron from the valence band to the conduction band. Thus, for analysis of the valence band, it is useful to adopt the *concept of a hole* as a new quasiparticle; i.e., by a hole we refer to an electron missing from the valence band. These quasiparticles can be introduced and described on the basis of simple considerations. If the valence band is full, the total wavevector of all electrons in the valence band is zero:

$$\mathbf{k}_v = \sum_i \mathbf{k}_i = 0, \tag{4.13}$$

where the sum accounts for all occupied valence states. Let us assume that one of the electrons with wavevector $\mathbf{k}_e$ is removed from the valence band. The total wavevector of the valence electrons becomes

$$\mathbf{k}_v = \sum_i \mathbf{k}_i = -\mathbf{k}_e. \tag{4.14}$$

On the other hand, removing this electron is identical to the creation of a hole in the valence band. One can attribute the wavevector of Eq. (4.14) to this hole: $\mathbf{k}_h = -\mathbf{k}_e$. Then, the energy of the valence electrons decreases by the term $E_v(\mathbf{k}_e)$, and, thus, one can also attribute the energy $E_h(\mathbf{k}_h) = -E_v(\mathbf{k}_e)$ to this hole. If the energy band is symmetric; i.e., $E_v(\mathbf{k}) = E_v(-\mathbf{k})$, we can write for the hole energy

$$E_h(\mathbf{k}_h) = -E_v(\mathbf{k}_e) = -E_v(-\mathbf{k}_e) = -E_v(\mathbf{k}_h). \tag{4.15}$$

Thus, we can characterize the hole by a wavevector, $\mathbf{k}_h$, and an energy, $E_h(\mathbf{k}_h)$, and consider the hole as a new quasiparticle created when the electron is removed from the valence band. In the conduction band, the electron energy, $E_c(\mathbf{k})$, increases as the wavevector, $\mathbf{k}$, increases. Conversely, in a valence band, near the maximum energy of the band the electron energy, $E_v(\mathbf{k})$, decreases as $\mathbf{k}$ increases. However, according to Eq. (4.15), the hole energy increases with the hole wavevector, $\mathbf{k}_h$. That is, the hole behaves as a usual particle. Thus, one can introduce the velocity of the hole, $v_h = \partial E_h(\mathbf{k}_h)/\partial \mathbf{k}_h$, and then employ Newton's laws, etc. The absence of a negative charge in the valence band brought about when an electron is removed makes it possible to characterize a hole by a positive elementary charge; that is, *the holes carry positive electrical charge*.

It is worth emphasizing that the similarity between the electrons and holes is not complete: the holes exist as quasiparticles only in a crystal, whereas the electrons exist also in other physical media, as well as in the vacuum.

Symmetry of crystals and properties of electron spectra

Usually, the energy dispersion relations, $E_\alpha(\mathbf{k})$, are very complex and can be obtained only numerically in the context of approximate methods.

Fortunately, the Brillouin zone possesses a symmetry which directly reflects the symmetry of the unit cell of the crystal in coordinate space. If a crystal is mapped into itself due to transformations in the form of certain rotations around the crystalline axes and of mirror reflections, one can speak about the *point symmetry* of directions in the crystal. In the Brillouin zone, this symmetry generates several points with high symmetry with respect to the transformations of the zone in $\mathbf{k}$-space. The extrema of the energy dispersion $E_\alpha(\mathbf{k})$ always coincide with these high-symmetry points. In particular, this fact allows one to simplify and solve the problem of obtaining electron spectra. Near extrema, the energy spectra can be approximated by expansion of $E_\alpha(\mathbf{k})$ in series with respect to deviations from the symmetry points. Such an expansion can be characterized by several constants which define the reciprocal effective mass tensor:

$$\left(\frac{1}{m^*}\right)_{ij} = \begin{pmatrix} m_{xx}^{-1} & m_{xy}^{-1} & m_{xz}^{-1} \\ m_{yx}^{-1} & m_{yy}^{-1} & m_{yz}^{-1} \\ m_{zx}^{-1} & m_{zy}^{-1} & m_{zz}^{-1} \end{pmatrix}, \tag{4.16}$$

where i and j denote x-, y-, and z-coordinates.

Table 4.3 Symmetry points in group IV semiconductors and III–V compounds

Symmetry point	Position of extremum in k-space	Degeneracy
Γ	0	1
L	$\pm(\pi/a)[111]$, $\pm(\pi/a)[\bar{1}11]$, $\pm(\pi/a)[1\bar{1}1]$, $\pm(\pi/a)[11\bar{1}]$	4
Δ	$\pm\gamma(2\pi/a)[100]$, $\pm\gamma(2\pi/a)[010]$, $\pm\gamma(2\pi/a)[001]$, $|\gamma| < 1$	6
X	$\pm(2\pi/a)[100]$, $\pm(2\pi/a)[010]$, $\pm(2\pi/a)[001]$	3

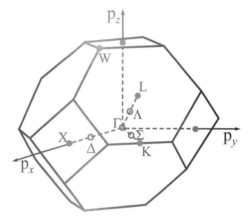

Figure 4.9 The first Brillouin zone of cubic crystals. Points of high symmetry, Γ, L, and X points, are shown.

Thus, due to the crystal symmetry, the problem of finding an electron energy spectrum $E_\alpha(\mathbf{k})$ is reduced to the following steps: (a) determination of the high-symmetry $\mathbf{k}$ points of the Brillouin zone; (b) calculation of energy positions of extrema; and (c) analysis of effective masses or other parameters of an expansion of $E_\alpha(\mathbf{k})$ within extrema.

The structure and symmetry of the Brillouin zone for cubic crystals of group IV semiconductors and of III–V compounds are very similar. Figure 4.9 shows the Brillouin zone of these semiconductor materials. The symmetry points are shown in Fig. 4.9 and presented in Table 4.3. Evidently, as a result of crystal symmetry, several points have the same symmetry; indeed, they are mapped into themselves under proper symmetry transformations. Such a degeneracy of the symmetry points is indicated in Table 4.3. In particular, the Γ, L, X, and Δ points are of central importance. They give the positions of the extrema of the electron energy in III–V compounds, Ge, and Si.

The bandstructures for GaAs and Si are presented in Fig. 4.10. The energy dispersions along two symmetric directions of the wavevectors [111] (from Γ to L) and [100] (from Γ to X) are shown. In each case, the energy, E, is taken to be zero at the top of the highest valence band, which is located at the Γ point for both of these materials. Since

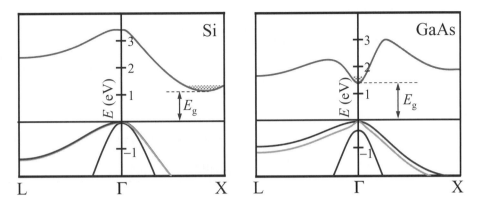

Figure 4.10 Bandstructures of Si and GaAs: the bandgap of Si is $E_g = 1.12$ eV; and the bandgap of GaAs is $E_g = 1.42$ eV.

electrons tend to be near the energy minima, one can think of electrons as being located inside marked regions of **k**-space. Frequently, these regions in **k**-space are referred to as *energy valleys*, or simply *valleys*. Materials with several valleys are called *many-valley semiconductors*. For III–V compounds there is only one valley around the point **k** = 0 (the Γ valley); however, in the case of Si there exist six Δ valleys in accordance with the degeneracy of the Δ points. It is worth emphasizing that also other symmetry points can play a considerable role in processes occurring far from equilibrium. The highest valence band and the lowest conduction band are separated by *energy bandgaps*, E_g. Let us postpone the analysis of valence bands for a while and focus on conduction bands. Conduction bands have different structures for the groups of materials considered in this section. The main difference is that for Si and Ge the lowest minima are located at Δ and L points, respectively, whereas for most III–V compounds there is only one lowest minimum, which is at the Γ point. The difference is not simply quantitative; indeed, it is qualitative and leads to a series of important consequences in the behavior of electrons.

As discussed previously, within the Γ, Δ, X, and L points the electron dispersion curves, $E_\alpha(\mathbf{k})$, can be expanded in series with respect to deviations from the minima, $\mathbf{k}_\beta$:

$$E_\alpha(k) = E(k_\beta) + \frac{1}{2}\left(\frac{\hbar^2}{m_\beta^*}\right)_{ij}(k_i - k_{i,\beta})(k_j - k_{j,\beta}). \qquad (4.17)$$

For the Γ point, we have the simplest case of the so-called *isotropic effective mass*, m^*:

$$\left(\frac{1}{m_\Gamma^*}\right)_{ij} = \frac{1}{m^*}\delta_{ij}. \qquad (4.18)$$

For Δ, X, and L points, i.e., for Δ, X, and L valleys, the reciprocal effective-mass tensor has only two independent components, corresponding to the longitudinal, m_l, and transverse, m_t, masses. For example, in the case of Si, for a coordinate system with the

Table 4.4 Energy-band parameters for Si and Ge

Group IV	Si	Ge
Type of bandgap	Indirect	Indirect
Lowest minima	Δ points	L points
Degeneracy	6	4
E_g (eV)	1.12	0.664
Electrons		
m_l/m_0	0.98	1.64
m_t/m_0	0.19	0.082
Holes		
m_{hh}/m_0	0.50	0.44
m_{lh}/m_0	0.16	0.28
Δ_{so} (eV)	0.044	0.29

z-axis along the axis where the Δ valley under consideration is located and the two other axes perpendicular to the first, one can obtain

$$\left(\frac{1}{m^*}\right)_{ij} = \begin{pmatrix} m_t^{-1} & 0 & 0 \\ 0 & m_t^{-1} & 0 \\ 0 & 0 & m_l^{-1} \end{pmatrix}, \tag{4.19}$$

where m_l and m_t are called the *longitudinal* and *transverse effective masses*, respectively. For group IV semiconductors, Si and Ge, the energy parameters are presented in Table 4.4. The degeneracy, given in Table 4.4, indicates the existence of six and four equivalent energy valleys for Si and Ge, respectively. According to Fig. 4.9, the L points are located at the edges of the Brillouin zone, i.e., only half of each energy valley lies inside the first Brillouin zone. This reduces the effective number of the valleys to four. In Table 4.4 the bandgaps, E_g, and the electron longitudinal, m_l, and transverse, m_t, effective masses are also presented.

For III–V compounds, different situations occur for different materials: some of these compounds are direct-bandgap crystals and others are indirect materials. Thus, for the conduction bands of these materials the conduction band edge can be found at the Γ point, at the L point, or at the X point. In Table 4.5, energy-band parameters for three sets of III–V compounds are shown; the direct or indirect nature of the crystals is indicated. Now we shall consider valence bands. Their structure is more complicated. For group IV semiconductors and for III–V compounds, the top of the valence bands at $\mathbf{k} = 0$ has high degeneracy, i.e., several valence bands have the same energy at this point. The degeneracy occurs because these bands originate from the bonding of three p orbitals of the atoms composing the crystals. Thus, if one neglects the interaction between the spin of the electrons and their motion (so-called *spin–orbit interaction*), one obtains three degenerate valence bands, each of which is also doubly degenerate as a result of electron spin. In fact, the spin–orbit interaction causes a splitting of these six-fold degenerate states. At $\mathbf{k} = 0$ they are split into (i) a quadruplet of states with degeneracy equal to 4 and (ii) a doublet of states with degeneracy equal to 2. This splitting of valence bands,

Table 4.5 Energy-band parameters for III–V compounds in the As family

	InAs	GaAs	AlAs
Type of gap	Direct	Direct	Indirect
Lowest minima	Γ point	Γ point	X points
E_g (eV)	0.354	1.42	2.16
Electrons			
m^*/m_0	0.025	0.067	0.124
Holes			
m_{hh}/m_0	0.41	0.50	0.50
m_{lh}/m_0	0.26	0.07	0.26
Δ_{so} (eV)	0.38	0.34	0.28

Δ_{so}, at $\mathbf{k} = 0$ is shown in Fig. 4.10. One often refers to the lower valence band as the *split-off valence band*. At finite $\mathbf{k}$, the spin–orbit interaction leads to further splitting of the upper valence band into two branches: *the heavy-* and *light-hole bands*. Parameters characterizing these bands are the *heavy-hole*, m_{hh}, and *light-hole*, m_{lh}, masses. The effective masses of the light hole, m_{lh}, and heavy hole, m_{hh}, as well as the distance to the split-off band, Δ_{so}, for Si, Ge, and some III–V compounds are presented in Tables 4.4 and 4.5.

It is worth mentioning that, despite the relative complexity of the picture presented for energy bands, the description of electron properties in terms of the electron and hole quasiparticles by using several $E_\alpha(\mathbf{k})$ dependences is *incommensurably* simpler than operating with the enormous number of valence electrons in the crystal ($\sim 10^{23}$ cm^{-3}).

Direct-bandgap and indirect-bandgap semiconductors

One of the important conclusions, which can be drawn from the energy-band picture described above is related to the optical properties of the crystals.

In Section 2.4, following M. Planck and A. Einstein, we found that electromagnetic radiation (light) can be thought of as a flux of photons with certain energy and momentum values given by Eqs. (2.39) and (2.40). The absorption and emission of light can be interpreted as the absorption and emission of discrete "portions" of the light or photons with a specific energy and momentum. Let us apply these findings to consider light–crystal interactions. In a crystal absorbing or emitting a photon of sufficient energy, an electron may be transferred between the valence and conduction bands. For a given frequency of light, ω, such a transition is possible if the energy and momentum conversation laws are satisfied

$$E_c(\mathbf{k}_1) - E_v(\mathbf{k}_2) = \hbar\omega,$$

$$\mathbf{k}_1 - \mathbf{k}_2 = \pm\mathbf{q},$$

where $\mathbf{k}_1$ and $\mathbf{k}_2$ are the wavevectors of the electrons participating in phototransition; here, $\mathbf{q}$ is the photon wavevector. The sign, $+ (-)$, in the second equation stands for photon emission (absorption). Throughout the whole optical spectral region (from infrared to ultraviolet light), wavelengths λ are much greater than electron de Broglie wavelengths, as estimated in Chapters 2 and 3. The photon wavevectors $q = 2\pi/\lambda$, in turn, are much smaller than the electron wavevectors ($|\mathbf{k}_1|, |\mathbf{k}_2| \gg |\mathbf{q}|$). This property reduces the above equations to

$$\mathbf{k}_1 \approx \mathbf{k}_2 = \mathbf{k} \qquad \text{and} \qquad E_c(\mathbf{k}) - E_v(\mathbf{k}) = \hbar\omega.$$

In other words, under absorption and emission of the light electrons transferred between the valence and conduction bands practically preserve their wavevectors. In other words, the electron wavevector changes very little. In an energy scheme like that presented in Fig. 4.10, the processes of light–crystal interaction can be interpreted as *vertical* electron interband transitions. Another conclusion following from this analysis is that the interband phototransitions are possible only for light with energy quanta exceeding the bandgap, $\hbar\omega \geq E_g$. This finding implies that a pure semiconductor crystal is optically transparent for light with $\hbar\omega < E_g$ ($\lambda > 2\pi c\hbar/E_g$, with c being the velocity of light in vacuum). For some optoelectronic applications, the spectral range near the onset of absorption/emission of light is critically important.

On combining this analysis with the previously described bandstructures of different materials we can see that phototransitions induced by light with energy near the bandgap are possible in semiconductors for which the conduction and valence bands have a minimum and a maximum, respectively, at the same Γ point. For example, in GaAs one can transfer an electron from the valence band to the conduction band directly without a change in its momentum. Crystals of this type are called *direct-bandgap semiconductors*.

In contrast, in order to move an electron from the valence band to the conduction band in Si and Ge, one needs not only to add an amount of energy – greater than the minimum energy difference between the conduction and valence bands – to excite an electron, but also to change its momentum by a large amount (comparable to the scale of the Brillouin zone). Such a semiconductor is called an *indirect-bandgap semiconductor*.

Summarizing, the bandstructure of a semiconductor material determines both electrical and optical properties. Manipulation of electrons using light, i.e., optoelectronic functions, is easier for a direct-bandgap semiconductor, such as GaAs. In contrast, silicon and other group IV materials are optically relatively inactive. The situation is changed in nanoscale Si and SiGe structures, where the momentum conservation law is no longer rigorously obeyed.

Bandstructures of semiconductor alloys

As emphasized in previous discussions, the energy bandstructure of a particular semiconductor determines its electrical and optical properties. For naturally existing semiconductor crystals such as monatomic Ge and Si, and binary GaAs, their fixed, and unalterable, energy bandstructures restrict their applications. One of the powerful tools for varying the bandstructure is based on alloying two or more semiconductor materials. Some alloys exhibit well-ordered crystal structures. Though an alloy always has a disorder of the constitutive atoms, contemporary technology facilitates partial control of

Table 4.6 Bandgaps for III–V alloys

Alloy	E_g (eV)
$Al_xGa_{1-x}As$	$1.42 + 1.247x$
$Al_xIn_{1-x}As$	$0.360 + 2.012x + 0.698x^2$
$Ga_xIn_{1-x}As$	$0.360 + 1.064x$
$Ga_xIn_{1-x}Sb$	$0.172 + 0.139x + 0.415x^2$
$Al_xGa_{1-x}Sb$	$0.726 + 1.129x + 0.368x^2$
$Al_xIn_{1-x}Sb$	$0.172 + 1.621x + 0.430x^2$

Table 4.7 Effective masses for the alloy $Al_xGa_{1-x}As$

Type of minimum	Effective mass, m_α^*/m_0
Γ point	$0.067 + 0.083x$
X minima	$0.32 - 0.06x$
L minima	$0.11 + 0.03x$
Heavy hole	$0.62 + 0.14x$
Light hole	$0.087 + 0.063x$

this disorder and produces high-quality crystals. The properties of such materials can be interpreted in terms of nearly ideal periodic crystals.

Consider an alloy consisting of two components: A, with a fraction x, and B, with a fraction $1 - x$. If A and B have similar crystalline lattices, one can expect that the alloy A_xB_{1-x} has the same crystalline structure, with the lattice constant a_c given by a combination of lattice constants of materials A, a_A, and B, a_B. The simplest linear combination leads to the equation (Vegard's law)

$$a_c = a_A x + a_B(1 - x). \tag{4.20}$$

Then, the symmetry analysis can be extended to these types of alloys. For SiGe alloys and III–V compounds, this leads us to the previously discussed symmetry properties of the energy bands. Since the bandstructures are similar, one can characterize certain parameters of the alloy as functions of the fraction x. This approximation is often called the *virtual-crystal approximation*. For example, the bandgap of an alloy can be represented as $E_g^{alloy} = E_g(x)$. Such approximate dependences are given in Table 4.6 for III–V alloys. They correspond to the bandgaps, E_g, at Γ points.

As the composition of an alloy varies, the internal structure of energy bands changes significantly. For example, in the case of $Al_xGa_{1-x}As$ alloys, the lowest energy minimum of the Γ conduction band of GaAs is replaced by the six X minima of AlAs as the value of x is increased. Indeed, near the composition $x \approx 0.4$, the alloy transforms from a direct- to an indirect-bandgap material. The x-dependences of the effective masses for various electron energy minima as well as for heavy and light holes for $Al_xGa_{1-x}As$ are presented in Table 4.7.

Clearly, the established capability of fabricating a variety of high-quality materials provides an excellent tool for modifying the fundamental properties of materials.

4.5   Semiconductor heterostructures

Further modification and engineering of material properties is possible with the use of *heterostructures*. Heterostructures are structures with two or more abrupt interfaces at the boundaries between the regions of different materials. With modern techniques for growth of materials, it is possible to grow structures with transition regions between adjacent materials that have thicknesses of only one or two atomic monolayers.

Band offsets at heterojunctions

Let us consider a junction between two different semiconductor materials, which generates an abrupt change in the energy gap as well as an abrupt change in the conduction- and valence-band energies. These abrupt changes result in band-offset steps.

To understand the principal novel features brought about by an abrupt energy change in the energy bandstructure, we need to deviate from the approach of the previous sections where, while considering the energy bands of semiconductors, we analyzed energy structures in terms of *relative positions* of the bands in each of the semiconductors. In this approach, absolute values of the energies were not important and only relative positions of the bands were taken into account. However, if two different materials are brought together, the absolute values of energies become critically important. There is a simple way to compare energy bands of different materials. Let us introduce the *vacuum level* of the electron energy, which coincides with the energy of an electron "outside" of a material. It is obvious that the vacuum level may be taken to have the same value for any material. One can characterize the absolute energy position of the bottom of the conduction band with respect to this level, as shown in Fig. 4.11. The energy distance between the bottom of the conduction band and the vacuum level, χ, is called the *electron affinity*. In other words, the electron affinity is the energy required to remove an electron from the bottom of the conduction band to outside of a material, i.e., to the so-called *vacuum level*. Thus, if we know the electron affinities for different materials, we know the values of the conduction-band bottoms with respect to each other.

With this definition of electron affinity, one can calculate the discontinuity in the conduction band at an abrupt heterojunction of two materials, A and B:

$$\Delta E_c = E_{c,B} - E_{c,A} = \chi_A - \chi_B, \tag{4.21}$$

where $\chi_{A,B}$ are the electron affinities of materials A and B. Similarly, one can calculate the discontinuity of the valence band for the same heterojunction:

$$\Delta E_v = E_{v,B} - E_{v,A} = \chi_B - \chi_A + \Delta E_g, \tag{4.22}$$

where $\Delta E_g = E_{g,A} - E_{g,B}$ is the bandgap discontinuity for the heterojunction, with $E_{g,A}$ and $E_{g,B}$ being the bandgaps of materials A and B, respectively. Thus, if this simple

Vacuum level

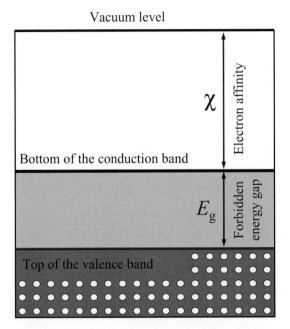

Figure 4.11 This diagram illustrates the electron affinity and vacuum level in a crystal.

approach – the *electron-affinity rule* – is applicable to a pair of semiconductor materials, one can calculate band offsets for an ideal heterojunction. Furthermore, if three materials, say A, B, and C, obey this rule, the following "transitivity" property is valid:

$$\Delta E_v(A/B) + \Delta E_v(B/C) + \Delta E_v(C/A) = 0,$$

where $\Delta E_v(A/B)$ is the valence-band discontinuity at the A/B interface. Hence, it is possible to calculate the band offset for one of three junctions if parameters for two of them are known.

Unfortunately, this rule fails for many semiconductor pairs. One reason for this failure is the dissimilar character of chemical bonds in adjacent materials. The formation of new chemical bonds at such a heterojunction results in charge transfer across this junction and the consequent reconstruction of energy bands, which leads to the breakdown of the electron-affinity rule. In real heterojunctions, band offsets can depend on the quality of the interface, conditions of growth, etc.

On combining different values of the electron affinity and the energy bandgap, we can expect different band line-ups at the interface between two semiconductor materials. In Fig. 4.12, sketches of the three possible types of band discontinuity are presented. The most common line-up is of the "straddling" type presented in Fig. 4.12(a), with conduction- and valence-band offsets of opposite signs, and the lowest conduction-band states occur in the same part of the structure as the highest valence-band states. This case is referred to as a *type I heterostructure*. The most widely studied heterojunction system, $GaAs/Al_xGa_{1-x}As$, is of this kind for $x < 0.4$. The next sketch, Fig. 4.12(b), depicts a heterostructure where the lowest conduction-band minimum occurs on one of the sides

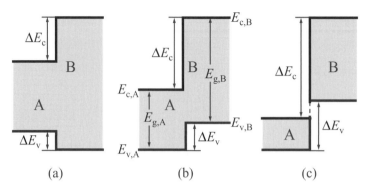

Figure 4.12 Three types of interfaces: type I (a); type II (b); and broken-gap lineup (c). The energy bandgaps of the materials A and B are indicated in (b).

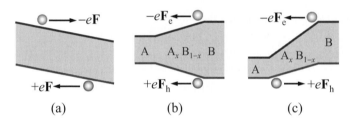

Figure 4.13 (a) A external electric field, $\mathbf{F}$, simply tilts the conduction and valence bands of a semiconductor. The forces $-e\mathbf{F}$ and $+e\mathbf{F}$ acting on the electron and the hole are equal in magnitude and opposite in direction. (b) The same direction of forces $-e\mathbf{F}_e$ and $+e\mathbf{F}_h$ on the electron and the hole caused by quasielectric field in the conduction band, $\mathbf{F}_e$, and in the valence band, $\mathbf{F}_h$. (c) Forces $-e\mathbf{F}_e$ and $+e\mathbf{F}_h$ of opposite directions for electrons and holes.

and the highest valence-band maximum on the other, with an energy separation between the two less than the lower of the two bulk bandgaps. This case represents a *type II heterostructure*. The combination AlAs/Al$_x$Ga$_{1-x}$As for $x > 0.4$ and some Si/Si$_x$Ge$_{1-x}$ structures are of this kind. Figure 4.12(c) illustrates a broken-gap line-up, in which the bottom of the conduction band on one side drops below the top of the valence band on the other. The example of this band line-up is given by InAs/GaSb, with a break in the forbidden gap at the interface of the order of 150 meV.

Graded semiconductors

Very often, graded semiconductors are used instead of abrupt heterointerfaces. To illustrate the idea, consider first a homogeneous piece of a semiconductor, say, a piece of uniformly doped silicon, but with an electric field applied. Then, the band diagram looks like that illustrated in Fig. 4.13(a) and is represented simply as two parallel tilted lines corresponding to the conduction- and valence-band edges. The separation between the two lines is the energy bandgap of the semiconductor; the slope of the two band edges is the elementary charge e multiplied by the electric field $\mathbf{F}$. When an electron or a hole

Table 4.8 Lattice constants for cubic
semiconductor materials ($T = 300$ K)

Semiconductor	Lattice constant (Å)
SiC	3.0806
C	3.5668
Si	5.4309
GaP	5.4495
GaAs	5.6419
Ge	5.6461
AlAs	5.6611
InP	5.8687
InAs	6.0584

is placed into this structure, a force $-e\mathbf{F}$ acts on the electron and $+e\mathbf{F}$ acts on the hole; the two forces are equal in magnitude and opposite in direction.

Slopes of conduction- and valence-band edges arise in the case of a graded transition from one material to another. Graded transitions from a narrow-bandgap to a wide-bandgap semiconductor that correspond to the abrupt heterojunctions of Figs. 4.12(a) and 4.12(b) are shown in Figs. 4.13(b) and 4.13(c). As is obvious from Figs. 4.13(b) and 4.13(c), in the case of graded heterostructures there is a built-in electric field that acts on electrons and holes. This field is called *quasielectric*. The quasielectric field does not exist in homogeneous crystals; that is why graded heterostructures can be used for new devices where the existence of a built-in electric field is required. Examples of materials used in graded nanostructure devices are Si_xGe_{1-x} and $Al_xGa_{1-x}As$, where x changes in the direction of growth. Graded structures and the accompanying quasi-electric forces introduce a new degree of freedom for the device designer and allow him to obtain effects that are basically impossible to obtain using only external (or *real*) electric fields.

4.6   Lattice-matched and pseudomorphic heterostructures

Now we shall consider some of the principal problems that arise in the fabrication of heterostructures. In general, one can grow any layer on almost any other material. In practice, however, the interfacial quality of such artificially grown structures can vary enormously. Even when one fabricates a structure from two materials of the same group or from compounds of the same family, the artificially grown materials of the heterostructure may be very different from the corresponding bulk materials. First of all, the quality of the materials near heterointerfaces depends strongly on the ratio of lattice constants for the two materials.

In Table 4.8, lattice constants for several group IV semiconductors and III–V compound semiconductors are presented; all of the cases presented represent cubic crystals. The lattice constants for some other materials can be found from Fig. 4.14. Depending on the structural similarity and lattice constants of the constituent materials, there exist two

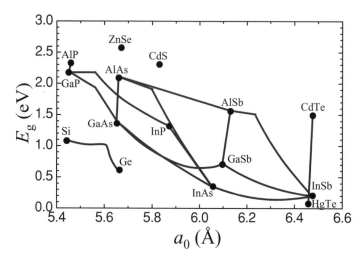

Figure 4.14 Room-temperature bandgaps, E_g, as functions of the lattice constant, a_0, for selected III–V and II–VI compounds and selected group IV materials and their alloys.

essentially different classes of heterointerfaces: lattice-matched and lattice-mismatched materials. Prior to an analysis of both classes, we highlight other factors affecting the quality and usefulness of heterointerfaces.

Valence matching

Since there are still no rigorous rules for how one can realize a given level of quality for heterojunctions, we consider a few examples, which illustrate the problems.

If lattice matching were the only obstacle, the Ge/GaAs system would be the ideal heterosystem, because, according to Table 4.8, it would allow one to realize the ideal combination of group IV semiconductors and III–V compounds. Indeed, on the basis solely of lattice-constant matching, the Ge/GaAs system appears to be the most promising candidate. However, it turns out that there is the problem of chemical compatibility for this heterostructure. Covalent bonds between Ge on the one hand and Ga or As on the other are readily formed, but they are what could be called *valence-mismatched bonds*, meaning that the number of electrons provided by the atoms is not equal to the canonical number of exactly two electrons per covalent bond. Hence, the bonds themselves are not electrically neutral. Consider a hypothetical idealized (001)-oriented interface between Ge and GaAs, with Ge to the left of a "mathematical plane" and GaAs to the right, as shown in Fig. 4.15. In GaAs, an As atom brings along five electrons (resulting in 5/4 electrons per bond) and is surrounded by four Ga atoms, each of which brings along three electrons (3/4 electron per bond), adding up to the correct number of two (8/4) electrons per Ga—As covalent bond. However, when, at a (001) interface, an As atom has two Ge atoms as bonding partners, each Ge atom brings along one electron per bond, which is half an electron more than is required for bonding. Loosely speaking, the As atom does not "know" whether it is a constituent of GaAs or a donor in Ge. As a result, each Ge—As bond acts as a donor with a fractional charge, and each Ge—Ga

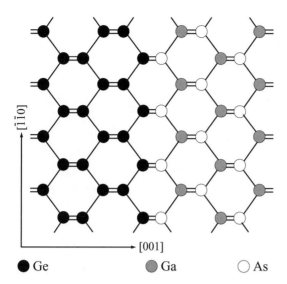

Figure 4.15 Departure from electrical neutrality at a "mathematically planar" (001)-oriented Ge/GaAs interface. The different atomic species, Ga or As atoms, and Ge atoms, do not bring along the correct number of electrons to form electrically neutral Ga—Ge or As—Ge covalent bonds of two electrons per bond. Reprinted with permission from W. A. Harrison, E. Kraut *et al.*, "Polar heterojunction interfaces" *Phys Rev.* B, **18**(8), 4402 (1978). © 1978 by the American Physical Society.

bond acts as an acceptor with the opposite fractional charge. To be electrically neutral, a Ge/GaAs interface would have to have equal numbers of both charges, averaged not only over large distances, but also locally. Given chemical bonding preferences, such an arrangement will not occur naturally during epitaxial growth. If only one kind of bond were present, as in Fig. 4.15, the interface charge would support a large electric field of 4×10^7 V cm^{-1}. Such a huge field would force atomic rearrangements during the growth, trying to equalize the numbers of Ge—As and Ge—Ga bonds. However, these rearrangements will never go to completion, but will leave behind locally fluctuating residual charges, with deleterious consequences for the electrical properties of materials and any device applications.

Interfaces with perfect bond-charge cancellation may readily be drawn on paper, but in practice there are always going to remain some local deviations from perfect charge compensation, leading to performance-degrading random potential fluctuations along the interface. This argument applies to other interfaces combining semiconductors from different columns of the Periodic Table of elements.

The above discussion pertains to the most widely used (001)-oriented interface. The interface charge at a valence-mismatched interface actually depends on the crystallographic orientation. It has been shown that an ideal (112) interface between group IV and III–V compounds exhibits no interface charge. An important example is the GaP-on-Si interface that has a sufficiently low defect density; as a result, it is used in various devices grown on Si. After these comments, we return to the discussion of the role and importance of lattice matching.

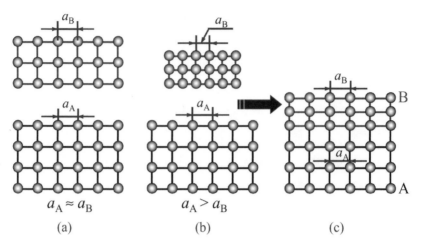

Figure 4.16 Lattice-matched materials (a) and lattice-mismatched materials (b); the resulting structure is strained (pseudomorphic) if the upper layer B adopts the lattice of substrate A (c).

Lattice-matched materials

For lattice-matched structures, the lattice constants of the constituent materials *are nearly matched*; i.e., the lattice constants are within a small fraction of a percent of each other. There is no problem, in general, in growing high-quality heterostructures with such lattice-matched pairs of materials. By "high-quality" we mean that the interface structure is free of lattice imperfections such as interface defects, etc. Such imperfections result in poor electrical and optical properties and may lead to fast and widespread degradation of the structure. Figure 4.16(a) illustrates a lattice-matched layer B on a substrate A. One can expect that the layer can be grown on the substrate if both materials are from the same group and the binding energies and crystal structures are very similar.

According to the data of Fig. 4.14 and Table 4.8, the AlGaAs/GaAs system is an example of a lattice-matched material. The system has a very small mismatch in the lattice constants of only about 0.1% over the entire range of possible Al-to-Ga ratios in the AlGaAs. As a result, such heterostructures can be grown free of mechanical strain and significant imperfections. Hence, these structures provide a practical way of tailoring bandstructures. In addition to these tailored electronic parameters, elastic and other lattice properties can be different in layers composing such a lattice-matched heterostructure.

Lattice-mismatched materials

The case of lattice-mismatched structures is characterized by a *finite lattice mismatch*. Figure 4.16(b) depicts this case. If one tries to match these lattices, strain in the plane of growth and a distortion along the growth axis arise. Thus, one obtains a strained layer with a lattice deformation. The lattice-mismatched structure can be characterized also by the *relative mismatch* of the lattice constants of the substrate, a_A, and the epilayer, a_B:

$$\epsilon = \frac{a_A - a_B}{a_A}. \tag{4.23}$$

Consider an elastic deformation of a lattice. It can be characterized by the vector of relative displacement $\mathbf{u}$. The displacement defines how any lattice point $\mathbf{r}$ moves to a new position, $\mathbf{r}' = \mathbf{r} + \mathbf{u}$, as a result of the deformation. Different regions of the crystal can be deformed differently. Thus, the displacement depends on coordinates: $\mathbf{u} = \mathbf{u}(\mathbf{r})$. In fact, only relative displacements are important. They are determined by the *strain tensor*:

$$u_{ij} = \frac{1}{2}\left(\frac{\partial u_i}{\partial \zeta_j} + \frac{\partial u_j}{\partial \zeta_i}\right), \tag{4.24}$$

where the tensor u_{ij} has the following components:

$$u_{ij} = \begin{pmatrix} u_{xx} & u_{xy} & u_{xz} \\ u_{yx} & u_{yy} & u_{yz} \\ u_{zx} & u_{zy} & u_{zz} \end{pmatrix}. \tag{4.25}$$

Here i and j denote x, y, and z components; $\zeta_i = x, y$, and z for $i = 1, 2$, and 3, respectively. In this discussion, we consider only *diagonal components* of u_{ij}. They determine a change in the crystal volume, from V to V', produced by the strain:

$$\delta = \frac{V' - V}{V} = u_{xx} + u_{yy} + u_{zz}. \tag{4.26}$$

The elastic energy density (elastic energy per unit volume) of a crystal may also be expressed in terms of the strain tensor. For cubic crystals, this energy is given by

$$U = \frac{1}{2}c_{11}\left(u_{xx}^2 + u_{yy}^2 + u_{zz}^2\right) + c_{44}\left(u_{xy}^2 + u_{xz}^2 + u_{yz}^2\right)$$
$$+ c_{12}\left(u_{xx}u_{yy} + u_{xx}u_{zz} + u_{yy}u_{zz}\right), \tag{4.27}$$

where c_{11}, c_{12}, and c_{44} are *elastic constants or elastic moduli* of the crystal. Equation (4.27) is a generalization of the expression for the potential energy of an elastic isotropic medium derived in Section 2.3. The elastic constants c_{12} and c_{44} are responsible for anisotropy of crystals, this is why they equal zero for an elastic isotropic medium. As a result, Eq. (2.17) is recovered in this case with $\Lambda = c_{11}$.

The *stress tensor* is defined in terms of derivatives of the elastic energy density with respect to strain-tensor components:

$$\sigma_{ij} = \frac{\partial U}{\partial u_{ij}}. \tag{4.28}$$

Boundary conditions at a surface or at an interface may be formulated in terms of the stress tensor:

$$\sigma_{ij}N_j = f_i, \tag{4.29}$$

where $\mathbf{N}$ is a vector perpendicular to the surface and $\mathbf{f}$ is an external force applied to the surface.

These equations are sufficient for calculations of the strain of a layer A grown on a mismatched substrate B. Let the lattice constants of these two materials be a_A and a_B, respectively. In this discussion, both materials are assumed to be cubic crystals and the direction of growth is along the [001] direction. If the layer A adopts the lattice periodicity of the substrate B, the in-plane strain of the layer is

$$u_{xx} = u_{yy} = u_{||} = 1 - \frac{a_B}{a_A}. \tag{4.30}$$

There should be no stress in the direction of growth. Thus, from Eq. (4.29) it follows that $\sigma_{zz} = 0$. On calculating σ_{zz} from Eqs. (4.27)) and (4.28), and from the obtained result, $\sigma_{zz} = c_{11}u_{zz} + c_{12}(u_{xx} + u_{yy})$, we find the strain in the direction perpendicular to the layer:

$$u_{zz} = -\frac{2c_{12}}{c_{11}}u_{||}. \tag{4.31}$$

Thus, the strain can be found through the mismatch of the lattice constants.

The strain results in two types of effects: (1) the strain can generate various imperfections and defects; and (2) the strain in the layer leads to a change in the symmetry of the crystal lattice, for example, from cubic to tetragonal or to rhombohedral, etc. Of course, the latter effect can modify the energy bandstructure of the layer.

Strained pseudomorphic heterostructures

Here, we consider imperfections generated by strain from a lattice mismatch. In order to understand the nature of the formation of imperfections in a layered structure, let us consider the characteristic energies of the structure. First of all, a layer grown on a substrate with a mismatched lattice should possess *extra elastic energy*, E_{el}, caused by the strain. This energy is a function of the thickness of the layer, d, and increases with increasing d. In the simplest case of uniform strain, the elastic energy can be calculated through its density U: $E_{el} = U \times d \times S$, where S is the area of the layer. On the other hand, the generation of misfit defects requires some energy. Let us denote this energy by E_{im}. If the extra elastic energy exceeds the energy associated with the imperfection, i.e., if $E_{el}(d) > E_{im}$, the system will relax to a new state with lower energy and imperfections will be generated. That is, the extra strain energy is the main physical reason for the instability and degradation of heterostructures fabricated from materials with a large mismatch of lattice constants.

Since the value of E_{im} remains finite even in thin layers, for certain thicknesses we may get $E_{el}(d) < E_{im}$. Thus, there is not sufficient strain energy and imperfections will not be generated. Such strained heterostructures can be of high quality. Hence, in some approximations, for each pair of materials there exists a critical thickness of the layers, d_{cr}; if $d < d_{cr}$ the lattice mismatch is accommodated by the layer strain without the generation of defects. The corresponding layered systems are called *pseudomorphic heterostructures*. In general, a pseudomorphic layer of material possesses some characteristics similar to those of the substrate and may possibly have the same lattice structure as the substrate material. In our case, a crystalline semiconductor layer grown on another semiconductor takes on the in-plane lattice periodicity of the substrate semiconductor. Figure 4.16(c) illustrates the case when the deposited layer adopts the lattice periodicity of a substrate material. Examples of such systems are the $Ga_{1-x}Al_xAs/Ga_{1-x}In_xAs$ and $GaAs/Ga_{1-x}In_xAs$ structures. In fact, these heterostructures are used to improve the characteristics of the so-called *heterojunction-field-effect transistors* which will be considered in Chapter 8. In spite of significant mismatches of lattice constants, these structures are virtually free of interface defects, because of the small, nanometer-scale, thicknesses of the pseudomorphic layers used in the fabrication of functioning heterojunction-field-effect transistors.

It is sometimes possible to grow defect-free systems with layer thicknesses exceeding the critical thickness. However, such systems are metastable and this may lead to device degradation as a result of the generation of misfit defects driven by temperature effects or other external perturbations. Central to the stability of pseudomorphic structures is the question of whether or not the strain energy leads to damage of the materials when the structures are subjected to various forms of external stress and processing. The experience accumulated in this field shows that in the case of small strain energy the heterostructures are stable. For example, in the case of the GaP/GaAsP layered system, the strain energy is about 10^{-3} eV per atom. Since this quantity is rather small in comparison with the energy required to remove the atom from its lattice site, this system can be stable for sufficiently thin layers.

The above-discussed strain states are shown in Fig. 4.17 as a function of x for Ge_xSi_{1-x} layers grown on Si substrates. The "phase diagram" – the critical thickness of the layer versus the Ge fraction – consists of three regions: strained layers with defects at large thicknesses, nonequilibrium (metastable) strained layers without defects at intermediate thicknesses, and equilibrium and stable layers without defects at small thicknesses. According to these results, a stable Ge layer on Si (the largest misfit) can not be grown with a thickness greater than 10 Å or so.

Let us consider the Si/Ge system in more detail. This system is very interesting and important because it has opened new horizons for silicon nanotechnology and Si-based applications. The data of Table 4.8 show that heterostructures based on Si and Ge materials should be designed so that they are *always* pseudomorphic.

First of all, the stability and quality of these Si/Ge pseudomorphic heterostructures depend strongly on the thicknesses of the strained layers, as discussed previously. In fabricating Si/Ge structures, one grows specific numbers of Si and Ge atomic monolayers. Thus, layer thicknesses can be characterized by the numbers of these monolayers. Let n and m be the numbers of Si and Ge monolayers, respectively. This system is known as the Si_n/Ge_m *superlattice*. The second important factor that determines the quality of these structures is the *material of the substrate* on which the superlattices are grown. We have discussed the case of Ge_xSi_{1-x} layers grown upon a Si substrate; see Fig. 4.17. For the fabrication of Si/Ge superlattices, the substrates of choice are frequently either Ge_xSi_{1-x} alloys or GaAs.

Let us consider Ge_xSi_{1-x} as a substrate. The elastic energy of a strained system depends on the alloy composition of the substrate. Figure 4.18 illustrates this dependence for various numbers of monolayers for the symmetric case, $n = m$. In accordance with the

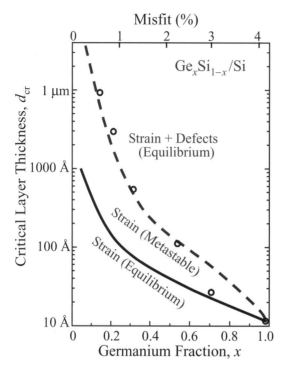

Figure 4.17 The stability–strain diagram for a Ge_xSi_{1-x} layer on Si substrate. After R. People and J. C. Bean, "Erratum: Calculation of critical layer thickness versus lattice mismatch for Ge_xSi_{1-x}/Si strained-layer heterostructures" *Appl. Phys. Lett.* **47**, 322 (1985), *Appl. Phys. Lett.* **49**, 229 (1986). Reprinted with permission from R. People and J. C. Bean, *Appl. Phys. Lett.*, **49** 229 (1986). © 1986 American Institute of Physics.

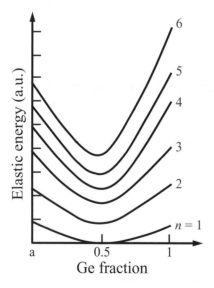

Figure 4.18 The elastic energy of strained Ge_n/Si_n superlattices on Ge_xSi_{1-x} substrate with various numbers of monolayers n as a function of the Ge fraction in the Ge_xSi_{1-x} substrate. The numbers on the curves indicate n.

previous discussion, the elastic energy increases with increasing thicknesses of strained layers for a given substrate material. Because of this, one employs superlattices with a few monolayers: $2 \leq (n, m) \leq 5$. Figure 4.18 also shows a nontrivial dependence of the strain energy on the alloy composition of the substrate; the minimal strain energy is expected for x in the range from 0.4 to 0.6.

Another important characteristic of pseudomorphic Si/Ge structures is the distribution of the elastic energy over the monolayers of the superlattice. It was shown that the most homogeneous distribution over layers occurs for the Si/Ge alloy with $x \approx 0.5$. From this point of view, $Si_{0.5}Ge_{0.5}$ substrates are preferable. However, these results depend strongly on the orientation of the substrate material. Often, the direction of growth on the substrate is chosen to be the [001] direction.

Although the technology of the growth and fabrication of Si/Ge structures is still in a developing stage, from our short analysis one can see that there is some qualitative and even quantitative knowledge concerning the behavior responsible for the stability and perfection of these structures.

Lattice-matched heterostructures

Let us return to Fig. 4.14 and discuss lattice-matched heterostructures in more detail. From this figure, we can determine the lattice constants of various compounds. First of all, one can see that the GaAs/AlAs system is really unique because the lattice constants have almost identical values. In order to achieve lattice matching for other cases, it is possible either to combine a binary compound and a ternary compound, or to use ternary–ternary compounds having appropriate ratios of atomic species within each layer. For example, in the case of GaInAs/InP structures, lattice matching is achieved exactly only for $Ga_{0.47}In_{0.53}As$, in which the ratio of Ga to In is 47 to 53 in the GaInAs layer; for the other ratios, the GaInAs layer is not lattice-matched with the InP. Moreover, the wide-bandgap $Ga_{0.51}In_{0.49}P$ material is compatible with the narrow-bandgap GaAs material.

In conclusion, the broad range of possibilities for controlling bandgaps and band offsets for both electrons and holes, as well as electron and hole effective masses, provides the basis for energy-band engineering. Through such energy-band engineering, it is possible to design and fabricate high-quality heterostructures with designated optical and electrical properties. If one can not achieve the desired properties using lattice-matched compositions, it is possible to employ strained pseudomorphic structures.

4.7   Organic semiconductors

In recent years, organic molecules have been shown to have properties that make them suited for novel electronic and optoelectronic devices. Such novel devices include organic light-emitting diodes, electronic circuits in mechanically flexible layers, crystal displays, novel molecular electronic devices, carbon-nanotube-based devices for data processing, and bioelectronic devices.

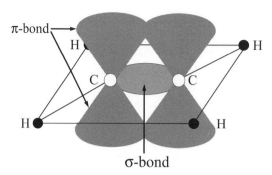

Figure 4.19 A schematic diagram of C_2H_4 (ethene) illustrating σ-bonds and π-bonds between carbon atoms.

Some organic molecules – including a variety of hydrocarbons – have been shown to have properties, such as relatively high conductivity and states with energy differences, $\Delta E = \hbar\omega$, corresponding to near-infrared, visible, and ultraviolet wavelengths, $\lambda = 2\pi c/\omega$, that make them suited for novel electronic and optoelectronic devices. These organic molecules may be deposited and bound on surfaces in specified patterns, using techniques described in Chapter 5, so that they form patterned layers with nanoscale dimensions. Thus, active layers in devices may be made very thin so that carrier transit time through a device is reduced, making possible very fast devices. Many organic polymers are flexible and it is thus possible to make electronic and optoelectronic devices on flexible thin films. Flexible displays – such as a computer display that may be rolled up into a pen – represent one of the novel applications of organic molecules in electronics and optoelectronics.

The importance of organic molecules which form conducting polymers is illustrated by the fact that Alan Heeger, Alan MacDiarmid, and Hideki Shirakawa won the Nobel Prize for Chemistry in 2000 for their pioneering research on conducting polymers. Working together, they learned how to make very pure samples of polyacetylene and found ways of "doping" them with bromine by exposing them to bromine vapor. They discovered that it was possible to increase the conductivity of polyacetylene by seven orders of magnitude! How does this work? To understand this, consider the σ- and π-bonds in a simple hydrocarbon such as ethene, C_2H_4, as depicted in Fig. 4.19. The localized σ-bond leads to the bonding of the two carbon atoms in a simple hydrocarbon such as ethene. The extended π-bonds are found in many hydrocarbons; these molecules are said to be π-conjugated. The extended π-bonds are delocalized in space along the length of the molecule and are thus similar to the extended energy bands that are formed when atoms are combined to form a crystal. Thus, these extended π-bonds open the way for charge transport along the polymer chain.

Polyacetylene, shown in Fig. 4.20, is a hydrocarbon that has alternating single and double bonds leading to slightly different C—C interatomic bond distances and therefore to energy levels separated in energy by different amounts, ΔE, as discussed in Chapter 3 for bonding and antibonding orbitals; see Fig. 3.17. The differences between the energy

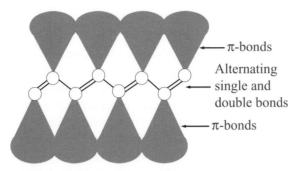

Figure 4.20 A polymer with carbon backbone having alternating single and double bonds between carbon atoms; such π-conjugated polymers have extended wavefunctions facilitating charge transport along the chain. Polyacetylene is a case in point.

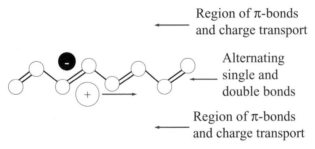

Figure 4.21 A π-conjugated polymer (without π-bonds shown) with an acceptor impurity. The acceptor atom – such as bromine or iodine – attracts and captures one of the electrons in the extended π-states and results in a region where there is a deficit of electron charge – a hole. The captured electron is relatively immobile compared with the hole, which moves along the polymer.

levels differ from one hydrocarbon polymer to another, making it possible to produce conducting polymers with different electrical and optical properties; π-conjugated polymers – like polyacetylene – have overlapping π-orbitals that conduct charge. The doping of a π-conjugated polymer with an acceptor impurity is illustrated in Fig. 4.21. The acceptor removes an electron from the polymer and leaves the polymer with a deficit of negative charge, i.e., with a positive charge – a hole (more detailed discussion about doping is given in Section 5.3). This positive charge attracts neighboring electrons and produces a local distortion of charge near the positive charge. This distorted charge density is known as a *polaron*. A polaron propagates along the molecule in a manner that bears some similarity to the propagation of a hole in a semiconductor. Indeed, the polaron behaves as a positive (screened) particle with an effective mass that is different from the mass of the electron. It is observed for polymers like polyacetylene that the mobility of the hole is high and resulting conductivities are close to those of metals. Thus, doping leads to conductivities in π-conjugated polymers that make them useful in electronic devices.

4.8   Carbon nanomaterials: nanotubes and fullerenes

In this section we will study new nanoscale objects – carbon-based nanomaterials. These include the so-called *carbon nanotubes* and *fullerenes*. These nanomaterials exhibit many unusual properties. In particular, the carbon nanotubes exhibit metallic, semiconductor, and dielectric properties and have great promise for applications in nanoelectronics.

We start with a brief review on carbon and ordinary carbon materials. Carbon, C, is a group IV element, as is silicon, Si. Carbon is known to be associated with rich and diverse chemistry and carbon atoms participate in the formation of a great number of molecules. Before the discovery of carbon-based nanomaterials, there were only two known forms of solid carbon: diamond and graphite. Diamond is a dielectric with a very large electron energy bandgap ($\sim$6 eV). Diamond can be p-doped, thus it should be considered as a wide-bandgap semiconductor. It is an indirect-bandgap semiconductor. In contrast, graphite is a semimetal. The structural and electronic properties of both of these carbon materials are not very promising for common electronic applications. However, the wide-bandgap properties of semiconducting diamond can be exploited in high-temperature electronics.

Carbon nanotubes

Actually, carbon nanotubes have the form of seamlessly rolled single sheets of carbon atoms. The cylindrical sheets may have diameters of only a few nanometers. They are very small objects that exhibit many different structures and properties. These nanotubes have no macroscopic analogues, such as the bulk semiconductors that served as analogues for traditional semiconductor nanostructures. Thus, to study the structure of the carbon nanotubes, one should use the most advanced microscopy techniques: *atomic force microscopy* and *scanning tunneling microscopy*, which we will consider in the next chapter.

A carbon nanotube is composed of carbon atoms. A defect-free *single-walled nanotube* consists of a single cylinder and is characterized by the tube diameter and a quantity known as the *helicity*. To understand the structure of carbon nanotubes, we will start with a two-dimensional graphite sheet, shown in Fig. 4.22(a). The single sheet of graphite, called *graphene*, has the form of a honeycomb-like lattice. Let $\mathbf{a}_1$ and $\mathbf{a}_2$ be the graphene lattice vectors and n and m be integers. The diameter and helicity of the nanotube are uniquely characterized by the vector $\mathbf{C} = n\mathbf{a}_1 + m\mathbf{a}_2 \equiv (n, m)$ that connects crystallographically equivalent sites on a two-dimensional graphene sheet. Here $\mathbf{a}_{1,2}$ are in units of $a_0\sqrt{3}$ with a_0 being the carbon–carbon distance. By using the vector $\mathbf{C}$ a carbon nanotube can be constructed by wrapping a graphene sheet in such a way that two equivalent sites of the hexagonal lattice coincide. The *wrapping vector* $\mathbf{C}$ defines the relative location of these two sites. In Fig. 4.22(a), the wrapping vector $\mathbf{C}$ connects the origin (0, 0) and the point with coordinates (11, 7). Thus, a nanotube denoted by indices (11, 7) is formed. A tube is called an *armchair* tube if n equals m, and a *zigzag* tube in the case of $m = 0$. Wrapping vectors along the dotted lines leads to tubes that are of zigzag or armchair form. All other wrapping angles lead to chiral tubes whose wrapping angle is specified relative either to

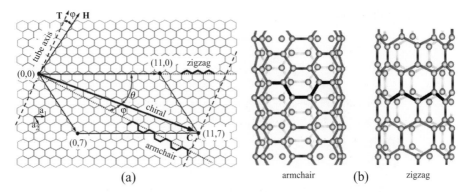

Figure 4.22 (a) The relation between the hexagonal carbon lattice and the chirality of carbon nanotubes. A carbon nanotube can be constructed from a single graphite sheet, called graphene, by rolling up the sheet along the wrapping vector **C**. (b) Fragments of "armchair" and "zigzag" carbon nanotubes.

the zigzag direction θ or to the armchair direction $\psi = 30° - \theta$. Both θ and the *wrapping angle* (*chiral angle*) ψ are in the range $(0, 30°)$ as a result of the hexagonal character of the carbon two-dimensional lattice of the graphene. Dashed lines are perpendicular to **C** and run in the direction of the tube axis indicated by the vector **T**. The solid vector **H** is perpendicular to the armchair direction and specifies the direction of nearest-neighbor hexagon rows. The angle between **T** and **H** is the chiral angle ψ. The unit cell of a nanotube can be constructed by finding the smallest lattice vector **T** which connects equivalent points of the lattice. It can be shown that this vector is given by

$$\mathbf{T} = [(n + 2m)\mathbf{a}_1 - (2n + m)\mathbf{a}_2]/q, \tag{4.32}$$

where $\mathbf{a}_1$ and $\mathbf{a}_2$ are the basis vectors of the graphene lattice; q is defined as follows:

$$q = \begin{cases} l, & \text{if } n - m \text{ is not a multiple of } 3l, \\ 3l, & \text{if } n - m \text{ is a multiple of } 3l, \end{cases}$$

where l is the greatest common divisor of n and m. The number of hexagons, $\mathcal{N}$, in the unit cell is

$$\mathcal{N} = \frac{2(n^2 + m^2 + nm)}{q}. \tag{4.33}$$

The tube radius, R, is given by

$$R = \frac{|\mathbf{C}|}{2\pi} = \frac{\sqrt{3}a_0}{2\pi}\sqrt{n^2 + m^2 + nm}. \tag{4.34}$$

The length of the carbon–carbon bond can be estimated as $a_0 \approx 0.14$ nm. It is easy to see that, for a general chiral nanotube, the number of atoms contained in the unit cell is very large. For example, an $(8, 7)$ tube has a radius of about 0.57 nm and contains 676 atoms in the unit cell. In determining the number of atoms, we have taken into account that each hexagon contains two carbon atoms: $n = 8, m = 7, n - m = 1, q = 1, \mathcal{N} = 338$, and the number of atoms is $2\mathcal{N} = 676$. In Fig. 4.22(b), the structures of two particular examples of the armchair and zigzag types of nanotubes are sketched.

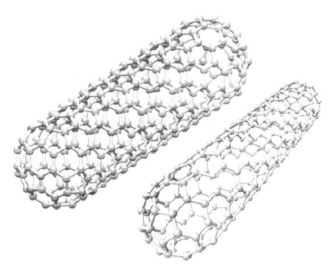

Figure 4.23 Single-walled carbon nanotubes of two different diameters.

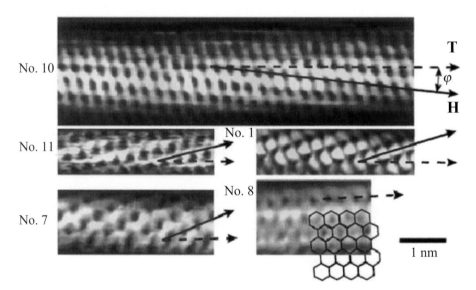

Figure 4.24 Atomically resolved scanning tunneling microscope images of individual single-walled carbon nanotubes. Reprinted with permission, from P. Moriarty, "Nanostructured materials," *Rep. Prog. Phys.*, **64**, 369 (2001). © IOP Publishing Limited.

In Figs. 4.23 and 4.24, atomically resolved images of individual single-walled carbon nanotubes are shown. The images were obtained by using scanning tunneling microscopy. On the surface of the cylinders (tube images), it is seen clearly that the lattice structure has a spacing between carbon atoms equal to $a_0 \approx 0.14$ nm. Such a lattice image makes it possible to identify the chirality of the tube. Dashed arrows in Fig. 4.24 represent the tube axis **T** and solid arrows indicate the direction of nearest-neighbor hexagon rows

H; compare this with Fig. 4.22(a). Tubes denoted by 10, 11, and 1 are chiral, whereas tubes 7 and 8 have a zigzag and an armchair structure, respectively. Tube 10 has a chiral angle $\psi = 7°$ and a diameter of 1.3 nm, which corresponds to the (11, 7) type shown in Fig. 4.22(a). A hexagonal lattice is plotted on top of image 8 to prove that the non-chiral armchair structure applies for this case.

In carbon nanotubes, an electron can propagate freely only in the direction along the tube axis. Thus, electron or hole transport in these structures has the properties of *one-dimensional* transport and the carriers can be characterized by a one-dimensional wavevector k. As for any periodic structure, we can think of the electron energies as a system of energy bands dependent on the wavevector k. The energy bands of a single-walled nanotube can be constructed from the energy bands of the graphene presented in Fig. 4.22(a). For the case of a multi-walled carbon nanotube, the energy bands may be constructed from the energy bands of a layer of graphite having a thickness equal to the thickness of the wall of the nanotube. Indeed, suppose that we find the energy bands of the graphene to be $E_N(\mathbf{K})$ with $\mathbf{K} = \{K_x, K_y\}$ being the two-dimensional wavevector. Instead of K_x and K_y, we can use projections of $\mathbf{K}$ onto the tube axis, $\mathbf{T}$, and onto the wrapping vector, $\mathbf{C}$, k and K_C, respectively; i.e., the energy is $E_N(K_C, k)$. However, along the $\mathbf{C}$ direction there is a periodicity. Thus, we can use the so-called *cyclic boundary conditions*, as we did in Section 4.4 to obtain the Brillouin zones. This results in "quantization" of the $\mathbf{K}$ projection onto the vector $\mathbf{C}$:

$$K_C = \frac{2\pi N_1}{a_0 \sqrt{3}} \tag{4.35}$$

with N_1 being an integer. For the electron energy, $E(k)$, we obtain

$$E(k) = E_N(K_C, k) = E_N \left(\frac{2\pi N_1}{a_0 \sqrt{3}}, k \right) \equiv E_{N,N_1}(k). \tag{4.36}$$

That is, each of the initial bands of the graphene N is split into a number of one-dimensional subbands N_1. In addition, the character of the splitting depends on the wrapping type, i.e., on the values of (n, m) for the nanotube. Then, classification as a metal, semiconductor, or dielectric can be made for these one-dimensional crystals. For different (n, m) values, distinct properties of the nanotube can be expected. A more detailed theory predicts that when $(n - m)$ is divisible by three the single-walled carbon nanotubes are of metallic nature; otherwise, they are semiconductors with bandgaps that depend on the nanotube diameter:

$$E_g \approx 2E_{C-C} \frac{a_0}{d}, \tag{4.37}$$

where E_{C-C} is the binding energy of two carbon atoms, a_0 is the carbon–carbon distance, and d is the diameter of the nanotube. Direct measurements support these considerations. Specifically, for semiconducting tubes, an inverse linear dependence of the bandgap on the nanotube diameter is confirmed, and the coefficient of proportionality is in good agreement with theoretical expectations. It turns out that both types of carriers can exist in those nanotubes, i.e., both the electron and the hole can be responsible for electric conductivity of the tubes.

Similarly to the electronic properties, the *mechanical properties* of nanotubes are related closely to those of a single graphite sheet. Since the graphite sheet is very stiff in the in-plane direction, one can expect nanotubes to have a similar stiffness constant along the tube axis.

Indeed, it was found that, in contrast to the electronic properties, the elastic moduli of nanotubes are almost insensitive to the tube diameter and the chiral angle. Their values are comparable to those of diamond or of a graphite sheet. In general, nanotubes are very stiff. For comparison, their elastic moduli are as much as five times larger than those of steel!

The response of the nanotubes to a large deformation is also remarkable. Indeed, most hard materials fail with a strain δ (see Eq. (4.26)) of about 1% or even less, because of generation of defects. The carbon nanotubes can sustain up to 15% tensile strain before fracture. Together with high stiffness, this provides a critical tensile strength of a single nanotube as much as 300 to 400 times larger than that for steel.

In conclusion, carbon nanotubes represent a new class of nanostructures, which differ from traditional solid-state device structures. The nanotubes can be fabricated with good control of their basic properties, including the electron bandstructure. They can be fabricated as semiconductors either with electron conductivity, or with hole conductivity. They can be contacted to metals and various types of processing techniques have been shown to be viable for the fabrication of electronic devices from carbon nanotubes.

Following the discovery of carbon nanotubes, it was recognized that this kind of perfectly organized nanostructure should not be limited to carbon. It was found that formation of nanotubes is a generic property of materials with *anisotropic two-dimensional layered structures*. The structures of this type are called *inorganic nanotube structures*; examples include WS_2, MoS_2, V_2O_5, and BN nanotubes. The study of these novel structures has led to the observation of a few interesting properties and promises potential applications, particularly in nanoelectronics.

Buckyball fullerenes

Now we consider the so-called *buckyball fullerenes* which represent the fourth major form of pure carbon. The previously discussed diamond and graphite and the well-studied carbon nanotubes are the other three.

The buckyball fullerene is a molecule with the chemical formula C_{60} and is one of the best-known nanoscale objects in nanoscience. Figure 4.25 depicts a buckyball. It consists of 60 carbon atoms occupying equivalent sites. Each atom is bonded to three other atoms. Single and double C—C bonds occur. The lengths of the two types of bond are 0.146 nm and 0.140 nm, respectively. That is, these bonds are practically identical. In Fig. 4.25 they are represented by light and dark-shaded lines. To form almost an ideal buckyball, nearest neighbors of carbon atoms are organized in pentagons and hexagons. Every pentagon in the case structure is surrounded by five hexagons. The truncated icosahedron structure shown in Fig. 4.25 has 12 pentagonal faces and 20 hexagonal faces at 60 vertices with a C atom at each vertex.

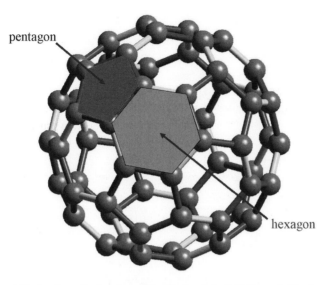

Figure 4.25 A schematic drawing of the C_{60} buckyball (fullerene). Single and double bonds are shown by light and dark shaded lines.

Because of the near-spherical shape of the truncated icosahedron, we can describe the electronic energy states of the buckyball fullerene by using the classification of quantum states developed for a spherically symmetric potential (Section 3.3) and the simplest atoms (Section 3.4). As was found there, the electron energy states can be classified by angular momentum quantum numbers (orbital numbers l). In the C_{60} fullerene, the total number of valence electrons is 240. Of them, 180 electrons are involved in the relatively deep-lying σ-bonding energy level, so that the electronic properties of the fullerene are determined primarily by the remaining 60 electrons involved in π-bonds. A total of 50 electrons may be accommodated in quantum states with orbital numbers from $l = 0$ to $l = 4$. With two-fold spin degeneracy, this gives the following level populations: $2 + 6 + 10 + 14 + 18$, as illustrated by Fig. 4.26. The remaining 10 electrons fill the energy level with the orbital number $l = 5$. Actually, the total number of these states is $2 \times (2 \times l + 1) = 22$. Now, we recall that strictly speaking a truncated icosahedron differs from a sphere and has a lower symmetry. On analyzing this lower symmetry, one finds splitting of l-states into three groups, as shown in Fig. 4.26 (h_{1u}, t_{1u}, and t_{1g} denote different types of symmetry). The lowest five-fold (excluding spin degeneracy) levels accommodate the remaining 10 electrons and can be designated as the highest occupied molecular orbitals. Two other groups of $l = 5$ states are lifted considerably; together with the $l = 6$ states they create the lowest unoccupied molecular orbitals.

Conceptually, the electronic properties of an individual fullerene molecule can be modified via replacement of a C atom with an atom having a greater or smaller number of valence electrons. Such a process can be thought of as a kind of doping. An example is azofullerene with the chemical formula $C_{59}N$, where one C atom is replaced by a N atom.

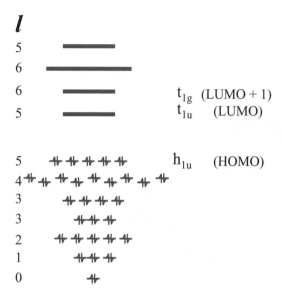

Figure 4.26 A diagram of the energy states of 60 valence electrons in the fullerene molecule. After P. Moriarty, "Nanostructured materials," *Rep. Prog. Phys.*, **64**, 355 (2001). © IOP Publishing Limited.

The existence of unoccupied molecular orbitals for C_{60} explains effects of intermolecular interaction of the buckyballs. For example, two C_{60} molecules, being neutral systems, attract each other through polarization of electron clouds. Such an attraction is caused by van der Waals forces. Owing to these forces, two fullerenes compose the dimer C_{60}–C_{60} with the coupling estimated to be about 0.27 eV. Similar van der Waals forces provide for interaction and adsorption of the fullerenes onto surfaces of various materials. Figure 4.27 depicts the fullerene molecules adsorbed onto a Si surface. Since the van der Waals forces are relatively weak, the fullerene molecules readily diffuse on the surface and, in large concentrations, they can form hexagonally close-packed islands, as shown in Fig. 4.28.

Then, the buckyballs themselves form face-centered cubic crystals (at room temperature) with large cohesive energy in this crystal ($\sim$1.6 eV per molecule). Interestingly, the lattice constant of the C_{60} crystal has the unusually large value of 1.42 nm.

Summarizing, the fullerenes are natural and very stable nano-objects manifesting a number of interesting physical and chemical properties, which can be controlled and used in nanoscience applications.

4.9   Closing remarks

We began this chapter with the definition of crystalline materials. We introduced two very important components of crystals, i.e., the electron subsystem and the crystalline lattice. The electronic applications of a material are determined primarily by its electronic

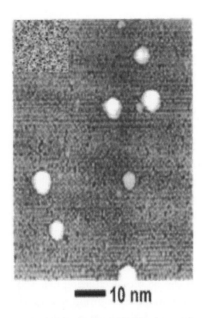

= 10 nm

Figure 4.27 An image of fullerene molecules adsorbed on a silicon surface (at low concentration). Reprinted with permission, from P. Moriarty, "Nanostructured materials," *Rep. Prog. Phys.*, **64**, 306 (2001). © IOP Publishing Limited.

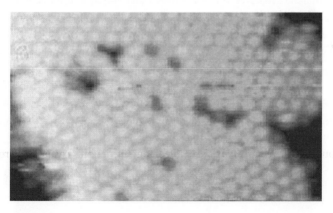

Figure 4.28 At large concentrations, the fullerenes form a hexagonal structure on a surface. Reprinted with permission, from P. Moriarty, "Nanostructured materials," *Rep. Prog. Phys.*, **64**, 358 (2001). © IOP Publishing Limited.

properties. As a result of the periodicity of crystals, one can interpret the electron energy in terms of the energy bands. Depending on the relative positions of the energy bands and their filling by electrons, a material can be described as a dielectric, semiconductor, or metal. We found that, despite the interaction of the electrons with the great number of atoms and ions composing a crystal, one can attribute a quasimomentum (a wavevector) to an electron and that, frequently, within the actual range of the wavevectors, the electron energy can be approximated by a simple dependence on wavevector. Several new notions that are necessary to describe electrons in these cases were introduced: (1) the positions of the energy minima, (2) the bandgap, and (3) the effective masses.

We analyzed semiconductor alloys and learned that for many cases of practical interest the alloy energy spectra can be described in a manner similar to that used to describe pure crystals. These alloys may be "engineered" to produce considerable variations in the electron parameters.

Next, we explained the formation of discontinuities in both valence and conduction bands when two materials are brought together to form a heterojunction. This effect provides the practical basis for modifying electron energy spectra, and it is particularly useful for spatial modulation of the potential profiles experienced by electrons and holes as well as for the creation of various artificial heterostructure-based nanostructures with energy barriers; these nanostructures include quantum wells, quantum wires, quantum boxes, and superlattices.

By analyzing the state of the art in the fabrication of heterostructures, it was established that high-quality heterostructures can be produced using materials with similar crystal properties, for example, materials from the same group. The ratio of lattice constants of the two materials is a critical parameter for such heterostructures. If these lattice constants almost coincide, as in the case of lattice-matched materials, one can produce a heterostructure without strain due to the absence of lattice mismatch or misfit imperfections. An example of such a lattice-matched system is the AlGaAs/GaAs heterostructure. For lattice-mismatched materials we have found that only thin layers can accommodate the lattice mismatch and retain near-perfect crystalline structure. The resultant structures are pseudomorphic strained layers. The strain in pseudomorphic heterostructures leads to a set of new phenomena. In particular, it affects the energy spectra in the strained layers. An example is the Si/Ge heterostructure.

Finally, we considered the structural and electronic properties of novel nanosized objects such as carbon nanotubes and carbon buckyballs.

More information on crystal symmetry, electron energy spectra, and the general theory of strain effects can be found in the following books:

C. Kittel, *Quantum Theory of Solids* (New York, John Wiley & Sons, Inc., 1963).

I. Ipatova and V. Mitin, *Introduction to Solid-State Electronics* (New York, Addison-Wesley Publishing Company, 1996).

A brief history of semiconductor heterostructures and clear motivation for their implementation is given by Professor Herbert Kroemer in his Nobel Lecture:

H. Kroemer, "Quasielectric fields and band offsets: teaching electrons new tricks," *Rev. Mod. Phys.*, **73**, 783 (2001).

Detailed treatments of semiconductor alloys and heterojunctions as well as reviews of methods of their fabrication and doping are presented in the following references:

J. Singh, *Physics of Semiconductors and Their Heterostructures* (New York, McGraw-Hill, 1993).

V. V. Mitin, V. A. Kochelap, and M. A. Stroscio, *Quantum Heterostructures* (New York, Cambridge University Press, 1999).

The following publications are devoted to calculations and discussions of electron energy spectra and wavefunctions in various quantum semiconductor structures:

T. Ando, A. B. Fowler, and F. Stern, "Electronic properties of two-dimensional systems," *Rev. Mod. Phys.*, **54**, 437 (1982).
G. Bastard, *Wave Mechanics Applied to Semiconductor Heterostructures* (New York, Halsted Press, 1988).
C. Weisbuch and B. Vinter, *Quantum Semiconductor Structures* (San Diego, CA, Academic Press, 1991).

A detailed description of carbon nanotubes and carbon buckyballs is presented in the following recent publications:

M. S. Dresselhaus, G. Dresselhaus, and P. C. Eklund, *Science of Fullerenes and Carbon Nanotubes* (San Diego, CA, Academic Press, 1996).
P. Moriarty, "Nanostructural materials," *Rep. Prog. Phys.*, **64**, 297 (2001).

4.10   Problems

1.   Using the lattice constant of silicon, $a = 5.43 \times 10^{-8}$ cm, and the fact that the number of Si atoms per unit volume, a^3, is eight, calculate the number of atoms per 1 cm^3 and the density of the crystalline silicon (silicon's atomic weight is 28.1 g mole^{-1}).

2.   Estimate the volume of the first Brillouin zone in **k**-space for a simple cubic lattice with the lattice constant $a = 5 \times 10^{-8}$ cm. Assume that the average energy of the electrons is $3k_B T/2$, where k_B and T are Boltzmann's constant ($k_B = 1.38 \times 10^{-23}$ J K^{-1}) and the ambient temperature, respectively. Estimate the volume occupied by electrons in **k**-space at $T = 300$ K with the effective mass $m = 0.1m_0$, with m_0 being the free-electron mass. Compare these two volumes and discuss whether Eq. (4.17) is valid for electrons with the parameters specified previously.

3.   Three electrons of the same energy are placed into three different energy valleys of silicon. The valleys are located at Δ points of the [100]-, [010]-, and [001]-axes. Assuming that all three electrons move along the same direction, say [100], and using Eqs. (4.17) and (4.19), find the ratio of the velocities for these electrons.

4.   Consider a valence band consisting of light- and heavy-hole branches. Assume that a heavy hole with energy E is transferred to a light-hole state with the same energy E. Find the ratios of the quasimomenta and the velocities of the hole in the initial and final states.

5.   For Al$_x$Ga$_{1-x}$As alloy, find the composition having an energy bandgap equal to 2 eV using Table 4.6. For this alloy, determine the effective masses in the Γ and X valleys using Table 4.7.

6. Assume that for some applications it is necessary to use a film of InGaAs of high quality, which can be grown on an InP substrate. By using Eq. (4.20) and data presented in Table 4.8, find the lattice-matched composition of this alloy, its lattice constant, the energy bandgap, and the wavelength of the light corresponding to this bandgap.

7. Assume that the conduction-band offset for an $Al_x Ga_{1-x} As$/GaAs heterojunction is 60% of the difference of the bandgaps of these materials. Find the composition of the AlGaAs layer necessary for the resulting heterojunction to have an energy barrier for the electrons equal to 0.3 eV. Calculate the energy barrier for the holes.

8. On the basis of the values of lattice constants given in Table 4.8, explain why it is feasible to grow stable $Al_x Ga_{1-x} As$/GaAs and $In_x Al_{1-x} As$/$In_y Ga_{1-y} As$ heterostructures; explain why it is difficult to grow stable heterostructures of GaP/SiC and InP/SiC.

9. Find the numbers of hexagons per unit cell in zigzag and armchair nanotubes. Calculate the radii of (7, 8) and (11, 10) nanotubes.

5 Growth, fabrication, and measurement techniques for nanostructures

5.1 Introduction

Having reviewed the basic properties of materials exploited in nanoelectronics, we shall now study the principal methods of high-quality material growth and nanodevice fabrication. Methods for the growth of perfect materials with controllable properties are critically important for nanostructure fabrication. Indeed, stringent requirements must be met for the growth of crystals for nanosize devices. These requirements include many factors and, first of all, ultra-high quality and purity, both controlled within extremely close limits. The following examples illustrate the term "ultra-high quality." For Si crystals used in nanodevices, concentrations of *controlled* impurities currently reach concentrations of less than one part in ten billion (1 in 10^{10}). For the case of Ge, this number is in the range of 1 in 10^{13}–10^{14}. The quality of a silicon crystal being used for nanoelectronics can be characterized in terms of the density of defects: they must be limited to several tens per 1 m^2 (!) of the Si wafer according to the Semiconductor Road Map, that was discussed in Chapter 1. The basic methods of growth of perfect crystalline materials and multilayered heterostructures we will discuss in Section 5.2.

To fabricate a nanostructure and a nanodevice two approaches can be undertaken. The first is based on a previously grown perfect material with further processing. This includes a number of fabrication stages and methods (nanolithography, etching, implantation, selective doping, etc.). In Section 5.3 we review these methods. The second approach exploits special regimes of material growth, when nanostructures are formed *spontaneously* due to the growth kinetics. Such a growth regime can control size, shape, and other properties of the nanostructures. Details of this approach are given in Section 5.4.

Progress in the refinement of fabrication techniques for making nanostructures depends on the great improvements made in characterization methods. In particular, composition and dopant distribution, lattice strain, and other parameters within nanostructures must be known with atomic-scale precision. Currently, the manipulation of a single atom (ion) in a solid is possible. In Section 5.5, we shall present the most important characterization techniques, such as atomic-force microscopy, scanning tunneling microscopy, transmission electron microscopy, and others.

The previously mentioned approaches for the production of nanostructures and nanoelectronic devices actually represent "evolutionary" improvements in the growth and processing methods applied previously in microelectronics. Nanoscale objects like carbon nanotubes and biomolecules require, in general, other techniques for production. These innovative techniques are highlighted in Sections 5.6 and 5.7.

Advances in nanotechnology require the utilization of methods and concepts from almost every area of science and engineering. Synthetic chemistry and even biology have much to offer for this emerging field. Some fundamental concepts coming from these fields can successfully be exploited. These include chemical and biological methods of surface nanopatterning, and preparing nanostructured materials with predefined, synthetically programmable properties from common inorganic building blocks with the help of DNA interconnect molecules. The basic ideas related to these chemical and biological approaches are discussed in Section 5.8.

The great technological advances brought about in mainstream microelectronics and nanoelectronics can be used for the fabrication of another class of nanodevices that employs both electrical and mechanical properties of nanostructures. This new generation of devices is commonly called *nanoelectromechanical systems* (NEMSs). Indeed, a strong enhancement of coupling between electronic and mechanical degrees of freedom appears on the nanometer length scale. This results in a new class of devices that includes nanomachines, novel sensors, and a variety of new devices functioning on the nanoscale. In Section 5.9 we study the methods of fabrication of this class of nanodevices.

5.2   Bulk crystal and heterostructure growth

Though technological methods and especially regimes of growing various types of crystals are generally different, they have a lot in common. Here we consider the common steps in growing pure materials using Si technology as an example.

Single-crystal growth

The following three steps are necessary to produce high-quality silicon crystals: (i) production of metallurgical-grade silicon (impurity level $\approx 5 \times 10^{16}$ cm^{-3}); (ii) improvement of the latter material up to electronic-grade silicon (the level of impurities is reduced to $\approx 5 \times 10^{13}$ cm^{-3} or less); and (iii) conversion to single-crystal Si ingots.

Metallurgical-grade silicon is typically produced via reaction of silicon dioxide (SiO$_2$) with C in the form of coke:

$$SiO_2 + 2C \rightarrow Si + 2CO, \qquad (5.1)$$

which requires very high temperature ($\approx 1800\,^\circ$C). Coke is a coal from which most of the gases have been removed. The silicon obtained at this step is not single-crystalline and is not pure enough for electronic applications, though it is good for some metallurgical applications such as the production of stainless steel.

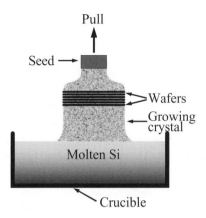

Figure 5.1 The Czochralski method for the growth of bulk semiconductors.

Further reductions in impurities can be achieved by carrying out the following reaction of the silicon with dry HCl:

$$Si + 3HCl \rightarrow HSiCl_3 + H_2. \qquad (5.2)$$

Now we obtain the trichlorosilane $HSiCl_3$, which is typically in a liquid state with a boiling point of $32\,^\circ C$. Simultaneously with $HSiCl_3$, other chlorides of impurities, such as $FeCl_3$, are formed. Since their boiling points are different, the simple fractional distillation technique can be applied: the mixture of $HSiCl_3$ and other impurity chlorides is heated and then condensed in a series of distillation towers at appropriate temperatures. By this technique $HSiCl_3$ is separated from impurities. The following reaction with H_2 then converts the trichlorosilane into highly pure electronic-grade silicon:

$$2HSiCl_3 + 2H_2 \rightarrow 2Si + 6HCl. \qquad (5.3)$$

The pure Si obtained by this process is still polycrystalline.

The final process, which converts polycrystalline silicon into single-crystal Si ingots, is based on the Czochralski method. In this method, a seed Si crystal provides a template for growth. First, this seed crystal is lowered into the molten Si material. (The melting point of Si is $1412\,^\circ C$.) Then it is raised very slowly so that the molten material touching the seed crystallizes as the seed is withdrawn from the molten material. Rotation of the seed crystal, stabilization of the temperature field, and other tricks are used to grow highly homogeneous ingots. The Czochralski method is illustrated in Fig. 5.1. Importantly, this technology facilitates doping in the course of crystal growth. Indeed, one can intentionally add precise quantities of impurities (dopants) into semiconductor melts to provide for regions of crystallization having the desired doping concentrations. This technique is used widely in growing silicon, germanium, and, with some modifications, compound semiconductors.

As the single-crystal ingot is grown, it is mechanically processed to obtain wafers of thicknesses of hundreds of micrometers, as shown schematically in Fig. 5.1. The wafers are used subsequently for producing individual devices, or integrated circuits, or for the

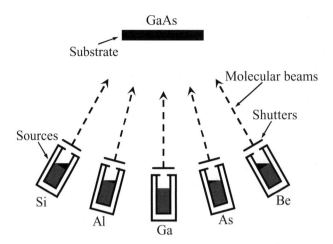

Figure 5.2 The MBE method for the growth of GaAs/AlGaAs heterostructures.

fabrication of much more complicated structures. In the following discussion, we focus
on the technique of growing single-crystal layers above a substrate wafer, which is called
epitaxial growth, or just epitaxy.

Epitaxial growth

Fabrication of a crystal layer upon a wafer of a compatible crystal makes it possible to
obtain very well-controlled growth regimes and to produce high-quality crystals with the
desired crystalline orientation at temperatures typically well below the melting point of
the substrate. During the epitaxial growth, several methods of delivering the necessary
atoms to the growing layer can be used. The most developed methods are *molecular-beam
epitaxy* (MBE), *chemical-vapor deposition* (CVD), and *liquid-phase epitaxy* (LPE). Here
we shall discuss MBE and CVD in more detail.

Molecular-beam epitaxy

The MBE method can be realized in a high vacuum, where molecular or atomic beams
deliver onto a substrate the necessary components for growing the desired crystalline
layer. For example, suppose that we want to grow an AlGaAs layer on GaAs. Then,
the substrate will be GaAs and the atomic beams are fluxes of the elements Al, Ga,
and As, as well as beams of dopants (typically, Si is used for n-doping and Be for
p-doping). Sources of the elements are contained in separately heated chambers. The
evaporated elements form beams, which are separately and closely controlled, collimated,
and directed onto the substrate surface, as illustrated by Figs. 5.2 and 5.3. Typical
flux densities in the beams are of 10^{14}–10^{16} atoms cm^{-2} s^{-1}. The substrate is held
at relatively low temperature ($\approx$600 °C for GaAs), while densities of the components
in the beams are large. This provides effective growth of the layer. A slow growth rate
($\approx$1 monolayer per second), which is often referred to as layer-by-layer growth, results in

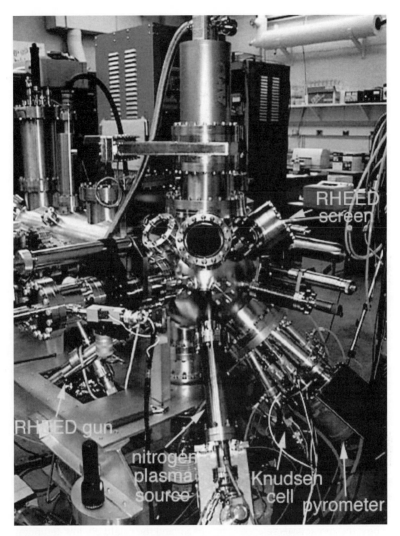

Figure 5.3 A typical MBE chamber. Used with permission from
www.ece.cmu.edu/~dwg/research/mbechamber.html. © 2007, ECE.

the growth of a high-quality layer. By controlling shutters for each beam, one can produce
abrupt changes in crystal compositions and doping concentrations on the scale of one
monolayer.

Chemical-vapor deposition

This epitaxial method allows one to realize a low-temperature growth regime and to
use high-purity chemicals for delivering the necessary atoms for growth of a crystalline
layer. The layers can be grown onto a seed crystal or substrate from mixtures of chemical
vapors containing both semiconductor elements and dopants.

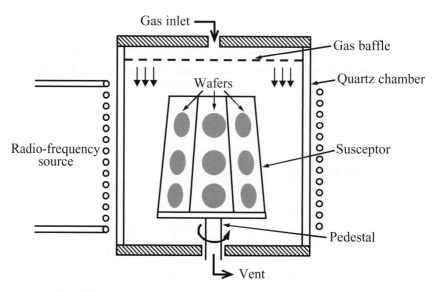

Figure 5.4 The CVD method for the growth of heterostructures.

The chemical-vapor deposition is conducted in a reaction chamber called a *reactor*, as illustrated by Fig. 5.4. In the reactor, a typical pressure of chemicals is $\approx 10^4$ Pa and heating is achieved by power from a microwave radio-frequency source.

In the case of growth of Si layers, several different gases containing Si atoms can be used. They include silicon tetrachloride ($SiCl_4$), silane (SiH_4), and dichlorosilane (SiH_2Cl_2). In the use of silicon tetrachloride, the following reaction with hydrogen occurs:

$$SiCl_4 + 2H_2 \rightarrow Si + 4HCl.$$

The reaction can be conducted at temperatures in the range of 1150–1250 °C. In the case of using silane and dichlorosilane, the reaction can be conducted at even lower temperatures (1000–1100 °C). These temperatures are well below the melting point of Si. Thus, these reactions release atoms of Si, and the relatively low-temperature regimes provide efficient crystal growth onto the seed.

For growth of A_{III}–B_V compounds, the following reactions are used:

$$A_{III}(CH_3)_3 + B_V H_3 \rightarrow A_{III}B_V + 3CH_4.$$

These reactions take place at temperatures of ≈ 600–700 °C. Dopants can be delivered, for example, by silane (n-doping) or diethylzinc (p-doping).

It is important that epitaxial methods can be applied to produce new materials that are difficult to grow by other methods. Examples are wide-bandgap nitrides of the group III elements. These include InGaN and AlGaN compounds.

In conclusion, epitaxial growth on the basis of CVD and MBE methods has become the dominant technique for the fabrication of perfect multilayered crystals of nanoscale thicknesses. Extremely high uniformity of layers has been demonstrated to be achieved by these epitaxial methods. Both group IV elements and III–V compounds are successfully grown with thickness control of the order of one monolayer. Different types of doping – uniform doping, modulation doping, and delta-doping – are realized with high accuracy. Since in the chemical reactor the partial pressures of chemicals are much higher than the pressure in the molecular beams of the MBE method, the rate of crystal growth realized in the CVD method is higher than that of MBE. The former may be used in industrial production, while the latter is rather well suited for research laboratories.

5.3 Nanolithography, etching, and other means for fabrication of nanostructures and nanodevices

In the previous section, we studied methods of growing high-quality single-crystal wafers and crystalline multilayered structures. We found that thicknesses of the layers may be on the nanometer scale. However, to produce an individual device or electric circuit scaled down to nanosize in two, or three, dimensions, one needs to exploit additional methods. First of all, on a wafer it is necessary to produce patterns corresponding to features of the nanodevice or circuitry. This may be done by using one of the so-called *nanolithography methods*.

Let us start with *photolithography*. This method involves the generation of a reticle, which is essentially a transparent quartz plate with the necessary pattern. Opaque regions on the reticle are made up of an iron-oxide layer, which absorbs ultraviolet (UV) light. A pattern on the reticle is usually made by a computer-controlled electron beam, which moves as prescribed by pattern-generation software. An electron-beam-sensitive material (the electron-beam resist) is placed onto the iron-oxide-covered quartz. The resist is exposed selectively by the electron beam and then the exposed material (positive resist) is removed. Finally, the iron-oxide layer can be selectively removed by etching to generate the desired pattern on the quartz plate.

During the next step, a thin uniform layer of photoresist is deposited onto the wafer surface. There are two types of photoresists: *positive* and *negative*. For positive resists, the resist is exposed with UV light wherever the underlying material is to be removed. In these resists, exposure to the UV light changes the chemical structure of the resist so that it becomes more soluble. The exposed resist is then washed away by developer solution, leaving windows of the bare underlying material. The mask, therefore, contains an exact copy of the pattern which is to remain on the wafer. Negative resists behave in just the opposite manner. Exposure to the UV light causes the negative resist to become polymerized, and thus more difficult to dissolve. Therefore, the negative resist remains on the surface wherever it is exposed, and the developer solution removes only the unexposed portions. Masks used for negative photoresists, therefore, contain the inverse (or photographic "negative") of the pattern to be transferred. Figure 5.5 shows the steps involved in photolithography, as well as the pattern differences generated by the use of positive and negative resists. Positive resists are now the dominant type of resists used in fabrication processes.

To transfer the patterns onto the wafer surface, the mask should be aligned with the wafer. Once the mask has been aligned accurately with the pattern on the wafer surface,

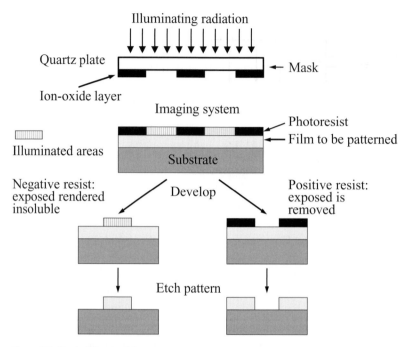

Figure 5.5 Basic lithographic processes.

the photoresist is exposed through the pattern on the mask to high-intensity UV light. One of the exposure methods, the contact method, is shown in Fig. 5.5. In this method, the photoresist is exposed to UV light while the wafer is in the contact position with the mask. As a result of the direct contact between resist and mask, very high resolution is possible in contact printing.

One of the last steps in the photolithography process is development. The results obtained after exposure and development are shown in Fig. 5.5 both for negative and for positive resists.

Actually, photolithography processes have been known for decades. To use them for nanostructures, we have to understand the limitations imposed by the wave nature of light. Indeed, as we discussed in Chapter 2, simple geometrical ray optics is applied for dimensions greater than the wavelength of light. The wave effect – diffraction of light – restricts the application of photolithography to minimum dimensional scales of about the wavelength of light. Thus, shorter wavelengths are preferable. Examples of some of the shorter wavelengths in use are 0.365 μm for UV mercury lamps and 0.193 μm for ArF excimer lasers. With these UV sources, it is possible to achieve linewidths of about 0.25 μm and 0.15 μm, respectively. Further penetration into the deep-UV region appears to be extremely difficult.

The previously mentioned diffraction limit is much smaller for X-rays, electron beams, and ion beams. Thus, advances in nanolithography are occurring as a result of the use of these short-wavelength beams. For example, electrons with an energy of 10 keV have a wavelength of about 0.1 Å; i.e., less than the lattice constants of any crystal. Now, the

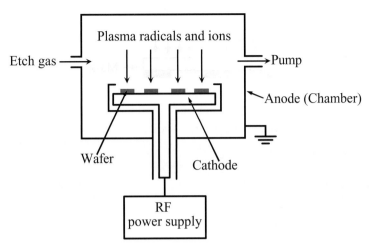

Figure 5.6 Reactive-ion etching.

ultimate linewidth is determined by interaction with the photoresist layer. In principle, it is possible to reach linewidths below 0.1 μm for direct writing with an electron beam onto the photoresist.

Etching

As soon as a resist pattern is formed, an *etching* method is generally used to produce the desired structures as shown in Fig. 5.5. There are many etching methods. A widely used method is wet chemical etching. For example, dilute HF is used to etch a SiO_2 layer covering silicon. The HF reacts with SiO_2 and does not affect the photoresist or silicon. That is, this wet chemical etch is highly selective. However, the rate of etching is the same for any direction, lateral or vertical, so the etching is isotropic. Using an isotropic etching technique is acceptable only for relatively large structures. For nanosize structures, anisotropic etching with faster vertical etching is preferable.

Anisotropic etching generally exploits a physical process, or some combination of both physical and chemical methods. The best-known method of anisotropic etching is *reactive-ion etching*. Reactive-ion etching is based on the use of plasma reactions. This method works as follows. An appropriate etching gas, for example a chlorofluorocarbon, fills the chamber with the wafers. The pressure is typically reduced, so that a radio-frequency (RF) voltage can produce a plasma. The wafer we want to etch is a cathode of this RF discharge, while the walls of the chamber are grounded and act as an anode. Figure 5.6 illustrates a principal scheme for the ion-etching method. The electric voltage heats the light electrons and they ionize gaseous molecules, creating positive ions and molecular fragments (so-called *chemical radicals*). Being accelerated in the electric field, the ions bombard the wafer normal to the surface. This normal incidence of bombarding ions contributes to the etching and makes the etching highly anisotropic. This process, unfortunately, is not selective. However, the chemical radicals present in the chamber give rise to chemical etching, which, as we discussed, is selective. Now we see that the method combines both isotropic and anisotropic components and can give good results for etching on the nanoscale.

Doping

As we discussed in Section 4.2, a perfect semiconductor is a dielectric at low temperatures: the valence band is completely filled with valence electrons, while the conduction band is totally empty. When the crystal temperature rises, some electrons can be excited into the conduction band, which results in the creation of a pair: an electron in the conduction band and a hole in the valence band. Thus, for the electron concentration, n, and the hole concentration, p, we obtain

$$n = p = n_i(T).$$

Here, $n_i(T)$ is the temperature-dependent *intrinsic density* of electrons and holes. Such a perfect semiconductor is itself called an *intrinsic material*. In practice, the intrinsic concentrations are small and can not be controlled effectively.

Thus, it is important to find a method to create the electron or/and hole concentrations needed for each particular device application. The most common method of varying electron and/or hole concentrations in a crystal is the process of adding impurities to the material; i.e., the *doping process*. The process is based on the following physical picture. When an impurity is added to a semiconductor, additional energy levels are contributed by the impurity to the energy structure of the crystal. Many of these additional energy levels fall near the bandgap energies. Here, we consider two especially important cases: (1) the impurity levels are close to the conduction band and (2) they are close to the valence band. In the former case the impurities can be *donors* of electrons to the conduction band, while in the latter case they can be *acceptors*; i.e., they will accept electrons from the valence band and thereby generate holes in the valence band. For example, impurities from group V of the periodic table of elements (P, As, Sb) are donors and impurities from group III (B, Al, Ga, In) are acceptors for group IV semiconductors. The energy "distances" of donor levels from the conduction band and of the acceptor level from the valence band are much smaller than the energy bandgap. This promotes thermally induced ionization of these impurities even at low temperatures and the creation of conducting electrons or holes. In Fig. 5.7, the population of energy states in semiconductors with donors and acceptors is illustrated for zero and finite temperatures.

We can perform the following simple estimates of energy levels of donors and acceptors. Consider first, for example, an atom from group V embedded in a semiconductor material of group IV. Such an atom has five valence electrons, of which four can participate in formation of covalent bonds with neighboring group IV atoms. But the fifth electron is an extra electron. We can consider this electron to be moving around a positive ion, i.e., the impurity can be thought of as a "hydrogen atom" embedded in a dielectric medium. If the radius of the electron state for this atom is large, we can suppose that the electron has the effective mass m^*, which is characteristic for the conduction band. By

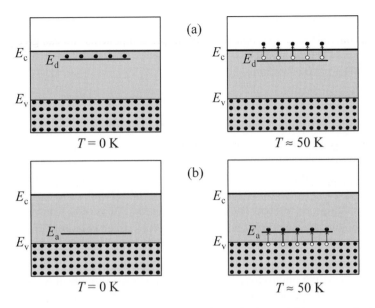

Figure 5.7 Doping of semiconductors: (a) donor doping and (b) acceptor doping.

applying the results of Section 3.4 that were obtained for the hydrogen atom, we find the ground-state energy of such a "hydrogen"-like state of the impurity:

$$E_1 = \frac{m^* e^4}{(4\pi\epsilon_0\epsilon)^2\hbar^2}, \tag{5.4}$$

with ϵ_0 and ϵ being the permittivity of free space and the dielectric constant of the material, respectively. For an acceptor atom, we may apply the same approach, considering a hole moving around a negatively charged ion. Equation (5.4) will be valid if the effective mass of the electron in the conduction band, m^*, is replaced by the effective mass of a hole, m_h, in the valence band: $m^* \rightarrow m_v$. Typical energies of donor and acceptor levels in silicon, for example, are about 30–60 meV from the conduction and valence bands, and the ionization occurs at temperatures above several tens of degrees Kelvin.

Similarly to the previously considered case of materials of group IV, in III–V compounds the atoms from group VI (S, Se, Te) are donors when they occupy sites of group V atoms. Atoms from group II (Be, Zn, Cd) act as acceptors when substituting group III atoms. A more complicated case occurs when doping is accomplished with atoms of group IV; for example, with Si or Ge. These impurities are called *amphoteric*, which means that they can act either as donors or as acceptors, depending on the sites they occupy. If such an atom occupies a group III site, it brings an additional electron and acts as a donor. When occupying a group V site, the atom accepts an electron and acts as an acceptor. In GaAs, typically, Si impurities occupy Ga sites; thus they are typically donors. However, if during the growth process there are As vacancies, Si atoms can fill these vacancies and serve as acceptors.

In conclusion, the doping of semiconductor materials provides a powerful tool for realizing the desired type of conductivity – electron conductivity or hole conductivity – and fabricating structures with the desired values of electric resistance. Doping methods provide a means for the control of the electronic properties over a wide range of values. These methods are well suited for application in the fabrication of semiconductor nanostructures.

5.4   Techniques for characterization of nanostructures

Progress in the fabrication, study, and use of nanostructures would not be possible without adequate techniques for the characterization of these structures. These techniques should allow one to determine the shape and geometrical parameters of nanostructures, the distribution of chemical composition, the strain fields, etc. Knowing all of these, one can predict the electronic and optical properties which will ultimately be relevant in applications. The questions of what the geometrical parameters of nanostructures are, and what their shapes are, can be addressed using atomic-force microscopy (AFM) or scanning tunneling microscopy (STM). Cross-sectional variants of STM and transmission electron microscopy (TEM) are sufficiently powerful tools for the investigation of buried structures. X-ray diffraction and some photoluminescence methods may also be used to determine size and shape, but in general AFM, STM, and TEM have better resolution. On the other hand, X-ray diffraction is a very powerful tool for measuring strain fields, defects, and imperfections. For the determination of statistically relevant properties, averaging over many individual nanostructures is necessary. STM and TEM often do not facilitate the study of large enough areas of a sample to yield quantitative data. Photoluminescence and diffraction techniques make possible the study of intrinsically large ensembles of nanostructures, so that a good statistical average is obtained automatically, whereas the quantification of fluctuations is sometimes difficult. The X-ray and optical methods are routinely applied for material characterization, and discussion about them can be found elsewhere.

In this section, we shall present a brief overview of new techniques based on AFM, STM, and TEM, which can be applied for a detailed characterization of nanostructures.

Scanning tunneling microscopy

This novel technique yields surface topographies and work-function profiles on an atomic scale directly in real space. In Section 4.5, we explained that the removal of an electron from the conduction band of a solid requires a certain amount of energy called the *electron affinity*. For a metal or a doped semiconductor, when the conduction band is partially filled, the energy needed to remove an electron is lower and it is called the *work function*. Let us consider two conducting solids separated by a space. In terms of classical physics, a transfer process of an electron from one solid into another can be thought of as an electron transfer over a vacuum barrier. The process requires additional energy and because of this it has a small probability. In Section 3.3 we found that, according to quantum mechanics, a particle can penetrate a classically forbidden spatial

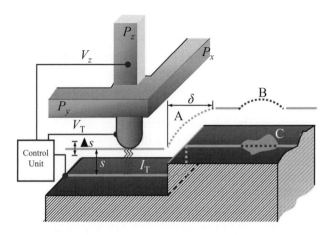

Figure 5.8 The principle of operation of the scanning tunneling microscope.

region under a potential barrier. This phenomenon was called *tunneling*. Thus, electron transfer between two solids can occur as a tunneling process through (under) the vacuum barrier. Tunneling experiments have been performed, for example, by using two metal films separated by vacuum or a solid-state insulator (a sandwich structure). Each of the metal films can be considered as an electrode and when a voltage bias is applied to these electrodes a so-called *tunneling electric current* is produced. This current can give information on electronic properties, but obviously the information will be averaged over the area of the metal-film surface. By appropriate shaping of one of the electrodes spatial resolution of far smaller scales than that of sandwich structures can be achieved. Since vacuum is conceptually a simple tunnel barrier, such experiments pertain directly to the properties of the electrodes and their bare surfaces. Clearly, vacuum tunneling offers fascinating and challenging possibilities to study surface physics and many other related topics.

The principle of STM, which is based on electron tunneling, is straightforward. It consists essentially of a scanning metal tip (one electrode of the tunnel junction) over the surface to be investigated (the second electrode), as depicted in Fig. 5.8. The metal tip is fixed to a rectangular piezodrive marked as P_x, P_y and P_z. The piezodrive is made of a piezoceramic material. The tunnel current, I_T, is a sensitive function of the gap between the tip and the surface, s; i.e., $I_T \propto V_T \exp(-A\phi^{1/2}s)$, where ϕ is the average barrier height; the numerical value of A is equal to unity if ϕ is measured in eV and s in Å. Obviously the barrier height, ϕ_0, is equal to the work function for a metal or doped semiconductor. With a typical value of ϕ of several eV, I_T changes by an order of magnitude for every 1 Å change of the gap, s. The control unit, CU, applies a DC voltage, V_z, to the piezodrive, P_z, such that I_T remains constant when scanning the tip with P_x and P_y over the surface. At constant function ϕ, $V_z(x, y)$ yields the topography of the surface, that is $z(x, y)$, directly, as illustrated at a surface step in Fig. 5.8. The curvature of the tip causes the abrupt step to appear to be smeared over a distance δ. For a constant tunneling current, changes in the work function, ϕ, are compensated by corresponding changes in the distance, s. Thus, a lower work function at a contamination spot C would be observed as an additional surface structure denoted by B in the Fig. 5.8. These "work-function-induced structures" and true structures can, however, be separated by the following method. Let the gap, s, be modulated by Δs while scanning at a frequency higher than the cut-off frequency of the control unit. Thus, the modulation is no longer compensated by the feedback loop of the control unit. As a result, the current will be modulated by ΔI_T. Then, the ratio $j_d \equiv \Delta(\ln I_T)/\Delta s \approx \phi^{1/2}$ directly gives the work function revealing the spot C, in the simple situation shown in Fig. 5.8. Since V_z, I_T, and j_d can be measured, the topography and the work function can be reconstructed.

This principal scheme of tunneling microscopy provides (i) stability of a vacuum gap in the sub-Å range and (ii) a lateral resolution in the Å range. This requires excellent vibration damping and very sharp tunnel tips. The first requirement is met by using highly developed and clever mechanical means. In fact, it is possible to use bungee cords if they are properly placed and have the desired elastic properties! Tunnel tips used nowadays are typically made of tungsten or molybdenum wires with tips of overall radii <1 μm. However, the rough macroscopic grinding process creates many rather sharp minitips. The tunnel current is extremely sensitive to the vacuum gap, s; this is why the *minitip closest to the sample* defines the whole current through the tip. Actually, the lateral resolution is given by the width of the *tunnel channel*, which is extremely narrow. Additionally, focussing of the tunneling current (in addition to the geometrical one) occurs due to a local lowering in height of the tunnel barrier at the apex of the tip. At present, the resolution of scanning tunneling microscopy reaches 0.05 Å vertically and well below 2 Å laterally.

Scanning tunneling microscopy is subject to some restrictions in application: only conductive samples can be investigated, and measurements usually have to be performed in ultra-high vacuum.

On the other hand, the tunnel current is sensitive to material composition and strain. Atomic resolution in both lateral and vertical directions makes STM an ideal tool for the investigation of growing surfaces and facets at this scale, which can give insight into growth mechanisms. STM systems *attached to a growth chamber* allow measurements to be made without breaking the vacuum after growth.

The tunnel current in STM is sensitive only to a thin layer at a sample surface, and therefore it might seem that buried structures are beyond the scope of STM studies. However, buried structures can be studied by STM. Indeed, after cleaving samples STM can be performed at the cleavage edge. Such a *cross-sectional STM* can reveal details on the inner structure of buried nano-objects. In Fig. 5.9(a), the measured and simulated cross-sectional STM profiles for a stack of InAs islands on GaAs substrate are shown. It is seen that the buried InAs islands have the shape of truncated pyramids. A compositional intermixing in the islands was found, with the GaAs composition decreasing linearly from 0.4 at the base to 0 at the top of the islands. The corresponding lattice-parameter distribution in the growth direction is shown in Fig. 5.9(c). This indicates directly an increase of compressive strain in the GaAs matrix above and below the islands.

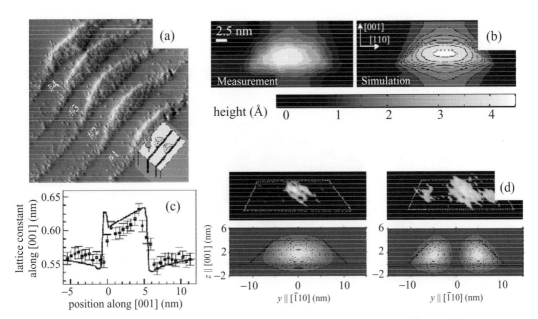

Figure 5.9 Cross-sectional scanning tunneling microscopy (STM): (a) an image of a stack of InAs islands in GaAs; (b) comparison between measured and simulated height profiles for a similar sample; (c) the lattice parameter in the growth direction in an InAs island (the experimental data were obtained from cross-sectional STM; the solid line is from a simulation assuming an In content increasing from island base to island apex); and (d) the electronic wavefunction measured at two different tip biases, compared with simulations for the ground and the first excited states. Reprinted with permission from J. Stangl, V. Holý *et al.*, "Structural properties of self-organized semiconductor nanostructures," Figs. 25 and 26, *Rev. Mod. Phys.* **76**, 725–783 (2004).

It is remarkable that, apart from providing structural information, low-temperature scanning tunneling spectroscopy has been used for *wavefunction mapping* of single-electron states in nanostructures. In Section 3.4, while studying the electron states in spherically symmetric potentials we introduced the states of different symmetries: s, p, etc. When they are applied to the InAs dots (islands) the STM methods directly reveal s-, p-, d-, and even f-type states as made visible by an asymmetry of the electronic structure, which can be attributed to a shape asymmetry of the islands. Simulation of the electron ground state and first excited state of an InAs island corresponds well to the STM image, showing that the wavefunctions in such islands are indeed atom-like; see Fig. 5.9(d). In the panel on the left, these are electrons in the ground state; in the panel on the right, electrons from both the ground state and the first excited state contribute to the measured electron distribution. These two measurements were performed at different voltages at the STM tip: at a low bias of 0.69 V, only s electrons contribute, whereas at a larger bias of 0.82 V, both s and p electrons contribute to the STM image.

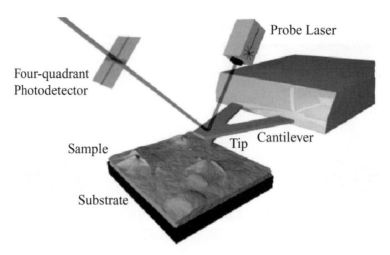

Figure 5.10 Atomic-force microscopy in the contact mode. The tip has dimensions of about 30–50 nm at the end.

Atomic-force microscopy

An atomic-force microscope measures the force between the sample surface and a very fine tip. The force is measured either by recording the bending of a cantilever on which the tip is mounted – the contact mode – or by measuring the change in resonance frequency due to the force – the tapping mode. For example, in Fig. 5.10, the contact-mode AFM technique is sketched. With a typical resolution of several nanometers laterally and several Å vertically, AFM is ideally suited to characterize the shapes of nanostructures. For large scan sizes up to 100 μm × 100 μm, the lateral arrangement can also be obtained. With AFM, any surface can be investigated; hardly any sample preparation is required. A drawback of AFM is that only structures on a surface can be investigated. Furthermore, most semiconductor materials oxidize under ambient conditions, so that, strictly speaking, the AFM images usually show the surface of this oxide. When obtaining quantitative data such as lateral sizes and heights of structures, this has to be kept in mind, as well as the fact that the image is actually a convolution of the sample's surface morphology with the shape of the microscope tip. Figure 5.10 is a schematic view of a contact-mode AFM. Essentially, a micrometer-size cantilever has an extremely sharp tip attached to it, which is sharpened to about 30–50 nm at the end. A low-power probe laser beam is reflected off the top of the cantilever and into a four-quadrant photodetector, which records the position of the reflected beam. Note that the probe beam need not be perfectly aligned (as long as some part of the beam is reflected into the detector, and the surface does not reflect too heavily into the detector), and need not even be smaller than the detector (since the difference between the quadrant signals allows the determination of the beam position). The photodetector measures the position of the reflected beam, which in turn gives information about the position of the cantilever and hence the tip. If the whole apparatus is raster-scanned across the surface (or the sample is scanned under the microscope), then an image of the surface relief can be generated.

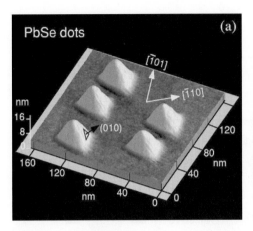
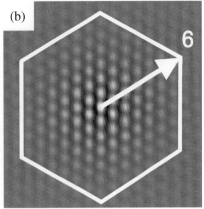

Figure 5.11 PbSe islands with [001]-type facets: (a) an AFM image of the top surface of a PbSe/PbEuTe island multilayer; (b) an AFM image of an area of 3 μm × 3 μm of the top surface of a PbSe/PbEuTe island multilayer. Islands are arranged in a regular array up to the sixth-nearest neighbor. Reprinted with permission, from J. Stangl, V. Holý *et al.*, "Structural properties of self-organized semiconductor nanostructures," Figs. 25 and 26, *Rev. Mod. Phys.* **76**, 725–783 (2004).

Examples of the quantitative analysis of AFM images are shown in Fig. 5.11. There, the top surface of PbSe/PbEuTe multilayers is shown. Both materials are semiconductors. From Fig. 5.11(a), one can see that PbSe forms triangular pyramids with [001]-type side facets. The lateral ordering can also be analyzed by AFM. In Fig. 5.11(b), a hexagonal in-plane arrangement of the pyramids is evident.

Transmission electron microscopy and scanning electron microscopy

Among the methods which allow one to "see" things at the nanometer scale, two types of electron microscopy play an important role. Transmission electron microscopy (TEM) makes possible the visualization of thin slices of material with nanometer resolution. This technique has subnanometer resolution, and, in principle, can resolve the electron densities of individual atoms. A TEM operates much like an optical microscope, but uses electrons instead of visible light, since the wavelength of electrons is much smaller than that of visible light. As we have already discussed, the resolution limitation of any microscopy is based on the wavelength of the probe radiation. As studied in Chapter 2, the electron wavelength is much smaller than that of visible light. Since electrons are used instead of light, glass lenses are no longer suitable. Instead, a TEM uses magnetic lenses to deflect electrons. Beyond this, a TEM is very similar to a conventional microscope, complete with condenser lenses, objective lenses, and projector lenses.

In a TEM, the electrons are collimated from the source and passed through the sample, and the resulting pattern of electron transmission and absorption is magnified onto a viewing screen. The image is typically recorded with a charge-coupled-device (CCD) camera, whose working element is a Si chip with the surface divided into a large array of pixels sensitive to charge carried by electrons. In scanning electron microscopy (SEM),

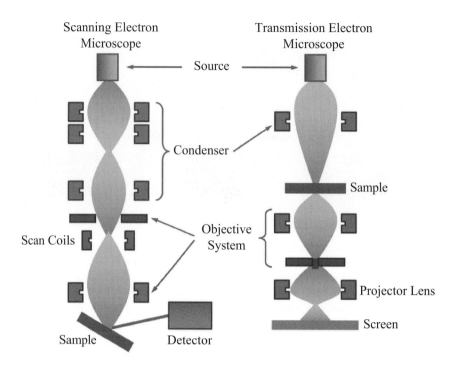

Figure 5.12 Schematic diagrams of a scanning electron microscope (SEM) and a transmission electron microscope (TEM).

the electron beam is not projected through the whole sample area. Instead, it is raster-scanned across the surface, and the secondary electrons, or X-rays, emitted from the surface are recorded. This generates a lower-resolution image, but allows the direct mapping of surface features, and can even be used for elemental analysis. Both types of electron microscope are depicted schematically in Fig. 5.12.

The electron microscopy techniques are widely used due to their very high spatial resolution and sensitivity to composition. TEM can be performed either on thin slices parallel to the sample surface (plane-view TEM) or on cross-sectional slices. Hence, buried nanostructures can be examined well by TEM, with some restrictions due to specimen preparation: in many cases, the lateral island diameter is comparable to the slice thickness. The image analysis is often not straightforward but requires elaborate image-analysis techniques and/or model calculations. Compared with other techniques, usually very small areas are investigated, so that no statistically averaged values can be obtained.

We present here only one example of atomically resolved transmission electron microscopy, from which the positions of unit cells, strain, and composition information were derived. The results obtained, after digital analysis of the lattice image, are given in Fig. 5.13: the strain distribution (i.e., distortion of atoms constituting the island from their regular positions) for an InGaAs island on a GaAs substrate is depicted. Remarkably, these techniques make it possible to visualize a detailed map of the strain for an object of size a few tens of nanometers. From Fig. 5.13, it is seen clearly how

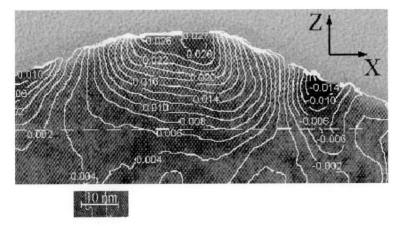

Figure 5.13 Strain distributions obtained from the TEM images of InGaAs islands in GaAs by using the method of digital analysis of lattice images. Reprinted with permission, from J. Stangl, V. Holý *et al.*, "Structural properties of self-organized semiconductor nanostructures," Fig. 27(a), *Rev. Mod. Phys.* **76**, 725–783 (2004).

the strain increases at the apex of the InGaAs island, while around the island the strain changes its sign.

In conclusion, powerful characterization techniques have been developed to study nanosize objects. The techniques give three-dimensional images in real space and on an atomic scale in all three dimensions. The methods are nondestructive. They provide means to perform structural and chemical analyses of the materials used in nanostructures. Moreover, these techniques make it possible to observe and measure directly the electron distributions inside the nanostructures; that is, it is possible to observe the electron probability densities!

5.5   Spontaneous formation and ordering of nanostructures

From the previous section, it is evident that crystal-growth and device-fabrication techniques are highly developed and are already having an impact on nanoscale semiconductor structures and devices. Every step toward realizing and perfecting artificial nanostructures involves a number of new physical and chemical processes and requires very serious efforts and technical innovations.

Importantly, Mother Nature points to another way to produce nanostructures. Indeed, the phenomenon of spontaneous formation of periodic domain structures in solids with a macroscopic periodicity has been known for several decades. Progress in TEM, STM, and AFM is facilitating the reliable and accurate observation, investigation, and control of surfaces and periodic structures with the characteristic periodicity of 1–100 nm. This opens a new way to use self-organizing growth processes as well as the formation of periodically ordered structures on semiconductor surfaces for the direct fabrication of quantum nanostructures and devices.

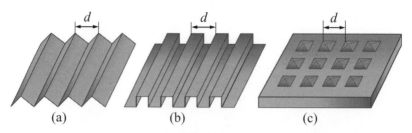

Figure 5.14 Three types of periodic structures: (a) periodically faceted surfaces; (b) planar domains; and (c) three-dimensional strained islands.

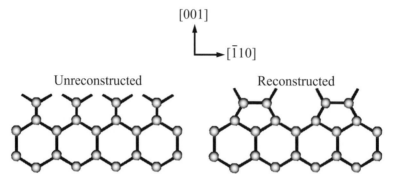

Figure 5.15 Reconstruction of Si surface.

There are three distinctive classes of spontaneously formed nanostructures on a surface, as illustrated in Fig. 5.14. These nanostructures are (1) periodically faceted surfaces (Fig. 5.14(a)), (2) periodic structures of planar domains (Fig. 5.14(b)), and (3) ordered arrays of three-dimensional coherently strained islands in lattice-mismatched heteroepitaxial systems (Fig. 5.14(c)). Despite the fact that the geometries of the three classes are different, there exist common features for all of these nanostructures. The main one is that the driving force of the periodic ordering is a long-range elastic interaction. The elastic-strain field is created due to the discontinuity of the intrinsic elastic properties on domain boundaries and/or by the lattice mismatch between two materials composing the heteroepitaxial system. The three classes depicted in Figs. 5.14(a)–(c) are equilibrium structures. In particular, they can be observed upon annealing of the crystal, or upon interruption of the crystal growth. Here we restrict our discussion to the formation of three-dimensional coherently strained islands.

We begin with a brief discussion of the strain arising at surfaces of solids. Since atoms in the surface layer of any material are in a different environment from that experienced by those in the bulk, the surface layer energetically favors a lattice parameter different from the bulk value in the directions parallel to the surface. Being adjusted to the bulk lattice, the surface layer is intrinsically *stretched* or *compressed*. Therefore, the surface is characterized by intrinsic surface stress. At the surface, even the symmetry of the crystal can be changed; this is known as *surface reconstruction*. Figure 5.15 illustrates such a reconstruction of the Si surface.

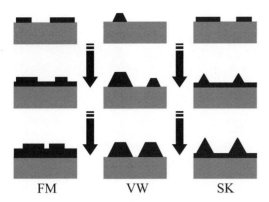

Figure 5.16 Three growth modes of heteroepitaxial systems: Frank–van der Merwe (FM), Volmer–Weber (VW), and Stranski–Krastanow (SK).

To some extent the intrinsic surface stress of a solid is analogous to the surface tension of a liquid. However, there is a fundamental difference between the properties of a liquid surface and those of a solid surface. Indeed, liquids are typically isotropic. The processes of formation and deformation of a liquid surface are identical and can be described by a single parameter that characterizes the energy of cutting of bonds on the surface. This is the so-called *surface energy*. However, in crystals the process of the formation of a surface is quite different: the distance between atoms changes and symmetry can change. This is quite different from the simple cutting of bonds. In addition, crystals are anisotropic and the energy needed to create a free surface of a given orientation depends on the orientation.

These features of the surfaces of solids give rise to different epitaxial growth regimes. Three growth modes are distinguished traditionally. They are the (1) Frank–van der Merwe (FM), (2) Volmer–Weber (VW), and (3) Stranski–Krastanow (SK) growth modes. Respectively, they can be described as (1) layer-by-layer growth, (2) island growth (three-dimensional), and (3) layer-by-layer plus island growth. These three modes are illustrated in Fig. 5.16. The particular growth mode for a given system depends on the surface energies and on the lattice mismatch between the material of the substrate and that of the grown layer. The growth regimes discussed in Section 5.2 are of the FM type.

In the following discussion we define the surface energy as the *excess energy of a very thin distorted layer* (one or two monolayers) at the free surface of a solid. The interface energy is the excess energy of a thin layer at the interface between two solids. In lattice-matched systems, the growth mode is governed only by interface and surface energies. Let γ_1 and γ_2 be the surface energies of the substrate and epitaxial layer (epilayer), respectively, and let γ_{12} be the interface energy. Then, if the sum of the epilayer surface energy and the interface energy is lower than the energy of the substrate surface, i.e., $\gamma_1 > \gamma_2 + \gamma_{12}$, the deposited material wets the substrate and the FM growth mode occurs. A change in $\gamma_2 + \gamma_{12}$ can drive a transition from the FM to the VW growth mode. These two modes of coexistence of crystalline materials are quite analogous to those of a liquid.

For a strained epilayer with small interface energy, γ_{12}, the initial growth may occur layer-by-layer; however, as the layer becomes thicker it may lower its increasing strain

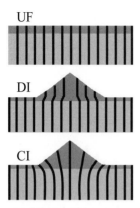

Figure 5.17 Elastic-strain relaxation during Stranski–Krastanow growth (schematic). Light gray areas denote the substrate, dark gray areas denote the lattice-mismatched epilayer. The lines symbolize lattice planes. Top, uniformly strained film (UF); middle, dislocated relaxed islands (DI); and bottom, coherently strained islands (CI).

energy by forming *isolated islands* in which strain is relaxed. This results in the SK growth mode. Thus, the SK mode depends critically on the lattice mismatch. Three scenarios for strain relaxation are sketched in Fig. 5.17. Figure 5.17(a) corresponds to uniform pseudomorphic growth without any stress relaxation, which can exist for a very thin strained layer, as discussed in Section 4.7. Figure 5.17(b) corresponds to a *dislocated* relaxed island. Finally, the island in Fig. 5.17(c) is *coherently* strained. From the point of view of the use of self-forming nanosize islands, coherently strained structures are of most importance, because of their high quality.

The scenarios presented in Fig. 5.17 occur as a result of the interplay of several parameters: (1) the ratio of the surface energy and the energy of the dislocated interface, (2) the amount of deposited material Q, and (3) the lattice mismatch, ϵ (see Eq. (4.23) and Fig. 4.16). As an example, a diagram of possible growth regimes is depicted in Fig. 5.18. In Fig. 5.18, the various growth regimes are identified in terms of the amount of deposited material, Q, versus lattice mismatch, ϵ.

For a dense system of islands, the elastic interaction between islands via deformation of the substrate is essential. The system of interacting islands is then a system of elastic domains where the energy minimum corresponds to a structure with periodic domains. Thus, there is a possibility of growing an ordered system of nano-islands.

Following this general analysis of formation of three-dimensional islands under pseudomorphic growth of crystalline materials, we consider particular examples of such self-growing nanosize heterostructures.

The first experimental evidence of the formation of the coherently strained islands was obtained by transmission electron microscopy of the InAs/GaAs system (GaAs is the substrate, while InAs is the epilayer). As follows from the data of Table 4.8, this is the lattice-mismatched system with mismatch parameter $\epsilon \approx 7\%$. Coherently strained and stable islands have been found for many systems: Ge/Si, GeSi/Si (Si substrate of [100] orientation), AlInAs/GaInAs, InAs/InP, CdSe/ZnSe, and others.

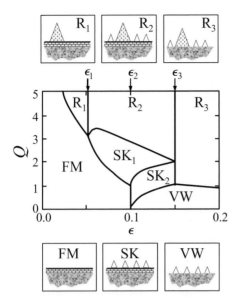

Figure 5.18 An equilibrium phase diagram of a lattice-mismatched heteroepitaxial system as a function of the total amount of deposited material Q and the lattice mismatch ϵ. The small panels at the top and bottom illustrate the morphology of the surface in the six growth modes described in the text. The small empty triangles indicate the presence of stable islands, while the large shaded ones refer to ripened islands. Reprinted with permission from I. Daruka and A. Barabasi, "Dislocation-free island formation in heteroepitaxial growth: a study at equilibrium," *Phys. Rev. Lett.*, **79**, 3708 (1997). © 1997 by the American Physical Society

   To study the formation of the islands, the growth-interruption method typically is applied. This method allows one to control the amount of the deposited material even below one deposited monolayer at the same growth temperature. For example, at a growth temperature of 480 °C the following features were revealed. When the average thickness of InAs deposition reaches a critical value of 1.6–1.7 monolayers, a morphological transition from an InAs layer to three-dimensional InAs islands occurs. After two monolayers of InAs have been deposited, an array of well-developed islands is observed. The average lateral size, the height, and the density of the islands are 100–140 Å, 50–80 Å, and 10^{10}–10^{11} cm^{-2}, respectively. Figure 5.19 depicts a single InAs island on GaAs substrate studied by scanning tunneling microscopy.

   In general, results of growth and post-growth annealing studies show that the islands grown in the SK mode are usually stable against ripening. Under given growth conditions they have well-defined sizes and shapes. For example, in SiGe/Si, essentially only four forms of islands are observed: (1) shallow pre-pyramids, (2) square pyramids, (3) "hut clusters" – elongated pyramids, and (4) large domes with facets in several directions. Figure 5.20 depicts a diagram that links possible shapes of islands with their volume and the mismatch parameters for Ge or GeSi on Si substrate. In the first stage of growth, shallow pre-pyramids appear that later convert to pyramids and then to domes. Pyramids and domes are clearly observed during growth at higher temperatures, whereas the much smaller hut clusters form at lower growth temperatures. The typical lateral sizes of the

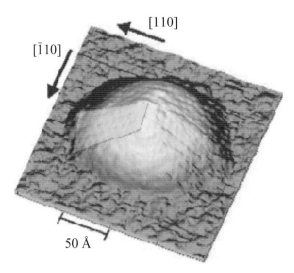

Figure 5.19 A single InAs island on GaAs substrate. Reprinted with permission, from J. Stangl, V. Holý *et al.*, "Structural properties of self-organized semiconductor nanostructures," Fig. 12(b), *Rev. Mod. Phys.* **76**, 725–783 (2004).

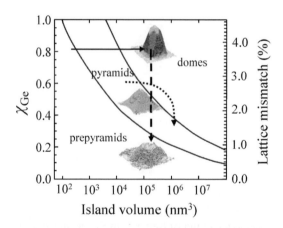

Figure 5.20 Shape transition of Ge or SiGe islands grown on Si [001] during growth (solid arrow), postgrowth annealing (dotted arrow), and Si capping (dashed arrow). The solid curves represent the critical volumes for pyramids and domes. Corresponding STM images are embedded in the diagram. Reprinted with permission, from J. Stangl, V. Holý *et al.*, "Structural properties of self-organized semiconductor nanostructures," Fig. 14, *Rev. Mod. Phys.* **76**, 725–783 (2004).

islands are 800–1,000 Å and their heights are 60–100 Å. All of these features of island growth can be explained by analyzing surface and interface energies, as well as the elastic energy of strained materials. Crystalline anisotropy is also an important factor in the formation of islands.

As for lateral correlation of the island, the mutual influence of islands on the same surface is rather weak and can be observed experimentally only for growth at very low rates; i.e., close to thermodynamic equilibrium. Typically, the lateral self-ordering

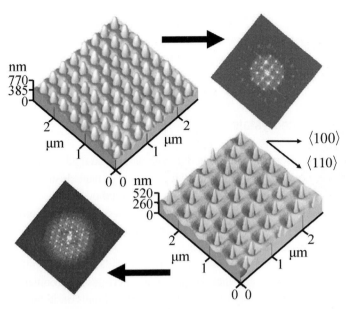

Figure 5.21 AFM images of the surface of a Ge layer grown on lithographically prepatterned Si [001] substrates. In the sample shown on the top left, the islands are arranged in a regular array along two orthogonal [110]-directions. In the sample shown on the bottom right, the unit vectors of the two-dimensional array of pits are oriented along the [100]- and [110]-directions, leading to a 45° island alignment. X-ray diffraction (top right and bottom left) demonstrates the high degree of ordering of islands. Reprinted with permission from J. Stangl, V. Holý *et al.*, "Structural properties of self-organized semiconductor nanostructures," Fig. 21, *Rev. Mod. Phys.* **76**, 725–783 (2004).

of islands can yield short-range ordered patterns with astonishing regularity; however, this process does not result in true long-range ordering. Ordering can be realized by using island nucleation on lithographically prepatterned substrates. Similarly to the case for nonpatterned substrates, for prepatterned substrates the nucleation of the islands is affected by the surface curvature and the surface stress. Thus, the substrate can be prepared to provide initial nucleation at selective locations on the substrate. Various methods can be used; e.g. shallow etching of trenches in Si and subsequent overgrowth with a GeSi multilayer, followed by Ge nucleation; and direct growth of Ge in two-dimensional periodically etched pits. The latter method results in long-range-ordered island structures, as shown in Fig. 5.21. Nucleation occurs at the intersection of the side facets within the pits. Thus, a combination of self-organizing processes in crystal growth with prepatterning methods may be used to achieve precise placement of individual islands, as well as high ordering throughout nanosize heterostructures.

In conclusion, for device applications, uniform islands with controlled positioning are required. This goal is difficult to achieve using self-organization alone, but a combination with other techniques may be successful. The following three routes are most promising: (1) a combination with conventional lithography will allow controlled positioning of self-organizing nanostructures without losing their inherent advantages; (2) seeded growth with catalytic nanoparticles facilitates the fabrication of structures significantly smaller than the SK islands; and (3) a combination of self-assembly with the techniques of organic chemistry and biochemistry. Nowadays, the controlled fabrication of extremely small structures of a few nanometers in all three dimensions and their selective addressing seems quite feasible.

5.6   Clusters and nanocrystals

The spontaneously formed and ordered nanostructures studied in the previous section are merely one type of a number of known examples of nanoclusters and nanocrystals, which can be grown by using various technological methods.

A nanocluster can be thought of as a size-dependent collection of atoms (from several atoms to several thousands of atoms). In an ideal case, the cluster is isolated. Isolation means the absence of "foreign" chemical species within the cluster volume, or on its surface. Obviously, the simplest way to realize isolation is to synthesize a cluster under vacuum conditions and then keep it in an inert-gas environment. Typically, an ideal cluster has a high density of unsatisfied *dangling* bonds on its surface. Such ideal unsupported clusters are not very useful for *functional* nanostructured materials and devices. Indeed, one needs to manipulate the clusters, to place them onto a surface in a certain order, to provide interaction with them, etc. However, if the total number of atoms in a cluster $\mathcal{N}$ is large, the fraction of atoms at the cluster surface which can have unsatisfied bonds is of the order of $\mathcal{N}^{\frac{2}{3}}$. The ratio of atoms on the surface to N decreases as $1/\mathcal{N}^{\frac{1}{3}}$ and a big cluster is almost an "ideal" cluster. Importantly, the dangling bonds have a high reactivity and thus a semiconductor cluster prepared under high-vacuum conditions will readily be oxidized on exposure to the atmosphere. Practically, cluster dangling bonds can be terminated artificially by using an organic, or inorganic, additive. Such a passivation of a cluster surface leads to an effective functionalization of the cluster.

The following two techniques of cluster and nanocrystal fabrication are different from the growth methods considered in the previous section: gas-phase and colloidal cluster syntheses.

Under gas-phase cluster synthesis, clusters are grown in a gas, prior to their passivation and deposition onto a surface. Such clusters are formed when the vapor pressure of atoms composing the clusters is much larger than it should be under equilibrium at a given temperature. Nonequilibrium atom vapors can be created by various methods, for example by laser evaporation of a solid, laser- or thermally induced decomposition of species containing necessary atoms, etc.

For example, vacuum laser-induced decomposition of silane leads to formation of Si clusters as ultra-small nanocrystals. Typically, the nanocrystal sizes range from 3 nm to 10 nm. The clusters can be deposited onto the surface of a metal, graphite, or silicon. In Fig. 5.22, an STM image of Si clusters is presented. One can see the atomic structure of the Si-(111) surface and several Si clusters of various sizes. On the left of the image, the detailed structure of a larger Si cluster is observable. Interestingly, the surface of the Si crystalline substrate oriented normally to the (111) direction has a large reactivity to

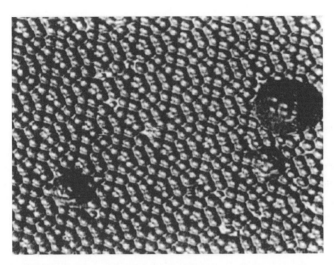

Figure 5.22 Silicon nanoclusters absorbed on a (111) surface of Si. Several clusters, of differing sizes, are seen on this STM image. Reprinted with permission from D. W. McComb, B. Collings *et al.* "An atom-resolved view of silicon nanoclusters," *Chem. Phys. Lett.* **251**, 8–12 (1996). Permission from Elsevier.

Si clusters. This explains the large sticking coefficient for Si clusters. As a result, on being deposited onto the surface, these clusters do not diffuse onto the substrate at room temperature.

Compound semiconductors can also be synthesized in the gas phase. For example, laser evaporation of materials containing Ga and As atoms in an inert gas produces GaAs clusters with dimensions of tens of nanometers. If these clusters are exposed to air, they are capped by an As-rich oxide shell. Then, the clusters can be deposited onto a surface (Si-(100), for instance) and manipulated by using an atomic-force microscope.

The colloidal synthesis of almost monodisperse nanoparticles is based on controlled nucleation and growth of clusters in a precursor-containing solution. Both metallic and semiconducting clusters can be grown. The most studied case is synthesis of III–V and II–VI compounds. The method involves injection of reagents containing the cluster constituents (for example, Cd and Se) into a hot solvent, where nucleation of CdSe occurs. Molecules of the solvent cap and thereby passivate the nucleated cluster. For the above example, *tri-n-octylphosphide* (TOPO) is used as a solvent. The reagents are Me_2Cd and TOPSe.

A hot solution of TOPO is favorable for nucleation of CdSe. TOPO capping the clusters significantly slows the cluster growth and makes the properties of the product more controllable. Careful control of the solution temperature enables production of nanoparticles with a small size dispersion. The size dispersion of nanocrystals can be within a few percent of the average diameter. The latter is a few to tens of nanometers. The shape of nanocrystals is close to spherical. On being deposited onto a surface, these nanocrystals form an ordered lateral structure. In Fig. 5.23, a tunneling electron microscope image is presented, which provides evidence of the formation of a

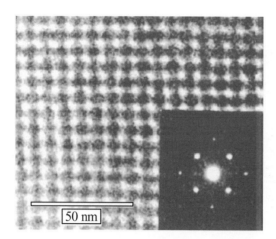

Figure 5.23 A transmission electron microscopy image of CdSe nanocrystals of size 4.8 nm grown by the colloidal method and then deposited on a surface. X-ray diffraction demonstrates the high ordering of these nanocrystals. Reprinted with permission, from P. Moriarty, "Nanostructured materials," Fig. 19(b), *Rep. Prog. Phys.* **64**, 324 (2001). © IOP Publishing Limited.

self-organized 10-nm CdSe nanocrystal superlattice. Interestingly, artificial passivation of these clusters modifies intercluster interaction. The latter is responsible for intercluster spacing. Thus, the ordering of CdSe and other clusters on a surface can be controlled by the regime of their fabrication in the colloid. In general, colloidal chemistry is convenient for "engineering" of basic properties of nanocrystals, including their shape, size, and state of surface. For example, more sophisticated nanocrystals prepared in a colloid can include a multilayered coating: the CdSe clusters can be covered by a layer of HgS followed by a layer of ZnSe and so on.

Clusters and nanocrystals possess a number of properties different from those of bulk materials. Indeed, they occupy an intermediate place between individual natural atoms/molecules and bulk crystals. For semiconducting materials, the clusters have electron energy bands practically the same as those of corresponding bulk materials. However, their small sizes restrict electron motion, giving rise to electron confinement and quantization in all three directions. As a result, the fundamental electrical, optical, and mechanical properties are modified. In the next chapter we will study some of these modifications.

5.7   Methods of nanotube growth

From the discussion of carbon nanotubes in Chapter 4, it is clear that the properties of these nano-objects are considerably different from the properties of other nanostructures and nanodevices fabricated on the basis of bulk-like materials. The same is valid for methods of growth of nanotubes. In this section, we consider several growth methods for carbon nanotubes. Basically, these methods are arc-discharge, laser-ablation, and chemical-vapor deposition. Arc-discharge and laser-ablation methods were

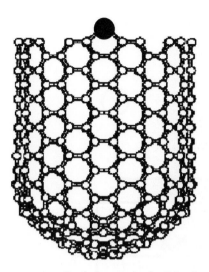

Figure 5.24 Catalytic growth of a (10, 10) armchair carbon nanotube using a metal atom (large black sphere) as a catalyst. Carbon atoms are depicted by white balls. Reprinted with permission, from J. Charlier and S. Iijima, "Growth mechanisms," in M. S. Dresselhaus, G. Dresselhaus, and Ph. Avouris (Eds.), *Carbon Nanotubes*, Fig. 11, Topics in Applied Physics vol. 80 (Berlin, Springer-Verlag, 2001), 55–79.

historically the first approaches used to fabricate nanotubes. Chemical-vapor-deposition growth methods are new and highly promising for scale-up of defect-free nanotube materials.

Carbon nanotubes are constituted solely of carbon atoms. Thus, any growth method for these nanostructures should first provide carbon atoms, and then condense the carbon vapor under certain nonequilibrium conditions at temperatures below the graphene melting point (≈ 4100 K).

Carbon-vapor condensation can result in several different forms of carbon-containing products: fullerenes, graphitic layers, nanotubes, etc. To obtain a considerable yield of the nanotubes, corresponding growth regimes should be used. For example, relatively high temperatures (1000–1300 K) are necessary to form single-walled nanotubes. Indeed, the rolling-up of a graphene sheet into a tube requires additional energy, especially for small diameters of single-walled tubes. Then, it was found that the key role in nanotube growth can be played by catalytic processes involving transition metals (iron, nickel, cobalt, etc.). Particular catalytic mechanisms are complicated.

Figure 5.24 illustrates one such mechanism for growing a (10, 10) armchair nanotube (white balls) with a Ni (or Co) atom (large black sphere) chemisorbed onto the open tube edge. Though Co and Ni atoms are strongly bound, they are still very mobile at the edge of the growing tube. Consequently, the metal catalyst keeps the tube open as a result of its mobility around the open edge, ensuring that any pentagons or other high-energy local structures are rearranged to hexagons. The latter occurs through the exchange mechanism: the metal catalyst assists two incoming carbon atoms (or a C_2 molecule) in the formation of carbon hexagons, thus increasing the tube's length.

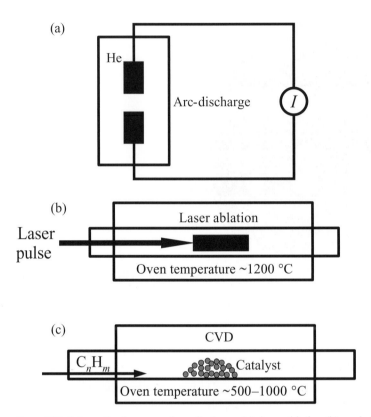

Figure 5.25 Schematic diagrams of arc-discharge (a); laser ablation (b); and chemical-vapor-deposition (c) methods of growth of carbon nanotubes.

Arc-discharge and laser ablation

In an arc-discharge, carbon atoms are evaporated by a plasma of helium gas ignited by high currents passed through opposing carbon anode and cathode as illustrated by Fig. 5.25(a). Evaporation of carbon atoms in an arc-discharge is a strongly nonequilibrium process, thus these atoms should undergo sequential condensation at temperatures below the melting point of graphite (3000 K). Carbon nanotubes arise as one of the products of this process. The synthesized multi-walled nanotubes have lengths of the order of 10 μm and diameters in the range 5–30 nm. The nanotubes are typically bound together by molecular van der Waals interactions and form tight bundles.

For the growth of single-walled tubes, a metal catalyst is needed in the arc-discharge system. For example, producing substantial amounts of single-walled nanotubes is possible by arc-discharge with the use of a carbon anode containing a small percentage of cobalt catalyst in the discharge camera. As a result, abundant single-walled nanotubes are generated in the soot material. Optimization of the growth of single-walled carbon nanotubes in an arc-discharge is achieved by using a carbon anode containing a large atomic percentage of a transition metal (for example, up to 4% nickel catalyst).

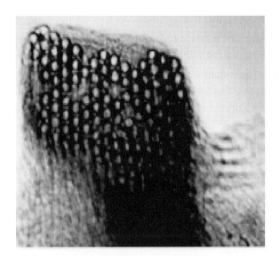

Figure 5.26 A bundle of carbon nanotubes grown by laser ablation. Reprinted with permission, from Hongjie Dai, "Nanotube growth and characterization," in M. S. Dresselhaus, G. Dresselhaus, and Ph. Avouris (Eds.), *Carbon Nanotubes*, Fig. 4, Topics in Applied Physics, vol. 80, pp. 29–53 (Berlin, Springer-Verlag, 2001).

The process of growth of high-quality single-walled nanotubes takes place also during the so-called *laser-ablation* (laser-oven) method illustrated by Fig. 5.25(b). The method utilizes intense laser pulses to ablate a carbon target containing 0.5 atomic percent of catalytic nickel and cobalt. The target is placed in a tube-furnace heated to 1200 °C. During laser ablation, a flow of inert gas is passed through the growth chamber to carry the grown nanotubes downstream to be collected on a cold finger, where carbon condensation occurs. The resulting tubes are primarily in the form of ropes consisting of tens of individual nanotubes close-packed into hexagonal crystals via the van der Waals interactions. Packaging of individual nanotubes into a bundle is clearly seen in Fig. 5.26.

Typically, arc-discharge and laser-ablation methods lead to a number of by-products: fullerenes (graphitic polyhedrons, some with enclosed metal particles), amorphous carbon, etc. Further purification is necessary to obtain the nanotubes. This process involves refluxing the nanotubes in a nitric acid solution for an extended period of time, which oxidizes away amorphous carbon particles and removes metal catalyst species. Production of single-walled nanotubes of high quality by laser ablation and arc-discharge is relatively simple and has already led to the wide availability of samples useful for studying their fundamental properties.

Chemical-vapor deposition

A schematic experimental setup for chemical-vapor-deposition growth is depicted in Fig. 5.25(c). The growth process involves heating a catalyst material to high temperatures in a tube furnace and flowing a hydrocarbon gas through the tube reactor for a period of time. Materials grown over the catalyst are collected upon cooling the system to room temperature. Parameters controlling nanotube growth are the hydrocarbons, catalysts, and growth temperature. The active catalytic species are typically transition-metal nanoparticles formed on a support material such as alumina. The growth mechanism involves the dissociation of hydrocarbon molecules catalyzed by the transition metal, and dissolution and saturation of carbon atoms in the metal nanoparticle. The precipitation of carbon from the saturated metal particle leads to the formation of tubular carbon solids. Tubule formation is favored over formation of other forms of carbon such as graphitic sheets with open edges. This is because a tube contains no dangling bonds.

Relatively high temperatures (1000–1300 K) are necessary to form single-walled nanotubes with small diameters and allow the production of nearly defect-free nanotube structures. Among all hydrocarbon molecules, methane is the most stable at high temperatures against self-decomposition. The methane chemical-vapor-deposition approach is promising for enabling scale-up of defect-free nanotube materials to the kilogram or even ton level.

Directed growth of single-walled nanotubes

Ordered, single-walled nanotube structures can be grown directly by methane chemical-vapor deposition on catalytically patterned substrates. Consider, for example, a method developed to grow suspended nanotube networks on substrates containing lithographically patterned silicon pillars. The growth starts with developing a liquid-phase catalyst precursor material that has the advantage over solid-state catalysts of allowing the formation of uniform catalyst layers for large-scale catalytic patterning on surfaces. The precursor material consists of a triblock copolymer, aluminum, iron and molybdenum chlorides in mixed ethanol and butanol solvents. The aluminum chloride provides an oxide framework when oxidized by hydrolysis and calcination in air. The triblock copolymer directs the structure of the oxide framework and leads to a porous catalyst structure upon calcination. The iron chloride also can lead to catalytic particles needed for the growth of nanotubes. The catalyst precursor material is first spun into a thin film on a polydimethyl siloxane stamp, followed by contact printing to transfer the catalyst precursor selectively onto the tops of pillars pre-fabricated on a silicon substrate. The stamped substrate is calcined and then used in chemical-vapor-deposition growth.

Remarkably, the nanotubes grown from the pillar tops tend to be directed from pillar to pillar. The directed growth of suspended single-walled nanotubes is presented in Fig. 5.27 for three different configurations of the pillars: (a) a nanotube power-line-like structure, (b) a square of nanotubes, and (c) an extensive network of suspended nanotubes. Such a directed growth can be understood as follows. Nanotubes are nucleated only on the tower-tops since the catalytic stamping method does not place any catalyst materials on the substrate below. As the nanotubes lengthen, the methane flow keeps the nanotubes floating and "waving in the wind" since the flow velocity near the bottom surface is substantially lower than that at the level of the "tower-tops." This prevents the nanotubes from being caught by the bottom surface. The nearby towers on the other hand provide

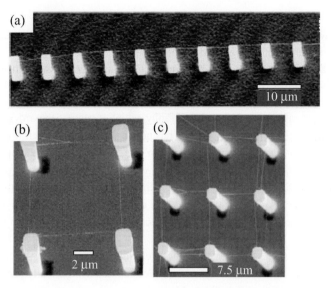

Figure 5.27 Pillar tops are connected by suspended single-walled nanotubes that form (a) a nanotube power-line-like structure; (b) a square of nanotubes; and (c) an extensive network of suspended nanotubes. Reprinted with permission, from Hongjie Dai, "Nanotube growth and characterization," in M. S. Dresselhaus, G. Dresselhaus, and Ph. Avouris (Eds.), *Carbon Nanotubes*, Fig. 9, Topics in Applied Physics, vol. 80 (Berlin, Springer-Verlag, 2001), pp. 29–53.

fixation points for the growing tubes. If a waving tube contacts an adjacent tower, the tube–tower van der Waals interactions will catch the nanotube and hold it aloft. By such a growth mode the suspended nanotubes can be made of macroscopic length. For example, tubes longer than 150 μm can be grown.

Growth of isolated nanotubes on a specific surface site

It is of importance that the chemical-vapor-deposition method allows one to grow individual nanotubes at specific sites on flat SiO_2 substrates. The approach involves methane chemical-vapor deposition onto substrates containing catalyst islands patterned by electron-beam lithography. Nanotube chips with isolated nanotubes grown from the islands have already been obtained. This growth approach readily leads to tubes originating from well-controlled surface sites, and makes possible the development of a controlled method to integrate nanotubes into addressable structures.

5.8   Chemical and biological methods for nanoscale fabrication

The technologies studied in previous sections are based primarily on physical processes and methods. The use of chemical and biological methods opens new ways for nanofabrication. It is obvious that chemical synthesis can produce a wide range of nanoparticles, including inorganic, organic, and biological nanosystems. In addition, chemistry and biology can be complementary to existing physical nanofabrication methods.

In particular, selected techniques of chemistry and biology portend alternatives to conventional photolithography. One approach is based on self-assembly phenomena. It is known that there are molecules that, due to physical, biological, or chemical processes, tend to congregate into long-range periodic structures. They can facilitate the realization of regular periodic masks on the surface of a metal or a semiconductor with a characteristic period of 10–50 nm and with internal pores of diameter 5–25 nm. Such resolution is superior to that of conventional photolithography, which is now of the order of 100 nm. In addition, the self-assembly methodology has potential for being much less expensive and less time-consuming than the electron-beam lithography techniques currently used for writing such small structures.

The chemistry methods can be used to tailor the chemical composition and structure of a surface. This can be achieved through a new direct-write tool – dip-pen nanolithography, which generates surface-patterned chemical functionality with a length scale of 1–100 nm.

In turn, achievements in biological science generate principally new methodologies for preparing nanostructured materials with predefined and synthetically programmable properties from given inorganic building blocks.

Chemical and biological innovations for nanotechnologies are briefly considered in this section.

Chemical self-assembly of nanoscale structures

Here we consider chemical systems based on *block copolymers*. They exhibit self-ordering, which can be used for nanopatterning.

A block copolymer is a macromolecule that consists of several polymer blocks, which typically can be grown in a few steps. In Fig. 5.28 an example of a *triblock polymer* is presented. The synthesis of this macromolecule starts with the polymerization of the styrene monomer. The amount of monomer supplied for the first step is only sufficient to create a *miniature styrene chain* (the first block) with an average degree of polymerization of 9. (Note: the degree of polymerization represents the number of monomeric units in a macromolecule or a block.) A second block with a similar average degree of polymerization grows when another chemical – isoprene – is added. In the third step, carbon dioxide (CO_2) installs *carboxyl end groups* at the top of the previous miniature diblock copolymers. One terminus of the triblock (the lower) is hydrophobic while the opposite terminus is hydrophilic. These triblock molecules have a rod–coil architecture with a *stiff rodlike segment* (the second block) covalently connected to more torsionally flexible segments. Chemical processes, typically, lead to a polydispersity of resulting triblock molecules; however, their dispersity is small: the ratio of the mass-averaged to number-averaged molecular masses ranges from 1.06 to 1.1.

Several triblock polymers can compose different clusters, as illustrated in Fig. 5.28. On being deposited onto a surface, triblock polymers form a macromolecular thin film with a high degree of self-ordering, as can be seen from the scheme in Fig. 5.29. The triblock polymer film has a typical "standing-up" structure. The shaded circle indicates a chemisorbing headgroup and the open circle an endgroup, which can be chosen from

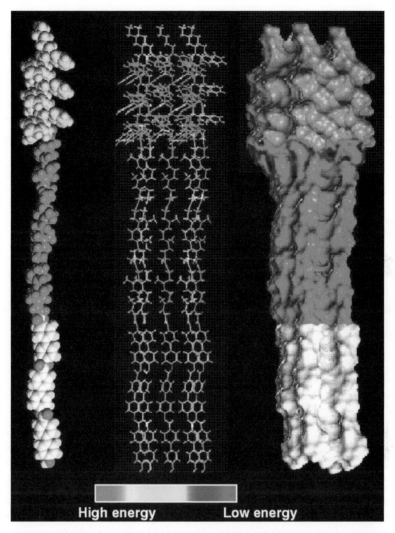

High energy   **Low energy**

Figure 5.28 Molecular graphics of the triblock self-assembling molecule (left) and a cluster of triblock macromolecules (center and right), shade-coded for energy in the various sectors of the cluster. Reprinted with permission from S., Stupp, V. LeBonheur *et al.*, "Supramolecular materials: self-organized nanostructures," *Science*, **276**, 384 (1997). © 1996 AAAS.

a variety of chemical functionalities. A top view of a real film obtained by transmission electron microscopy is depicted in Fig. 5.30. The evident ordering of macromolecules in the film is driven by the stiff rodlike segments of the triblocks. Specifically, a balance of attractive and repulsive forces mediates the formation of *macromolecular cells*, as for a crystal. In general, block copolymers easily cover the surfaces of various metals and semiconductor crystals and form two-dimensional periodic structures of various symmetries ranging from low-symmetry to square and hexagonal structures. Artificially built polymer films have thicknesses ranging from tens to hundreds of nanometers with pores of various sizes and various shapes.

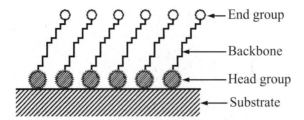

Figure 5.29 A schematic diagram of a triblock polymer film; the polymer is shown "standing-up." After F. Schreiber, "Self-assembled monolayers: from simple model systems to biofunctionalized interfaces," *J. Phys.: Condens. Matter*, **16**, R881 (2004). © IOP Publishing Limited.

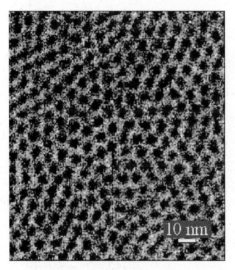

Figure 5.30 A transmission electron micrograph of a film formed by the triblock molecules, revealing regularly sized and shaped aggregates that self-organize into superlattice domains. Reprinted with permission from S. Stupp, V. LeBonheur *et al.*, "Supramolecular materials: self-organized nanostructures," *Science*, **276**, 384 (1997). © 1997 AAAS.

There are two ways to create nanostructures using self-assembly of block copolymers: (i) precipitation of metal or other inorganic crystals onto a polymer mask attached to a substrate; and (ii) using the polymer mask for the subsequent processing of the substrate, or as a support for additional auxiliary masks. After the mask has been formed on the surface of a semiconductor, the device-fabrication technology remains the same as with the conventional nanofabrication and employs dry ion or selective etching, oxidation, metallization, diffusion, selective growth, etc.

Biological methods

As examples of the use of biological methods as elements of nanoscale technologies, we consider nanopatterning and nanoassembly techniques that employ proteins and DNA,

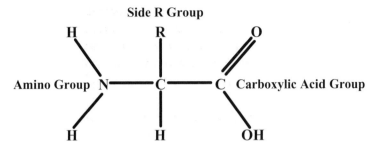

Figure 5.31 A generic amino acid.

as well as the use of biochemical molecular-recognition principles for the assembly of nanoscale inorganic building blocks into macroscopic functional materials.

First, let us consider some of the basic properties of proteins. Proteins are formed by binding together sequences of *amino acids* to form chain-like molecules whose "links" are amino acids. Proteins consisting of only a few amino acids are known as *peptides*. Important subgroups of proteins are *enzymes* and *antibodies*. Proteins constitute one of the most important classes of biomolecules and they are sometimes referred to as the "machines" of the body in view of the large number of diverse functions that they perform. As mentioned previously, amino acids are the building blocks of all proteins. *Amino acids* are among the most important molecules found in biological systems. In fact, by binding amino acids together to form chains of amino acids, it is possible to synthesize all known *peptides* (short chains of amino acids), all known *proteins* (long chains of amino acids), and all *antibodies* (selected proteins). Antibodies are proteins that have high affinities for binding to specific molecules known as *antigens*. Thus, an antibody in a cellular environment has a high probability of binding to its antigen if the antibody and antigen are in close proximity. Peptides, proteins, and antibodies are enormously important in biology. As just one example, proteins function as gates that control the flow of ions into and out of neurons.

Figure 5.31 illustrates a generic amino acid. All amino acids are derived from the generic structure shown in Fig. 5.31 by varying the side group, R. When two amino acids are in close proximity to each other, the carboxyl group on the end of one amino acid has an affinity to bind to the amino group on the end of the other amino acid. When this bonding occurs, the amino group gives up an H atom, and the carboxyl group gives up an OH moiety. The liberated OH and H form H_2O and an amide bond, CO—NH, remains as the bond linking two amino acids. In Fig. 5.32, three of the twenty common amino acids are depicted. Glycine (G) is the amino acid with the simple side group of a single hydrogen (H) atom. Arginine (R) and aspartic acid (D) are also shown in Fig. 5.32. As discussed previously, when amino acids like glycine (G), arginine (R), and aspartic acid (D) are in water, as is generally the case in biological systems, the amino (NH_2) group on the terminus of one amino acid has an affinity (tendency) for binding to the carboxyl (COOH) group on another amino acid; after such a binding event, the two amino acids are bound together by an amide bond (CO—NH bond), and a water molecule, H_2O, is produced that becomes part of the surrounding water environment.

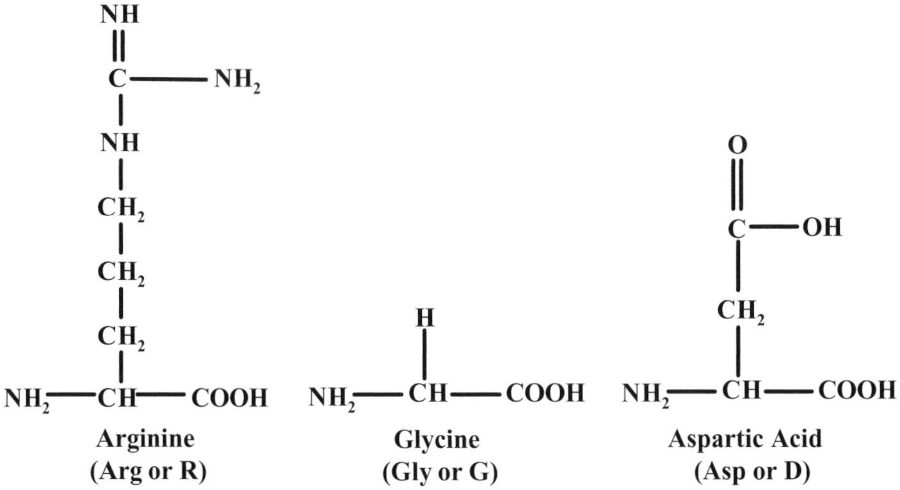

NH
‖
C————NH$_2$
|
NH
|
CH$_2$
|
CH$_2$
|
CH$_2$
|
NH$_2$——CH——COOH

Arginine
(Arg or R)

H
|
NH$_2$——CH——COOH

Glycine
(Gly or G)

O
‖
C——OH
|
CH$_2$
|
NH$_2$——CH——COOH

Aspartic Acid
(Asp or D)

Figure 5.32 Examples of some amino acids.

Such a process provides a very simple example of how chemistry may be used to self-assemble molecules and structures. By binding a few amino acids together a peptide is formed. From this discussion it is clear that amino acids are indeed the building blocks of many biomolecules, including peptides, proteins, and antibodies. As explained and illustrated previously, these biomolecules – peptides, proteins, and antibodies – may be formed through binding of amino and carboxylic groups at the two ends of the amino acids. The full set of twenty common amino acids has the following members: alanine (A), arginine (R), asparagine (N), aspartic acid (D), cysteine (C), glutamine (Q), glycine (G), glutamic acid (E), histidine (H), isoleucine (I), leucine (L), lysine (K), methionine (M), phenylalanine (F), proline (P), serine (S), threonine (T), tryptophan (W), tyrosine (Y), and valine (V).

Just as the binding of NH$_2$ and COOH groups plays a role in linking many biomolecules, they may be used to bind a COOH-functionalized (COOH-coated) quantum dot to the amino terminus of a peptide, protein, or antibody. In this way, we mimic a self-assembly technique found in nature as a key step in our arsenal of nanofabrication techniques. It is also possible to use another chemical bond found in nature – the *thiol bond* – to provide a frequently used technique for assembling nanodevice structures. Specifically, the amino acid cysteine, depicted in Fig. 5.33, and several other amino acids (leveine, valine, methionine, and serine), have special uses in chemical self-assembly since each of these amino acids contains a sulfur (S) atom in its side group, R. Specifically, the sulfur atom has an affinity for binding to a (111) surface of gold and it binds to quantum dots like CdS through an S—S or *thiol* bond. These thiol bonds are also instrumental in the phenomenon of protein folding, where they cause cysteine molecules at various points along a given protein to be attracted to each other, thus causing protein folding. Indeed, these S—S bonds are the strongest bonds found of all possible bonds among amino acids in protein systems.

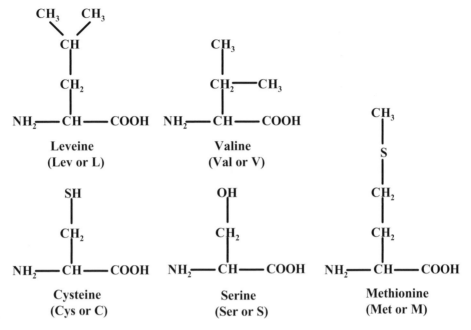

Figure 5.33 Additional examples of amino acids.

The affinity of sulfur to bind gold is sufficient to result in the wide use of S—Au bonds to bind molecules to Au surfaces. In the field of molecular electronics, there are efforts to use molecules as wires, and it is necessary to bind these wires to metal contacts. By chemically binding a sulfur atom to the end of such a molecular wire, it is possible to bind the S-functionalized ends of the wire to Au. The use of these S—Au bonds in molecular electronics is so prevalent that they are referred to as the *alligator clips* of molecular electronics.

Now that we have considered basic properties of proteins, we turn to their applications in nanofabrication. It is known that proteins form the external surface of many bacteria. Analogously to the case of block copolymer films analyzed previously, surface-layer proteins can be used as nanoscale biological masks. As shown schematically in Fig. 5.34, surface-layer proteins can be isolated from bacterial cells, and subsequently reassembled on the surfaces of a solid. The surface-layer proteins can be deposited upon various metals and semiconductor crystals. On their surfaces, surface-layer proteins form two-dimensional periodic structures of various symmetries. The sizes of the elementary cells of these surface crystals range from 3 nm to 30 nm. The thickness of an artificially built surface layer is about 2–8 nm, and it has pores of various sizes (2–8 nm) and shapes. In the upper part of Fig. 5.35, one such periodic structure is shown. The figure illustrates typical processing steps for nanopatterning: (a) deposition of surface-layer protein crystals onto the substrate; (b) metallization by deposition of metal; (c) dry etching, which allows the transfer of the pattern to the substrate; and (d) a plane view of the resulting nanostructure with highly ordered holes.

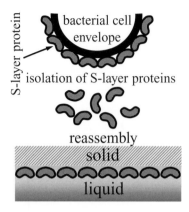

Figure 5.34 A schematic drawing of the isolation of surface-layer proteins from bacterial cells and their reassembly into crystalline arrays in suspension at a solid support.

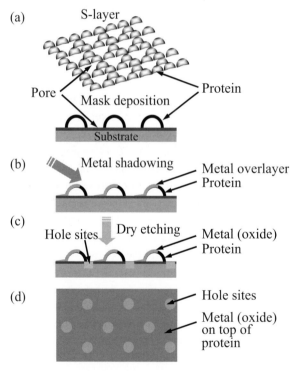

Figure 5.35 Processing steps for nanopattern transfer: (a) deposition of S-layer protein crystals onto the substrate; (b) shadow metallization by electron-beam deposition; (c) dry etching to transfer the pattern to the substrate; and (d) a plane view of the idealized nanostructure. After T. A. Winningham, S. G. Whipple, *et. al.*, "Pattern transfer from a biological nanomask to a substrate via an intermediate transfer layer," *J. Vac. Sci. Technol.*, **19**, 1796–1802 (2001). Reprinted with permission from Thomas A. Winningham, Steven G. Whipple, and Kenneth Douglas, *Journal of Vacuum Science & Technology* B, **19**, 1796 (2001). © 2001 AVS The Science & Technology Society.

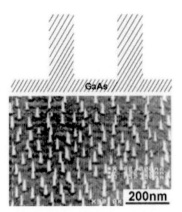

Figure 5.36 A scanning electron microscopy image of GaAs semiconductor wires obtained by means of the surface-layer protein technology. The wires have a length of 80 nm and a diameter of about 8 nm. Reprinted with permission, from M. Haupt, S. Miller *et. al.*, "Nanoporous gold films created using templates formed from self-assembled structures of inorganic-block copolymer micelles," *Advanced Materials*, **15**, 829–831 (2003), Fig. 3(a), p. 831.

Then, such a surface "superlattice" of holes can be used, for example, to fabricate semiconductor nanoscale free-standing wires. Figure 5.36 depicts a scanning electron microscopy image of semiconducting GaAs wires fabricated using surface-layer protein technology. The wires have a length of 80 nm and a diameter of about 8 nm at the top of each wire, while the base is thicker. High ordering of free-standing nanoscale wires is obtained using this biology-based method of nanopatterning.

Biological methods facilitate the realization of the assembly of nanoscale inorganic building blocks into functional materials. As an example, the biological macromolecule DNA may be used as a synthetically programmable assembler of nanoscale structures in order to fabricate a wide variety of different ensembles of selected nanocomponents. Such ensembles may be assembled in a variety of architectures. This method is based on the molecular-recognition properties associated with DNA. DNA may be prepared and functionalized with virtually any chromophore, acceptor, donor, or an active group in an automated fashion. It is known that the molecular-recognition properties arise due to *base pairing* in DNA strands. Specifically, the commonly encountered double-stranded DNA is composed of two single strands of DNA. These single strands are composed of linear sequences of the four DNA bases adenine (A), cytosine (C), guanine (G), and thymine (T). The two single strands bind together to form a single strand as a result of the high affinities for G and T to bind together and C and A to bind together. Thus, as a simple example, GTCAC and TGACA bind together to form a double-stranded DNA molecule. The two single strands, GTCAC and CAGTG, are *complementary* to each other. These molecular-recognition processes can guide the assembly of nanoparticles into extended structures. In principle, the method selects nanoparticles with certain chemical compositions and sizes, and it controls also the distance and coupling between the particles in the resulting nanostructured materials.

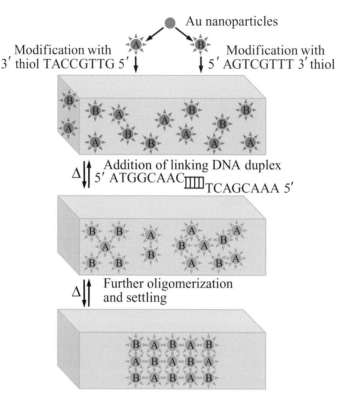

Figure 5.37 Formation of a biomolecule–inorganic cluster hybrid. Reprinted with permission, from C. A. Mirkin, "Programming the assembly of two- and three-dimensional architectures with DNA and nanoscale inorganic building blocks," *Inorg. Chem.*, **39**, 2258–2272 (2000). © American Chemical Society.

As nanoparticle building blocks, nanocrystals of metals, semiconductors, magnetic particles, fullerenes, etc. are currently available. The sizes of these nanocrystals/nanoparticles can be controlled precisely, in many cases from 1 nm to 20 nm in diameter.

The formation of a *biomolecule–inorganic cluster hybrid* is one of the important biological methods for creating nanoscale architectures. This process can be illustrated by the following example. Let two different sequences of *non-complementary* eight-base-pair DNA be synthesized with *alkanethiol endgroups*. Then, two batches of Au nanoparticles are functionalized with these DNA strands. In the scheme presented in Fig. 5.37 this functionalizing process is shown in the upper part of the figure. Two particular sequences of DNA molecules are indicated as well. If the two batches of Au particles were simply mixed, there would be no DNA recognition and hence no particle aggregation. However, the addition of *linker* DNA strands changes the situation. Such a linker molecule is composed of a DNA molecule that has three regions: a central region of double-stranded DNA and two end regions that are composed of unpaired single strands of DNA. These unpaired single-stranded ends are known as *sticky ends*. If the linker DNA molecules contain eight-base-pair sticky ends that are *complementary* to the base pairs attached

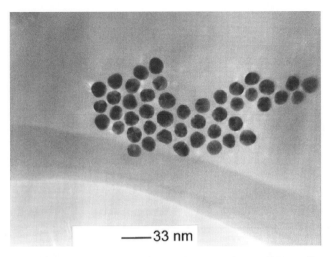

Figure 5.38 A transmission electron microscopy image of a two-dimensional structure of Au nanoparticles linked by DNA molecules. Reprinted with permission, from C. A. Mirkin, "Programming the assembly of two- and three-dimensional architectures with DNA and nanoscale inorganic building blocks," *Inorg. Chem.*, **39**, 2258–2272 (2000). © American Chemical Society.

to the Au particles, the particles start to aggregate, as shown in the lower part of the scheme in Fig. 5.37. The process of linking these nanoparticles with DNA molecules is known as "oligomerization" since DNA molecules are also known as *oligonucleotides*. In this example, the so-called sticky ends, 5′ATGGCAAC and TCAGCAAA5′ are single strands of DNA bound to opposite ends of a segment of double-stranded DNA. The double-stranded portion of this molecule is represented by the ladder symbol in Fig. 5.37. As is illustrated in Fig. 5.37, the 5′ATGGCAAC sticky end binds to the 3′thiolTACCGTTG complementary molecule which is bound to an Au quantum dot by this bond. Likewise, the TCAGCAAA5′ sticky end binds to the 5′AGTCGTTT3′thiol complementary molecule which is bound to an Au quantum dot by the thiol bond. Thus, an object can be built from nanosized building blocks linked by DNA molecules. By such a method, both two-dimensional and three-dimensional ordered nanosystems have been fabricated. Figure 5.38 depicts a transmission electron microscopy image of a two-dimensional structure of Au nanoparticles linked by DNA molecules.

Similarly, semiconductor nanocluster–DNA hybrids have been used to construct extended macroscopic structures of CdSe/ZnS quantum dots interconnected with DNA molecules. The wavelength of light emitted by a quantum dot is determined by its size as well as by the material. Table 5.1 summarizes the bandgaps of some direct-bandgap materials that are used to fabricate quantum dots. The materials represented in Table 5.1 have wavelengths spanning the ultraviolet, visible, and infrared regions of the electromagnetic spectrum.

When we integrate DNA with inorganic building blocks, we can exploit some of the properties of the latter. For example, semiconductor nanocrystals hybridized with DNA can be used as fluorescent biological labels, because of specific optical properties of

Table 5.1 The bandgaps of the direct-bandgap
bulk materials used for fabrication of quantum dots

Compound semiconductor	Bandgap (eV)
AlN	6.15
CdS hexagonal	2.4
CdS cubic	2.55
CdSe hexagonal	1.75
CdSe cubic	2.17
CdTe	1.49
PbS	0.41
PbSe	0.27
ZnS	3.68
GaN	3.36

these nanoclusters such as their emission at well-defined wavelengths. These optical properties will be discussed in the next chapters.

Although DNA arguably is the most tailorable and versatile molecule for organizing nanoscale materials into extended structures, its use has some limitations. Most notably, it is not a high-temperature material, and therefore the structures initially generated from DNA interconnects will not be stable at elevated temperatures.

As illustrated previously, NH_2 and COOH groups play a role in linking many biomolecules, and may be exploited further in interconnecting ensembles of quantum dots. Consider the case in which ZnS-coated CdSe quantum dots (QDs) coated with carboxyl groups are functionalized with GGGC peptides. These CdSe–ZnS QDs functionalized with GGGC peptides will be denoted simply as CdSe–ZnS–GGGC. This commonly used notation does not specify the number of peptides bound to each quantum dot. In practical applications, the number of peptides may vary from one to twenty or more depending upon the application. In the present case, where the outermost amino acid is a cysteine molecule with its sulfur-containing side group, the C amino acids may be used for chemical self-assembly. Indeed, if, as the first step in the fabrication process, an Au wire is immersed in a beaker containing a typical density of 10^{16} CdSe–ZnS–GGGC complexes per cm^3, many of the CdSe–ZnS–GGGC complexes will bind to (111) surfaces of the Au wire since the S atoms in the cysteine side groups have affinities for binding to (111) surfaces of Au. In the next fabrication step, the CdSe–ZnS–GGGC-coated Au wire is immersed in a beaker containing 10^{16} CdS nanocrystals per cm^3, and there is subsequent binding of CdS nanocrystals to the C amino acids of the GGGC biomolecules that are attached to the CdSe–ZnS nanocrystals that are bound to the Au wire. Upon repeated alternating immersions of the quantum-dot functionalized wire in the two beakers of CdSe–ZnS–GGGC and CdS, the nanocrystals are assembled as illustrated in Fig. 5.39. The nanocrystals assembled in this manner have densities in excess of 10^{17} per cm^3! These densities are orders of magnitude higher than those achievable by current commercially used semiconductor-device-fabrication techniques. This example illustrates the use of chemical self-assembly to integrate ensembles of semiconductor nanostructures. To create electrically functional integrated semiconductor nanocrystal

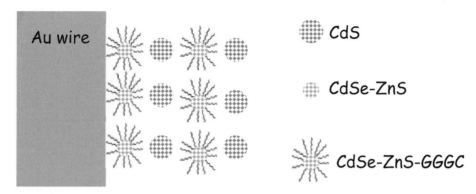

Figure 5.39 A quantum-dot network assembled using chemically directed assembly techniques.

networks using chemically directed assembly with molecular links, it is necessary to use molecular links that conduct electricity.

Since proteins (and short proteins – peptides) constitute one of the major classes of biomolecules, an understanding of the electronic properties of these general classes of molecules is of potential importance in bioelectronics. Charge transport in peptides has recently been studied by inserting amino acids containing a side group containing a natural chromophore, which is a molecule that produces charge carriers when exposed to light. In this way, charge can be introduced into the peptide by illuminating the chromophore in the side group. In addition, phenylalanine, tryptophan, and tyrosine have been considered theoretically as possible conducting elements of peptides since their side groups are rich in conductive π-bonds. As we discussed for conducting polymers, such π-bonds provide charge conduction. Results from the chromophore-based studies of charge transport indicate that charges move along these peptide-based wires at speeds of about 1.5×10^5 cm s^{-1}, a relatively slow speed from the standpoint of Si- and GaAs-based electronic devices. This relatively slow transport implies that electronic systems of integrated nanocrystals with peptide-based interconnects must be designed with architectures that overcome the limitations imposed by carrier-transport speeds in peptides; perhaps, the design of massively parallel architectures offers a possible resolution of the difficulties imposed by the limited propagation speed in the peptide-based interconnects.

Our previous discussion of the use of DNA as a self-assembling agent makes it clear that DNA may be used to link metallic and semiconducting nanostructures. There are also current research efforts aimed at using DNA as molecular wires that conduct electric current. Of course, these efforts stem from the growing interest in the use of DNA as a charge-carrying element in bioelectronic devices. In these efforts, a key parameter determining the change of transport properties of DNA is the highest occupied molecular orbital (HOMO). As shown in Fig. 5.40, the commonly encountered type of DNA is a double-stranded helix-shaped molecule with each of the two strands being composed of sequences of the following four bases: adenine (A), cytosine (C), guanine (G), and thymine (T). As discussed previously, these sequences may contain the A, C, G, and T bases in any order, and the adjacent bases on the two strands are pairs of either G–C or

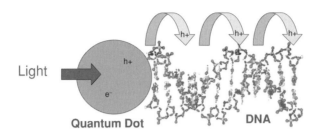

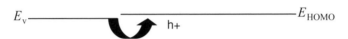

Figure 5.40 Transition of a hole from a quantum dot to DNA.

A–T. For a simple double strand of DNA that contains only G–C pairs, the energy levels of the DNA exhibit a gap of the approximate magnitude of 2.0 eV between the HOMO and the lowest unoccupied molecular orbital (LUMO).

In molecules, energy states may be filled, partially filled, or empty just as in semiconductors. As discussed earlier in this book, the valence band of an undoped semiconductor is nearly full of electrons at low temperatures, and the conduction band is nearly empty under these conditions. For molecules, it is traditional to refer to the HOMO simply as the HOMO "level" or the HOMO band. The highest energy in the HOMO band, therefore, plays a role analogous to the high-energy edge, or top, of the valence band, E_v, of a semiconductor. Similarly, there is an energy gap between the top of the HOMO band and the low-energy edge, or bottom, of the next (empty) molecular orbital, known as the LUMO, which is referred to frequently as the LUMO band. The bottom of the LUMO band plays a role analogous to the low-energy edge of the conduction band, E_c.

One of the current approaches to studying charge transport in DNA is to bind DNA to a semiconductor quantum dot and to inject charge into the quantum dot as shown in Fig. 5.40. As shown in the lower portion of Fig. 5.40, it is desirable in such experiments to pick a semiconducting material for the quantum dot that has some of its energy states aligned with some of the orbitals of the DNA. In the case illustrated in Fig. 5.40, the valence-band energy is selected so that it falls just below the HOMO of the DNA. For a DNA molecule that is composed of a pure strand of G bases bound to a pure strand of C bases, the top of the HOMO band has an energy of -7.34 eV as measured relative to the vacuum level. Two readily available colloidal semiconductor quantum dots with valence bands having energies close to the HOMO band energy are TiO_2 and ZnO. TiO_2 has conduction- and valence-band energies of -4.21 eV and -7.41 eV relative to the vacuum level, respectively. ZnO has conduction- and valence-band energies of -4.19 eV and -7.39 eV relative to the vacuum level, respectively. For either of these semiconductors, the energy alignment of the valence-band edge with the HOMO band is as shown in Fig. 5.40. With such a band alignment, holes in the semiconductor may be injected into the DNA.

In practical experiments on charge transport using these QD–DNA structures, light with energy greater than the bandgap of ZnO (3.2 eV) or TiO_2 (also 3.2 eV) is used to create electron–hole pairs in the quantum dot. Since ZnO and TiO_2 are both indirect-bandgap materials, these electron–hole pairs do not recombine as rapidly as they do in direct-bandgap materials. (This is one of the reasons why TiO_2 is used in solar cells, where there is a need to maximize the production of electric currents from photo-produced carriers.) As shown in Fig. 5.40, the photo-produced holes may escape the quantum dot and move into the DNA wire. Alternatively, we may view this process as an electron near the top of the HOMO band making a transition into the quantum dot, where it recombines with a hole in the valence band. Experiments based upon these techniques are currently being conducted by several research groups to assess the current-carrying capability of DNA. One of the interesting findings emerging from such studies is that charges (holes) tend to become trapped near guanine-rich regions of DNA molecules. This is not surprising, since the ionization potentials of GC and AT base pairs are -7.34 eV and -7.99 eV, respectively. Since these ionization potentials correspond to the energies required for an electron to transition from the top of the HOMO band to the (unbound) continuum, it follows that the HOMO band-edge energy has a local maximum near guanine-rich sites along the DNA. Accordingly, a hole that is propagating along the DNA wire and dissipating its energy as it propagates – by producing vibrational modes in the DNA, as an example – may become trapped at the guanine-rich sites. The charge-transport properties of DNA are still not understood fully. However, if conducting DNA wires may be engineered by selecting appropriate base sequences, it may be possible to use DNA not just as a self-assembling agent of complex ensembles of nanocomponents, but also as an electrically active linking elements in these ensembles of nanostructures. It may even be possible to design novel types of ultra-high-density information-processing systems that greatly exceed the maximum information-processing capabilities realizable by the downscaling of today's integrated circuits! At present, it is reasonable to expect that years of research will be required to determine whether such nanostructure-based chemically self-assembled systems are useful for advanced information processing.

Dip-pen nanolithography

Since the invention of the scanning tunneling microscope and similar techniques, there have been attempts to develop new nanolithography methods. In particular, STM- and AFM-based methods were applied to oxidize, scrape, or etch nanostructures on surfaces. Some of these methods were mentioned in previous sections. However, proposed methods are generally limited to the growth of thin oxides on selected metal and semiconductor surfaces, or to multistep etch procedures that cannot be generalized to parallel and productive patterning of nanostructures.

Finally, dip-pen nanolithography (DPN) was introduced as a direct-write scanning-probe-based lithography in which an AFM tip is used to deliver chemical reagents directly to nanoscopic regions of a target substrate, as shown in Fig. 5.41.

The DPN technique is a type of soft lithography, where the word "soft" refers to the chemical composition of the nanostructures which can be fabricated. They are made

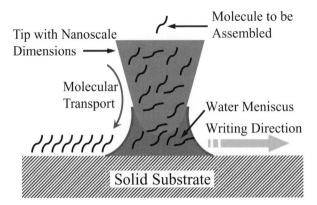

Figure 5.41 The principal "writing" element of a dip-pen lithographic system is a nanoscale structure having a tip with nanoscale dimensions. Fluids flowing over the surface of the tip are deposited onto the surface of a substrate in the region where a meniscus is formed between the tip and the surface. Reprinted with permission, from C. A. Mirkin, "Programming the assembly of two- and three-dimensional architectures with DNA and nanoscale inorganic building blocks," *Inorg. Chem.*, **39**, 2258–2272 (2000). © American Chemical Society.

of organic ligands rather than solid-state materials. Though it is unlikely that dip-pen technologies can displace conventional solid-state fabrication methods, they are highly complementary. Many interesting scientific and important practical issues pertaining to miniaturization can be addressed as a result of having these soft lithography methods, whereby molecules can be patterned in a controlled fashion on the sub-100-nm length scale. For example, such a method can generate nanoscale molecule-based conducting structures and provide their contacting with macroscopically addressable electrodes prepared via conventional microfabrication methods.

In inventing DPN, the idea was to miniaturize a 4000-year-old technology: the technology of the quill or dip-pen. The largest difference is that we wanted to do on a nanoscopic scale what a quill pen can do on a macroscopic scale.

Considering the fundamentals of DPN, first we should mention that the simple idea of transporting any "ink" through a nanoscale AFM tip to a surface via the well-known capillary action does not work properly. Instead, the basic idea is to design a system with an ink that would *chemically react* with a substrate of interest. The scheme in Fig. 5.41 represents the main DPN elements: the moving AFM tip is coated with ink molecules; a water meniscus forms between the tip and the solid substrate; ink molecules react with the substrate. Such an effect of reaction of a chemical with a surface is called *chemisorption*. Thus chemisorption acts as a driving force for moving the molecules from the tip to the substrate and then results in formation of stable one-molecule-thick nanostructures. The key to making this approach successful is to select organic molecules with low water solubilities so that the chemisorption driving force facilitates the control of tip-to-substrate transport properties. This would prevent the uncontrolled, nonspecific adsorption and accumulation of multilayers of molecules on the surface.

Figure 5.42 illustrates the results of chemisorption. The panel on the left shows the coating – "painting" – of a 1-μm^2 area of the gold surface with octadecanethiol via the use of DPN. The formation of a one-monolayer structure is proved by the more detailed

(a)

(b)

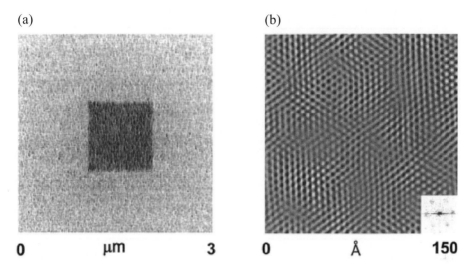

0   μm   3   0   Å   150

Figure 5.42 (a) The painting via DPN by octadecanethiol on gold. (b) A lattice-resolved image of an octadecanethiol nanostructure deposited on a single crystal of gold. Reprinted with permission, from C. A. Mirkin, "Programming the assembly of two- and three-dimensional architectures with DNA and nanoscale inorganic building blocks," *Inorg. Chem.*, **39**, 2258–2272 (2000). © American Chemical Society.

lattice-resolved image of the transported octadecanethiol, as depicted in the panel on the right of Fig. 5.42. The observed lattice is hexagonal with an intermolecular spacing of 5 Å, the known lattice constant for an octadecanethiol monolayer on gold formed by a number of other methods.

The application of the method is shown in Fig. 5.43. The panel on the left shows a molecular grid of octadecanethiol on gold. The grid is a single molecule thick, with 100-nm-wide lines. The panel on the right demonstrates an array of uniformly deposited 450-nm-diameter dots. Note that attaining this type of shape regularity and reproducibility would not be possible with a conventional pen, even over macroscopic dimensions.

Currently, writing with DPN generates a nanostructure with approximately 60–70 nm linewidths. The smallest fabricated structures are 15-nm-diameter dots spaced 5 nm apart. DPN has been developed to pattern a variety of ink–substrate combinations. The method is compatible with many inks, from small organic molecules to organic and biological polymers and from colloidal particles to metal ions and sols; patterned surfaces range from metals to insulators and semiconductors. In Table 5.2, examples of combinations of chemical inks and surfaces are listed.

In conclusion, chemistry and biological methods can be applied for fabrication of nanosystems. Importantly, these new methods can be complementary with the conventional semiconductor nanotechnologies.

5.9   Fabrication of nanoelectromechanical systems

In previous sections, we studied state-of-the-art semiconductor technologies that may be used to produce nanoscale structures and devices for electronics. High-quality structures

Table 5.2 Examples of ink–substrate combinations used in DPN

Ink	Substrate	Notes
Alkylthiols	Au	15 nm resolution on single-crystalline surfaces, $\leq$50 nm on polycrystalline surfaces
Ferrocenylthiols	Au	Redox-active nanostructures
Silazanes	SiO_x GaAs	Patterning on oxides
Proteins	Au, SiO_x	Both direct write and indirect assembly
Conjugated polymers	SiO_x	Polymer deposition verified spectroscopically and electrochemically
DNA	Au, SiO_x	Sensitive to humidity and tip-silanization conditions
Fluorescent dyes	SiO_x	Luminescence patterns
Metal salts	Si, Ge	Electrochemical and electrolytic deposition
Colloidal particles	SiO_x	Viscous solution patterned from tip

(a)   (b)

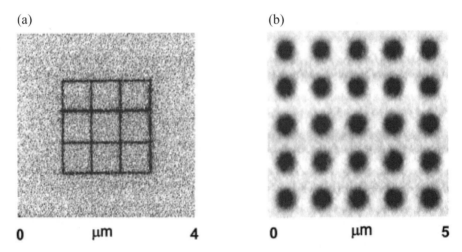

0   μm   4   0   μm   5

Figure 5.43 The nanoscale grid (left) and the dot array (right) generated via DPN. Reprinted with permission, from C. A. Mirkin, "Programming the assembly of two- and three-dimensional architectures with DNA and nanoscale inorganic building blocks," *Inorg. Chem.*, **39**, 2258–2272 (2000). © American Chemical Society.

were necessary in order to obtain superior electronic properties. The great achievements and the advances brought about in mainstream electronics by these technologies can be used for the exploration of other nanodevices commonly called *nanoelectromechanical systems* (NEMSs). This class of devices includes nanomachines, novel sensors, and a variety of new devices that function on the nanoscale.

Nanomechanical devices promise to revolutionize measurements of extremely small displacements and extremely weak forces, particularly at the molecular scale. Indeed, with surface and bulk nanomachining techniques, NEMSs can now be built with masses approaching a few attograms (1 attogram $= 10^{-18}$ g) and with cross-sections of about

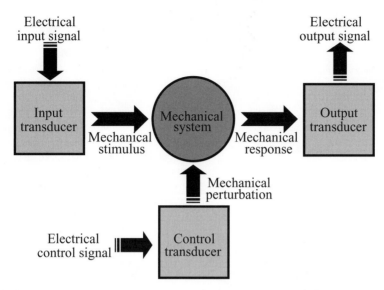

Figure 5.44 A schematic representation of a three-terminal electromechanical device.

10 nm. The small mass and size of NEMSs give them a number of unique attributes that offer immense potential for new applications and fundamental measurements. In general, the potential applications of NEMSs are likely to be enormous and could benefit a diverse range of fields, ranging from nanoelectronics to medicine and biotechnology. In this section, we study the basic concepts of NEMSs, the technology for NEMS fabrication, and the challenges arising in this field.

An electromechanical device can be thought of as a two-, three-, or, generally, multi-terminal transducer that provides input stimuli (i.e., signal forces), and reads out a mechanical response (i.e., output displacement). At additional control terminals, electric signals can be applied and subsequently converted by the control transducers into varying forces to perturb the properties of the mechanical element in a controlled and useful manner. The generic picture of a NEMS is shown in Fig. 5.44, where the input, output, and control transducers/terminals are presented schematically. The basic mechanical element of a NEMS is a nanosize suspended film, a membrane, or a beam; in the following discussion, we will use the term "beam." Easy *flexural deformations* of such mechanical elements provide high mechanical responsivity for a NEMS. Electronic devices to which the beam is coupled are assumed to be of comparable dimensions.

Let us consider the fabrication of the basic mechanical component of a NEMS. The crystal and heterostructure growth, and processing techniques, which have been studied previously are used widely to produce suspended semiconductor structures. These techniques can be applied to bulk silicon, epitaxial silicon, and systems based on III–V compounds and to other materials.

The procedure for fabricating a suspended structure is illustrated in Fig. 5.45. In its simplest form, the procedure starts with a heterostructure that contains structural

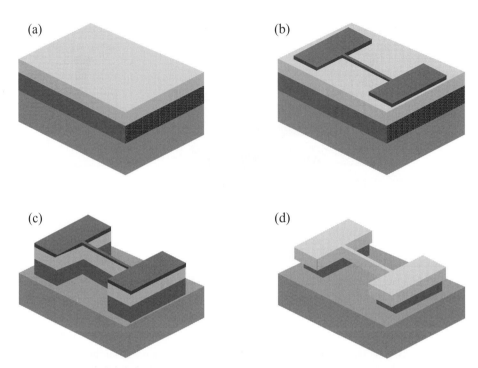

Figure 5.45 A general scheme for the fabrication of NEMS suspended structures using structural (gray) and sacrificial (dark) layers on a substrate (dark gray). (a) Three-layer base heterostructure, (b) etching-mask deposition, (c) anisotropic etching, and (d) selective wet etching of the sacrificial layer.

(gray) and sacrificial (dark) layers on a substrate (dark gray), as in Fig. 5.45(a). Masks on the top of this heterostructure can be patterned by a combination of optical and electron-beam lithography, followed by thin-film-deposition processes. The resulting mask protects the material from beneath during the next stage; see Fig. 5.45(b). Unprotected material around the mask is then etched away using a plasma etching process as in Fig. 5.45(c). Finally, a local chemically selective etching step removes the sacrificial layer from the specific regions to create a freely suspended nanostructure that is both thermally and mechanically isolated; see Fig. 5.45(d). This procedure can be repeated several times and combined with various deposition processes to produce the necessary mechanical nanostructure for a particular device. The flexibility of the process allows one to apply this general scheme to diverse materials and to fabricate fully suspended structures with lateral dimensions of approximately a few tens of nanometers.

Consider, for example, the important case of silicon nanomachining by the use of the so-called SIMOX (Separation by IMplantation of OXygen) process. The procedure starts from a Si wafer, which is processed by a large dose of oxygen-ion implantation. The implanted wafer is annealed at high temperature to form a SiO_2 layer of 0.05–1 μm. Above the SiO_2 layer formed, a single Si crystal layer is then grown. This top layer is 0.1–0.2 μm thick. As the result, one obtains a $Si/SiO_2/Si$ heterostructure that corresponds

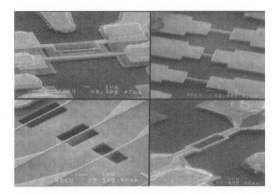

Figure 5.46 Four different NEMSs fabricated by SIMOX method. From A. N. Cleland, *Foundations of Nanomechanics*, Fig. 11.4 (Berlin, Springer-Verlag, 2003).

to the gray/dark/dark gray structure of Fig. 5.45. All stages are already illustrated in Fig. 5.45: the top Si layer is patterned and an etching mask is created as in Fig. 5.45(b); to the top Si layer anisotropic etching is applied as shown in Fig. 5.45(c); then, the oxide is subjected to selective wet etching to create finally a suspended structure as depicted in Fig. 5.45(d). The SIMOX process can be carried out with wafers of large area (4 to 6 inches in diameter) and facilitates the integration of a number of NEMSs and other electronic devices on a chip. Figure 5.46 illustrates some suspended structures fabricated by this method.

In conclusion, due to advanced technologies, fabrication of new nanoelectromechanical systems becomes possible. These systems have dimensions so small that their mechanical motion (vibrations) are coupled to the electrons much more strongly than in the case of bulk-like, massive samples. A number of nanodevices can be built on the basis of NEMSs. Chapter 8 provides additional discussion of these devices.

5.10   Closing remarks

In this chapter, we focussed on methods used for high-quality material growth and for nanodevice fabrication. We studied the growth of perfect crystals and multilayered heterostructures. We found that there has been a considerable and persistent improvement of traditional semiconductor technologies for material processing. We also found that these advances facilitate the fabrication of nanostructures and nanodevices with excellent precision and high reproducibility, and with the necessary electrical, optical, and mechanical properties.

Moreover, we analyzed novel approaches to producing nanostructures based on special regimes of material growth (the Stranski–Krastanow regime), in which nanostructures are formed spontaneously due to the growth kinetics. The processes of self-organization and self-ordering of nanostructures (quantum dots, quantum wires, etc.) give rise to a new and unique way to produce high-density ensembles of nanodevices with desired properties. Apart from the traditional methods of fabrication, those adapted for microelectronics and nanoelectronics, we studied fabrication techniques for such "original" nano-objects as nanotubes.

We described both improved and novel characterization and growth methods that facilitate the control of nanostructure geometries with atomic-scale precision as well as the manipulation of a single atom or ion in a crystal.

The latest achievements in the chemical and biological sciences have generated new approaches to nanofabrication that are complementary to conventional semiconductor technologies. We presented several examples of the new chemical and biological methods.

Finally, we studied in detail the fabrication of a new class of nanodevices, nanoelectromechanical systems. Such nanosized systems are so small that the motion of a few electrons may strongly affect the mechanical vibrations of nanostructures.

More information on crystal growth and device fabrication can be found in the following books and papers:

R. H. Hendel, S. S. Pei *et al.*, "Molecular-beam epitaxy and the technology of selectively-doped heterostructure transistors," in *Gallium Arsenide Technology*, ed. D. K. Ferry (Indianapolis, IN, Howard W. Sams, 1985).

K. Ploog, "Delta-doping in MBE grown GaAs: concept and device application," *J. Cryst. Growth*, **81**, 304 (1987).

D. Bimberg, M. Grundman, and N. N. Ledentsov, *Quantum Dot Heterostructures* (Chichester, John Wiley & Sons, 1999).

The following publications are devoted to self-organized growth of nanostructures:

R. Nötzel, "Self-organizing growth of quantum dot structures," *Semicond. Sci. Technol.* **11**, 1365 (1996).

V. A. Shchukin and D. Bimberg, "Spontaneous ordering of nanostructures on crystal surfaces," *Rev. Mod. Phys.*, **71**, 1125 (1999).

J. Stangl, V. Hol, and G. Bauer, "Structural properties of self-organized semiconductor nanostructures," *Rev. Mod. Phys.*, **76**, 725 (2004).

Scanning tunneling and atomic force microscopy techniques are described in the book

C. J. Chen, *Introduction to Scanning Tunneling Microscopy* (New York, Oxford University Press, 1993).

Detailed analysis of fabrication methods of carbon and other nanotubes is presented in

M. S. Dressellhaus, G. Dressellhaus, and P. C. Eklund, *Science of Fullerenes and Carbon Nanotubes* (San Diego, CA, Academic Press, 1996).

In the following two publications the chemical and biological nanofabrication methods are reviewed:

C. A. Mirkin, "Programming the assembly of two- and three-dimensional architectures with DNA and nanoscale inorganic building blocks," *Inorg. Chem.*, **39**, 2258–2272 (2000).

D. S. Ginger, H. Zhang, and C. A. Mirkin, "The evolution of dip-pen nanolithography," *Angewandte Chem.*, **43**, 30–45 (2004).

A description of fabrication of various types of NEMSs is presented in

A. N. Cleland, *Foundations of Nanomechanics* (Berlin, Springer-Verlag, 2003).

5.11 Problems

1. Describe the main differences between the Czochralski method of crystal growth and epitaxial growth. Which of these approaches can be applied to grow multilayered crystalline structures?

2. In the molecular-beam epitaxy method, the rate of crystal growth is characterized by the flux density, J, of atoms, constituting the growing film. For binary crystals AB, like GaAs, SiGe, etc., the fluxes of deposited components A and B are to be equal: $J_A = J_B = J/2$. Let the crystal density ρ_{AB} be given. Using given J, ρ_{AB}, and masses of atoms A and B, calculate the time necessary to grow a film of thickness d. Estimate numerically the growth time for GaAs film of thickness 100 nm at $J = 10^{15}$ atoms cm^{-2} s^{-1}; ($\rho_{GaAs} = 5.316$ g cm^{-3}).

3. When a photolithographic method is applied, the effect of diffraction of light restricts the minimal scale of an illuminated pattern necessary for further processing to fabricate a nanostructure. Explain the advantages in the use of short-wavelength light sources in nanolithography.

Assume that the minimal thickness of a light line d_{min} is related to the wavelength of illumination λ as $d_{min} \approx \lambda/2$. Calculate and compare the minimal scales of devices fabricated by exploitation of three laser sources: a red He–Ne laser ($\lambda = 0.63$ μm), a UV KrF laser ($\lambda = 0.243$ μm), and a UV ArF laser ($\lambda = 0.19$ μm).

4. Apply the "hydrogen model" of Eq. (5.4) for energy levels of donors in GaAs and InAs. Use the effective masses according to Table 4.5 and set the dielectric constants, ϵ, equal to 12.8 and 15.5, respectively. Calculate the ionization energies and estimate the radius of ground donor states for these materials. Find the average number of primitive cells "covered" by a single donor electron.

5. Explain the role of lattice mismatch for regimes of self-organizing growth of nanostructures. For which of the following heterostructures is the formation of nano-islands possible: GaAs/AlAs (the lattice constants are $a_0 = 5.64$ and 5.66 Å, respectively), GaAs/InAs ($a_0 = 5.64$ and 6.06 Å), and Si/Ge ($a_0 = 5.43$ and 5.65 Å).

6. Discuss the differences between the working principles of the nanoscale techniques of scanning tunneling microscopy and atomic-force microscopy.

7. Scanning tunneling microscopy (STM) is a new and important technique used to probe and characterize nanostructures. Explain why atomic-scale precision measurements are possible when a macroscopic tip of dimension about 1 μm is used. What are the limitations for applications of STM?

8.   Some chemical and biological methods of formation of surface nanopatterns can be complementary to standard semiconductor technologies. Which of these methods can be used for creation of periodic patterns and which are for arbitrary surface patterning?

6 Electron transport in semiconductors and nanostructures

6.1 Introduction

In previous chapters we studied advances in materials growth and nanostructure fabrication. In the case of electrons, we paid primary attention to the quantization of their energy in nanostructures. In fact, electronics relies upon electric signals, i.e., it deals with measurements of the electric current and voltage. Controlling and processing electric signals are the major functions of electronic devices. Correspondingly, our next task will be the study of transport of charge carriers, which are responsible for electric currents through nanostructures.

The possible transport regimes of the electrons are dependent on many parameters and factors. Some important aspects of these regimes can be elucidated by comparing the time and length scales of the carriers with device dimensions and device temporal phenomena related to operating frequencies. Such an analysis is carried out in Section 6.2. In Sections 6.3 and 6.4 we discuss the role of electron statistics in transport effects. Then, we consider the behavior of the electrons in high electric field, including so-called *hot-electron effects*. Analyzing very short devices, we describe dissipative transport and the velocity-overshoot effect. Finally, we consider semiclassical ballistic motion of the electrons and present ideas on quantum transport in nanoscale devices in Section 6.5.

6.2 Time and length scales of the electrons in solids

We start with an analysis of possible transport regimes of the electrons in nanostructures. Since there is a large number of transport regimes, we introduce their classification in terms of characteristic times and lengths fundamentally inherent to electron motion.

Electron fundamental lengths in solids

As we already noted in Section 4.3, the characteristic length in a crystalline solid is the lattice constant a_0. However, the scales which are relevant to the charge carriers are typically much larger than a_0. This fact, as stressed in Chapter 4, allows one to neglect fine crystalline structure and consider an electron as an almost free particle, in particular by assigning to the electron an effective mass that may differ from the mass of an electron in vacuum.

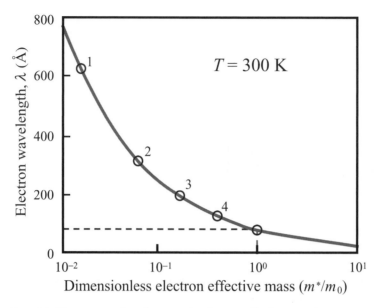

Figure 6.1 Electron wavelength versus the electron effective mass for room temperature ($T = 300$ K). Points 1 through 4 correspond to InSb, GaAs, GaN, and SiC, respectively.

The first fundamentally important length is the *de Broglie wavelength* of an electron in a solid. For a free particle this length was introduced in Chapter 2; see Eq. (2.45). For an electron in a semiconductor nanostructure with the effective mass m^* the de Broglie wavelength λ is typically greater than that of a free electron λ_0:

$$\lambda = \frac{2\pi\hbar}{p} = \frac{2\pi\hbar}{\sqrt{2m^*E}} = \lambda_0\sqrt{\frac{m_0}{m^*}}, \quad \lambda_0 = \frac{2\pi\hbar}{\sqrt{2m_0E}}, \tag{6.1}$$

where E is the electron energy and m_0 is the mass of the electron in vacuum. In Fig. 6.1 the value λ is shown as a function of m^*/m_0. Points 1–4 on the curve indicate wavelengths for electrons in InSb, GaAs, GaN, and SiC, respectively. We have used effective masses m^*/m_0 equal to 0.014, 0.067, 0.172, and 0.41, respectively, for these materials and we have assumed that the electron energy is $E = k_B T$. Here $T = 300$ K is the ambient temperature and k_B is Boltzmann's constant. We see that the de Broglie wavelength of an electron in typical semiconductors with m^* in the range $(0.01$–$1)m_0$ is of the order of 730–73 Å; i.e., it is really much larger than the lattice constants for the materials presented in Table 4.8. As the temperature decreases to 3 K, the de Broglie wavelength increases by one order of magnitude. Thus the wavelength becomes comparable to the sizes of semiconductor structures and devices fabricated by modern nanofabrication technology.

Size of a device and electron spectrum quantization

Let us introduce a geometrical size of a semiconductor sample $L_x \times L_y \times L_z$, as shown schematically in Fig. 6.2. Without loss of generality we assume that $L_z < L_y < L_x$. If

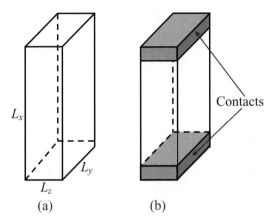

Figure 6.2 (a) Geometrical sizes of a semiconductor sample ($L_z < L_y < L_x$) and (b) a sample with contacts; the electron transport occurs along the x-direction.

the system is free of randomness and other scattering mechanisms are sufficiently weak, the electron motion is *quasiballistic* and the only length with which the geometrical sizes need be compared is the electron de Broglie wavelength λ. Since only an integer number of half-waves of the electrons can fit into any finite system, instead of a continuous energy spectrum and a continuous number of the electron states, one obtains a set of discrete electron states and energy levels, each of which is characterized by the corresponding number of half-wavelengths. This is frequently referred to as *quantization of electron motion*. Depending on the dimensions of the system, one can distinguish the following cases.

(a) The three-dimensional or bulk-like case, when the electron spectrum quantization is not important at all,

$$\lambda \ll L_x, L_y, L_z, \tag{6.2}$$

and an electron behaves like a free particle characterized by the effective mass m^*.
(b) The two-dimensional or quantum-well case, when the quantization of the electron motion occurs in one direction while in the other two directions electron motion is free:

$$\lambda \simeq L_z \ll L_y, L_x. \tag{6.3}$$

Such a case was discussed in Chapter 3 for the example of a potential energy dependent on a single coordinate. The electron energy for this case is given in the form of two-dimensional subbands, as described by Eq. (3.49).
(c) The one-dimensional or quantum-wire case, when the quantization occurs in two directions, so that the electron moves freely only in one direction – along the wire:

$$L_z \simeq L_y \simeq \lambda \ll L_x. \tag{6.4}$$

Such a case was discussed in Chapter 3 for the example of a potential energy dependent on two coordinates. The electron energy for this case is given in the form of one-dimensional subbands, as in Eq. (3.50).

(d) The zero-dimensional or quantum-box (quantum-dot) case, when the quantization occurs in all three directions and the electron can not move freely in any direction:

$$L_x \simeq L_y \simeq L_z \simeq \lambda. \tag{6.5}$$

The simplified models for this case also were analyzed in Chapter 3. The energy spectrum is discrete.

The last three cases also illustrate the *quantum size effects* in one, two, and three dimensions, respectively. If at least one geometrical size of a device is comparable to the electron wavelength, a quantum-mechanical treatment of the problem is strictly required.

Let us analyze the conditions when and the reasons why carriers lose their wave-like behavior so that they can be considered as classical particles. There are two major reasons. The first is non-ideality of the system, which leads to electron scattering. The second is related to finite temperature and electron statistics.

Electrons in solid-state devices are subjected to *scattering* by crystal imperfections, impurities, lattice vibrations, interface roughness, etc. These scattering processes are divided into two groups: *elastic* and *inelastic*. In classical physics, an elastic collision leads to a change only in the particle momentum (wavevector), whereas in an inelastic collision both the momentum and the energy change. An essential property of an *elastic collision* is that it *does not destroy the phase of the electron*. Indeed, after an elastic scattering the energy remains unchanged and the electron wavefunction $\Psi(\mathbf{r}, t)$ consists of different components, which are of type $e^{-i\Omega t}e^{i\mathbf{k}_j\mathbf{r}}$. All components have the same time-dependent phase $e^{-i\Omega t}$. Thus, the spatial distribution of the electron density $|\Psi(\mathbf{r}, t)|^2 = |\psi(\mathbf{r})|^2$ remains independent of time. In other words, elastic scattering does not destroy the *coherence* of electron motion. The same is true for the case of two or more impurities: the spatial wave pattern is generally complex, but it remains coherent.

Using the semiclassical language, if τ_e refers to the mean time between two elastic scattering events, we can define the *mean free path* of the electrons between elastic scattering events as $l_e = v\tau_e$, where v is the average electron velocity. Therefore, even for distances exceeding l_e the wave-like properties of electrons are coherent.

Inelastic scattering leads to a new result. This scattering produces electron waves with different energies and the resulting wavefunction has a complex dependence on both position and time; the beating of different wave components in time washes out the coherence effects. Let τ_E be the mean time between two inelastic collisions. The distance the electron propagates between these collisions is frequently called the *inelastic scattering length*, L_E. The electron preserves its quantum coherence for distances less than L_E and it loses coherence for larger distances. Generally, $L_E > \lambda$ unless extremely nonequilibrium conditions exist. Often L_E far exceeds the mean free path, l_e. In this case, the electron undergoes many elastic collisions before losing its energy. This process is known as *diffusion* and its displacement during τ_E is known to be

$$L_E = \sqrt{D\tau_E} \quad (\tau_E \gg \tau_e), \tag{6.6}$$

where the diffusion coefficient D is given by $D = v^2\tau_e/\alpha$, with $\alpha = 3$ for a three-dimensional electron gas, $\alpha = 2$ for a two-dimensional electron gas, and $\alpha = 1$ for a one-dimensional electron gas. Usually, τ_E and L_E decrease as the temperature of the system increases.

The second reason for averaging out the quantum behavior is the *temperature effect* on the electron statistics. Indeed, at finite temperatures, there are electrons with significantly different energies and this leads to a large spreading of the wavefunction phases and the coherence in the electron system is destroyed. One can estimate the characteristic length L_T associated with such a temperature-related smearing of the coherence. Electron energy broadening of the order of $k_B T$ leads to a spreading of the phases with time, t, that goes as $\Delta\phi = t \times (k_B T/\hbar)$. Thus, the time of spreading, t, can be estimated as the instant τ_T, for which $\Delta\phi \approx 1$, i.e., $\tau_T = \hbar/(k_B T)$. If the only scattering is elastic scattering, an electron diffuses in space over a distance of about $\sqrt{Dt}$ during the time t exceeding the mean-free-flight time τ_e. Therefore, during the time interval τ_T one obtains the *thermal diffusion length* $L_T = \sqrt{D\tau_T} = \sqrt{D\hbar/(k_B T)}$. At distances exceeding L_T, the coherence of the electron will be lost.

In fact, the dephasing effects caused by inelastic collisions and the temperature spreading of phases exist simultaneously. The spatial scale associated with the loss of quantum-mechanical coherence should be determined by the smaller of these two lengths:

$$l_\phi \simeq \min\{L_E, L_T\}. \tag{6.7}$$

Electron transport is determined by the wavefunction, i.e., by the superposition of the scattered electron waves. From the above considerations we can conclude that the *coherence length*, l_ϕ, defines the limit below which electron transport has a quantum character. Devices with geometrical sizes of the order of the coherence length are no longer characterized by macroscopic material parameters such as conductivity, average velocity, etc. Such systems are called *mesoscopic systems*. The proper theory for the description of mesoscopic devices is therefore quantum theory; the properties of such mesoscopic systems are determined by wave-like phenomena and are thus strongly dependent on the geometry of the sample, contacts, positions of scatterers, etc. For cases in which the transport distance, L_z, is long compared with l_ϕ, the device can be described within the framework of classical physics.

Quantum and classical regimes of electron transport

Let us compare the previous discussion of fundamental lengths with characteristic device sizes to illustrate and explain possible electron-transport regimes. For simplicity, we suppose that transport occurs along one dimension, say the x direction. The total current in each of the other two directions is zero, but these transverse sizes of the device can be important too.

Quantum and mesoscopic regimes of transport

We can define two non-classical regimes. If the de Broglie wavelength exceeds the length of a device L_x,

$$\lambda \geq L_x, \tag{6.8}$$

and $l_e \gg \lambda$, electron-transport is described in terms of the *quantum ballistic transport regime*. If the coherence length, l_ϕ (also referred to frequently as the *dephasing length*), exceeds L_x and λ,

$$l_\phi > L_x, \lambda, \tag{6.9}$$

electron transport is described in terms of the *mesoscopic transport regime*.

The classical transport regime

In the case when the size L_x exceeds the dephasing length,

$$L_x > l_\phi, \tag{6.10}$$

electron transport is described in terms of the *classical regime*. If the dimension L_x is less than the mean free path,

$$l_e > L_x, \tag{6.11}$$

electron transport is described in terms of the *classical ballistic regime*, which means that electrons can move through the device along classical trajectories without collisions.

If the dimension L_x is greater than the mean free path,

$$L_x \gg l_e, \tag{6.12}$$

electron transport is of a diffusive nature. If $L_x \sim l_E \gg l_e$, electrons do not lose their energy in moving across the device. Such transport is called *quasiballistic* transport. In the absence of an electric field the electrons preserve their energy under the quasiballistic regime. On combining the above-discussed inequalities with inequalities (6.2)–(6.4), one can see that there are three classical transport regimes for one, two, and three-dimensional electrons.

If the transverse dimensions, L_z and L_y, are both greater than the de Broglie wavelength but they are comparable to one of the characteristic classical lengths, the transport regime is characterized by *transverse classical size effects*. In this case, collisions with the device boundaries affect electron transport through the device. For example, if one or both of the transverse dimensions are of the order of the mean free path,

$$L_z, L_y \sim l_e,$$

the resistance of the device depends strongly on the properties of the side boundaries. Roughness of the boundaries increases the resistance and entirely controls it if $L_z, L_y \ll l_e$.

If the transverse dimensions become comparable to one of the diffusion lengths, we are dealing with another kind of classical size effect, namely, *diffusive classical size effects*. For example, if L_z or L_y is of the order of the energy relaxation length L_E the device boundaries provide an additional energy-relaxation channel. This diffusive size effect controls the mean energy of nonequilibrium electrons. For convenience the classification of the possible transport regimes is presented in Table 6.1.

Table 6.1 Classification of transport regimes

Quantum regime	Intercontact distance, L_x, is comparable to the electron wavelength, $L_x \leq \lambda$
Mesoscopic regime	Intercontact distance is less than the dephasing length, $L_x \leq l_\phi$
Classical regime (one-, two-, and three-dimensional electron transport)	Intercontact distance exceeds the dephasing length, $L_x > l_\phi$: classical ballistic regime, $l_e \geq L_x$ quasiballistic regime (energy-conserving): $L_E \geq L_x \geq l_e, l_\phi$ transverse size effects: effect related to the mean free path, $L_z, L_y \sim l_e$ diffusion effects, $L_z, L_y \sim L_E$

Time scales and temporal (frequency) regimes

The time scales which characterize transport phenomena determine the temporal and frequency properties of materials and devices.

There are two fundamental times defining the character of electron-transport behavior: the time between two successive scattering events, or the free-flight (scattering) time, τ_e; and the time which characterizes the duration of a scattering event, τ_s. Under ordinary conditions $\tau_e \gg \tau_s$. In fact, it is usually assumed that the scattering event is instantaneous, i.e., $\tau_s \to 0$. In this case either classical or quantum theory can be applied for the description of electron behavior, depending on the length scales. If, however, τ_e is comparable to or smaller than τ_s, which may happen under extremely strong scattering of nonequilibrium electrons, the quantum description of electron behavior is required regardless of the size of the system.

In classical transport regimes, the characteristic times and their relationships to the device sizes are of critical importance. They determine temporal and frequency regimes of device operation. For example, the *transit time* $t_{tr} = L_x/v$ determines the duration of signal propagation through a device; here v is the electron velocity. Therefore, t_{tr} defines the ultimate speed limit of the device: the device cannot effectively operate in the time range less than t_{tr} or at frequencies greater than t_{tr}^{-1}. This explains one of the trends of modern electronics: scaling down the device sizes.

The times related to transverse dimensions, $t_b = L_{z,y}/v$ (near the ballistic regimes), or $t_D = (L_{z,y})^2/D$ (for diffusive size effects), determine features of electron transport at frequencies of the order of t_b^{-1}, or t_D^{-1}.

In quantum mechanics, as described in Chapter 3, if external potentials are time-independent, the electrons are in *stationary states*. In this case, despite a possible complex dependence of the wavefunction on position, the temporal evolution of a stationary state is always determined by an exponential factor $\exp[-i(E/\hbar)t]$. If the alternating external field of an angular frequency ω is applied to the stationary electron system the response of the electron system may be referred to one of the following three different regimes depending on the frequency of the external field.

Ultra-high (quantum) frequencies

If $\hbar\omega$ is comparable to the characteristic stationary electron energy state, E, the nature of the electron's response will be essentially quantum-mechanical. Only transitions between states with the energy difference $\Delta E = \hbar\omega$ are allowed. If E is quantized, the interaction is possible only at resonance frequencies. By varying the device size, one can vary the energy spectrum and, as a result, change the frequency properties over a wide range. The kinetic times τ_e, τ_E, etc. lead to a broadening of these resonances. If this broadening exceeds the energy separations between quantized levels, the discrete quantum behavior changes to a continuum-like behavior reminiscent of classical mechanics.

If $\hbar\omega \ll E$ the electron's response to an alternating field is classical (quantization of transitions can be neglected). In the classical picture, the external alternating field will cause periodic electron acceleration and deceleration. The scattering interrupts these accelerations and decelerations. Depending on how many scattering events occur during the one period, we distinguish two different regimes of electron behavior.

High (classical) frequencies

If $\omega\tau_e \gg 1$ the electron motion during one period is not interrupted by scattering. In accordance with classical mechanics, the electron momentum oscillates with a phase opposite to that of the field.

Low frequencies

If $\omega\tau_e \ll 1$ the electron undergoes many scattering events during one period of the external field. Multiple scattering during the period brings the electron into a quasi-stationary state, which follows the oscillations of the external field. In other words, the electron momentum oscillates in phase with the field.

In closing this section we conclude that depending on the device dimensions, the temperature, and other conditions there is a variety of transport regimes. Each of these regimes demonstrates peculiar properties and requires a physical description suited to the relevant conditions.

6.3   Statistics of the electrons in solids and nanostructures

For further analysis of nanostructures, we need to review briefly the basic properties of many-electron systems. Indeed, any semiconductor material consists of a vast number of electrons, i.e., it is essentially a *many-electron system*. For such a system, the key question is how particles are distributed over the different energy states which characterize these particles. For example, if the particles move freely, we may be interested in information about their distribution over the velocities; if the motion of the particles is quantized, knowledge of their distribution over the energy levels is necessary, etc. Knowing such distributions, we may find all average characteristics of the many-particle systems.

The rules and principles according to which particles occupy the energy states in many-particle systems constitute the so-called *physical statistics*. In the physical statistics, to describe the occupation of states by particles, one uses a *distribution function* of particles. Under equilibrium, the distribution over the energy levels determines entirely the properties of many-particle systems. *The distribution function has the meaning of the probability of finding particles with a given energy, E.* Let E_l be the energy level of particles in a many-particle system, the index l numbers the energy levels. Then, the distribution function can be thought of as a function of the energy, $\mathcal{F}(E_l)$. Obviously,

$$\sum_l \mathcal{F}(E_l) = N, \tag{6.13}$$

where N is the total number of the particles. It turns out that the statistical principles in classical and quantum physics are different. The quantum physics brings statistical features that are absent in the classical description. These features are associated with the fact that elementary particles, including electrons, are identical and it is impossible, in principle, to specify their coordinates and trace a given electron. In addition, an "internal characteristic" of a particle, the spin, plays a very important role in many-particle physics. Although a definition of spin is absent in classical physics, we start with a brief review of classical statistics.

Classical statistics

In classical physics, under equilibrium the distribution function has an exponential form. This is the so-called Boltzmann distribution

$$\mathcal{F}_{\mathrm{B}}(E) = C e^{-E/(k_{\mathrm{B}}T)}, \tag{6.14}$$

with C and T being the normalization constant and the ambient temperature, respectively. The normalization constant can be found from Eq. (6.13). In general, statistics in classical physics does not restrict the number of particles occupying an energy level E.

As an example of the application of the Boltzmann distribution, consider the distribution of free particles with mass m over the velocities. First, we suppose that the system under consideration is uniform, i.e., external forces are absent. Then, the energy of these particles coincides with the kinetic energy $E = E_{\mathrm{kin}} = m\mathbf{v}^2/2$, where $\mathbf{v}$ is the particle velocity. For three-dimensional particles, we have $\mathbf{v} = \{v_x, v_y, v_z\}$ and each of the velocity components can vary in the interval $[-\infty, +\infty]$. Thus, by rewriting Eq. (6.14) we can introduce the distribution function as

$$\mathcal{F}_v(\mathbf{v}) = C_v e^{-m\mathbf{v}^2/(2k_{\mathrm{B}}T)}. \tag{6.15}$$

Since the velocity is a continuous value, the meaning of the distribution $\mathcal{F}_v$ is the following. If we define an infinitesimally small volume $\mathrm{d}^3 v \equiv \mathrm{d}v_x \times \mathrm{d}v_y \times \mathrm{d}v_z$ around a given velocity $\mathbf{v}$ in the "velocity space," then the number of particles whose velocity is inside this volume is $\mathcal{F}_v(\mathbf{v})\mathrm{d}^3 v$. That is, actually, $\mathcal{F}_v$ is the *density* of the distribution over the velocity. To find the normalization constant, we can integrate $\mathcal{F}_v$ over all possible **v**, and then we obtain the total number of the particles N. As a result, the distribution function, $\mathcal{F}_v(\mathbf{v})$, can be written in the form

$$\mathcal{F}_v(\mathbf{v}) = N \left(\frac{m}{2\pi k_{\mathrm{B}} T} \right)^{3/2} \mathrm{e}^{-m\mathbf{v}^2/(2k_{\mathrm{B}} T)} = N \left(\frac{m}{2\pi k_{\mathrm{B}} T} \right)^{3/2} \mathrm{e}^{-m\left(v_x^2 + v_y^2 + v_z^2\right)/(2k_{\mathrm{B}} T)}.$$
(6.16)

This is the so-called *Maxwellian distribution* of the particles. By using the Maxwellian distribution, we can calculate average characteristics of the so-called *ideal* gas of particles. For example, the average energy per particle is

$$\overline{E} = \frac{1}{N} \int \mathrm{d}^3 v \, \frac{m\mathbf{v}^2}{2} \mathcal{F}_v(\mathbf{v}) = \frac{3}{2} k_{\mathrm{B}} T,$$
(6.17)

the average velocity at equilibrium is $\overline{\mathbf{v}} = 0$, etc.

Fermi statistics for electrons

Now we return to the more general case of quantum statistics. As we stressed at the beginning of this section, the spin of the particle plays the crucial role for quantum statistics.

The definition of the spin of a particle was introduced in Section 3.4 as an additional "internal" degree of freedom. Though one can compare spin with classical rotation, in fact, spin is strictly a quantum-mechanical quantity and differs substantially from its classical analogue. The principal quantitative characteristic of spin is a dimensionless quantity called the *spin number s*. It is well established experimentally that an electron has a spin number equal to $\frac{1}{2}$. If one fixes an axis in space, the projection of the electron spin on this axis can be either $+\frac{1}{2}$ or $-\frac{1}{2}$. A complete description of an electron state requires a set of quantum numbers: three of them correspond to motion of a particle in space, say $l = \{l_1, l_2, l_3\}$, and one more corresponds to a spin, s. According to the classification discussed in Section 3.4, this case corresponds to a two-fold degeneracy for each energy level.

For many actual cases the electron spin is not important in altering the energy spectra, or the spatial dependence of wavefunctions, etc. There is one crucially important consequence of the fact that the electron spin number is a half integer. Indeed, the particles with half-integer spin numbers obey the Pauli exclusion principle, which we introduced in Section 3.4. It reads as follows: any quantum state $\{l, s\}$ can be occupied by a single particle only. In other words, two electrons in a system can not be simultaneously in the same quantum state. Stated in another way, two electrons may be in the same energy state if their spin quantum numbers are different (the so-called *degenerate state*); if one spin quantum number is $+\frac{1}{2}$, the other must be $-\frac{1}{2}$. The degeneracy is removed if there is interaction between electron spin and electron translational (orbital) motion, which is known as *spin–orbital interaction*. In this case, the electron spin affects the electron's spatial properties, and electrons with spin $+\frac{1}{2}$ and $-\frac{1}{2}$ have different energies. In Fig. 6.3, we illustrate possible populations of the energy levels by the electrons for two cases, degenerate and nondegenerate energy levels.

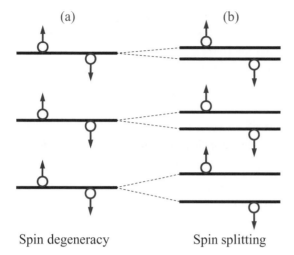

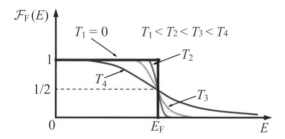

Spin degeneracy   Spin splitting

Figure 6.3 Occupation of energy levels by the electrons: (a) spin-degenerate levels and (b) nondegenerate (spin-split) levels.

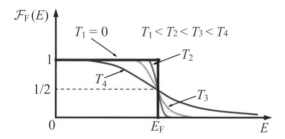

Figure 6.4 How the Fermi distribution function varies with crystal temperature.

It is clear that the Pauli exclusion principle leads to a new, non-classical statistics of electrons. Such statistics is called the *Fermi statistics*. Under equilibrium the occupation of the energy levels is described by the *Fermi distribution function*:

$$\mathcal{F}_{\mathrm{F}}(E_{l,s}) = \frac{1}{1 + e^{(E_{l,s} - E_{\mathrm{F}})/(k_{\mathrm{B}}T)}}, \qquad (6.18)$$

where T is the temperature of the system, $E_{l,s}$ is the energy of the quantum state characterized by the set of quantum numbers l and s, and E_{F} is the so-called *Fermi energy* or *Fermi level*. The evolution of the Fermi function $\mathcal{F}_{\mathrm{F}}(E_{l,s})$ with the temperature is shown in Fig. 6.4, where we use the temperature parameters $T_4 > T_3 > T_2 > T_1$ with $T_1 = 0$. Importantly, the Fermi energy can be related to the total number of electrons through the normalization condition of Eq. (6.13). Taking into account explicitly the summation over the spin, we obtain

$$\sum_{l,s} \mathcal{F}_{\mathrm{F}}(E_{l,s}) = N, \qquad (6.19)$$

which gives $E_{\mathrm{F}} = E_{\mathrm{F}}(N, T)$.

From Eq. (6.18), one can see that in accordance with the Pauli principle the occupation of any energy state, $\{l, s\}$, defined by Eq. (6.18), is always less than or equal to 1.

At high temperatures the second term in the denominator of Eq. (6.18) is substantially larger than unity and the Fermi distribution is close to the Boltzmann distribution:

$$\mathcal{F}_{\mathrm{F}}(E) \approx e^{(E_{\mathrm{F}}-E)/(k_{\mathrm{B}}T)}. \tag{6.20}$$

Equations (6.20) and (6.14) are the same when the normalization constant C is equal to

$$C = \exp\left(\frac{E_{\mathrm{F}}}{k_{\mathrm{B}}T}\right). \tag{6.21}$$

The corresponding curve is shown schematically in Fig. 6.4 at $T = T_4$.

In the limit of low temperatures, $T \to 0$, the function $\mathcal{F}_{\mathrm{F}}$ transforms into a step function:

$$\mathcal{F}_{\mathrm{F}}(E) = \begin{cases} 1, & E_{\mathrm{F}} > E, \\ 0, & E_{\mathrm{F}} < E, \end{cases} \tag{6.22}$$

i.e., $\mathcal{F}_{\mathrm{F}}(E) = 1$ for the energy levels below the Fermi energy E_{F} since all levels with $E < E_{\mathrm{F}}$ are occupied and $\mathcal{F}_{\mathrm{F}}(E) = 0$ for energies above E_{F} since these levels are empty. In this limit, the electron system is frequently referred to as a *highly degenerate electron gas*.

Now we can apply Fermi statistics to the electrons in a conduction band. Let n be the concentration of electrons in the conduction band with energy dispersion $E(\mathbf{k})$. We accept that the energy spectrum is independent of the spin. Thus, the set of the quantum numbers, l, is identical to the set of the electron wavevectors, $\mathbf{k}$. According to the Fermi distribution, the probability of finding an electron with the wavevector $\mathbf{k}$ is

$$\mathcal{F}_{\mathrm{F}}(E(\mathbf{k})) = 2\frac{1}{1 + e^{(E(\mathbf{k})-E_{\mathrm{F}})/(k_{\mathrm{B}}T)}}, \tag{6.23}$$

where the factor 2 comes from the spin degeneracy. The Fermi energy, E_{F}, and the electron concentration, n, are related through the following equation:

$$n = \frac{N}{V} = \frac{2}{V}\sum_{\mathbf{k}} \frac{1}{1 + e^{(E(\mathbf{k})-E_{\mathrm{F}})/(k_{\mathrm{B}}T)}}, \tag{6.24}$$

with V being the volume of the crystal. The summation in the latter formula can be converted to an integration. Indeed, according to the analysis given in Section 4.4 and Eq. (4.10), the electron wavevector takes the following values: $k_x L_x = 2\pi l_1$, $k_y L_y = 2\pi l_2$, and $k_z L_z = 2\pi l_3$, with l_1, l_2, and l_3 being integers. Here we introduced the crystal dimensions L_x, L_y, and L_z. (The crystal dimensions are related to the basis vectors of the lattice a_i and the number of primitive cells N_i as $L_x = a_x N_x$, etc.) Thus, summation over $\mathbf{k}$ is equivalent to summation over l_i. The latter can be approximately calculated via the integral: $\sum_{l_1,l_2,l_3}(\ldots) \approx \int\int\int dl_1\, dl_2\, dl_3(\ldots)$. Since the distribution function depends on $E(\mathbf{k})$, it is convenient to express the latter integral in terms of integration over $\mathbf{k}$. We can use the relationships

$$\Delta l_1 = \frac{L_x}{2\pi}\Delta k_x, \qquad \Delta l_2 = \frac{L_y}{2\pi}\Delta k_y, \qquad \Delta l_3 = \frac{L_z}{2\pi}\Delta k_z.$$

Since $L_x \times L_y \times L_z = V$, we can finally write

$$\sum_{\mathbf{k}}(\ldots) = \frac{V}{(2\pi)^3} \int \int \int \mathrm{d}k_x\, \mathrm{d}k_y\, \mathrm{d}k_z(\ldots). \qquad (6.25)$$

This procedure of replacement of the summation over discrete $\mathbf{k}$ by integration over continuous $\mathbf{k}$ is useful for the calculation of average quantities.

As an example, let us calculate the Fermi energy E_F of the electron system at low temperatures ($T \to 0$). The energy spectrum of the electrons is supposed to be isotropic, i.e., it depends only on the modulus $|\mathbf{k}| = k$: $E(\mathbf{k}) = E(k) = \hbar^2 k^2/(2m^*)$, where m^* is the effective mass of the electron. Let the electron concentration be n. According to the Fermi statistics, electrons will occupy all energy states below the Fermi energy. Since $E(k)$ is an increasing function of k, it follows, for $T \to 0$, that all states with $k \le k_F$ are occupied, where k_F is the so-called *Fermi wavevector* defined through $E(k_F) = E_F$. When calculating the concentration according to Eq. (6.24), we have to perform the summation (integration) over all occupied states, i.e., $k \le k_F$. For these k we have $\mathcal{F}_F = 1$, and it follows that

$$n = \frac{2}{V} \sum_{k \le k_F} 1 = \frac{2}{V} \times \frac{V}{(2\pi)^3} \int \int \int_{|\mathbf{k}| \le k_F} \mathrm{d}^3 k.$$

Evaluation of this integral gives the volume of a sphere of radius k_F, i.e., $4\pi k_F^3/3$. Then, we obtain a relationship between the Fermi wavevector k_F and the electron concentration n: $k_F = (3\pi^2 n)^{1/3}$. Finally, the Fermi energy of the degenerate electrons in a bulk crystal is

$$E_F = (3\pi^2)^{2/3}\frac{\hbar^2 n^{2/3}}{2m^*}, \quad \text{at } T \to 0. \qquad (6.26)$$

In this case, E_F increases as the $2/3$ power of the electron concentration n. Because the Fermi function contains the exponential factor, the low-temperature limit corresponds to the condition $E_F \gg k_B T$. In metals and heavily doped semiconductors, the electron gas remains degenerate up to room temperature. For example, in the case of a GaAs crystal with the effective electron mass $M^* = 0.067m_0$, where m_0 is the mass of the free electron, at the concentration $n = 10^{17}$ cm^{-3}, we find $k_F = 1.43 \times 10^6$ cm^{-1} and $E_F = 11.6$ meV. This energy corresponds to a temperature of 135 K. Thus, at $T < 135$ K the electron gas concentration $n = 10^{17}$ cm^{-3} in GaAs can be considered as degenerate and it is possible to use the above estimates for the Fermi energy, E_F, and the Fermi wavevector, k_F.

The degenerate electron gas is an interesting and very important physical system. This limiting case facilitates understanding a number of complex phenomena in a simple way. Indeed, as emphasized previously, in a degenerate electron gas all states below E_F are occupied. Let us imagine that a small external perturbation is applied to such a many-electron system. The perturbation first of all will cause a redistribution of electrons between the energy states. However, below the Fermi level all states are completely filled and no redistribution is possible. Instead, only those electrons that are at the Fermi level, i.e., that have energy just equal to E_F, can be affected by the perturbation. This results in the fact that only a small portion of the electrons can participate in the crystal's response to the perturbation. One can say that these "active" electrons are on the so-called *Fermi surface* in $\mathbf{k}$ space and the size of the Fermi surface determines the basic properties

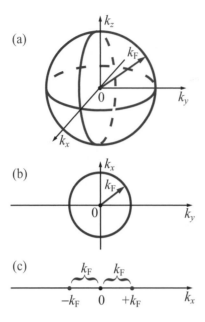

(a)

(b)

(c)

Figure 6.5 The Fermi surfaces for (a) three-, (b) two-, and (c) one-dimensional electron gases.

of the degenerate electron gas. For bulk-like crystals with the simple energy spectrum considered above, the Fermi surface is just a sphere of radius k_F. The Fermi surface for a three-dimensional electron gas is presented in Fig. 6.5(a). Using Eq. (2.4) it is easy to show that an electron on the Fermi surface has the velocity $v_F = \hbar k_F/m^*$.

We can consider low-dimensional electron systems quite similarly. As discussed in Section 4.5, by using heterostructures it is possible to fabricate artificial potential wells, which confine the electrons from the conduction band in such narrow layers that electron motion across the layers becomes quantized. This results in electron energies in the form of low-dimensional subbands given by Eq. (3.49) for the so-called quantum wells:

$$E_{l_3}(\mathbf{k}_{\parallel}) = \varepsilon_{l_3} + \frac{\hbar^2 k_{\parallel}^2}{2m^*},\tag{6.27}$$

where l_3 ($l_3 = 1, 2, \dots$) and $k_{\parallel}$, the two-dimensional wavevector, determine the electron motion in the plane of the layer. The Fermi distribution function in the form of Eq. (6.18) can be used to calculate the *two-dimensional electron concentration (sheet concentration of electrons)*, n_{2D}:

$$n_{2D} = n \times d = \frac{2}{S} \sum_{l_3,\mathbf{k}_{\parallel}} \frac{1}{1 + e^{(E_{l_3}(\mathbf{k}_{\parallel})-E_F)/(k_B T)}},\tag{6.28}$$

where we have introduced the sheet concentration of electrons, n_{2D} (the number of electrons per unit area), the thickness d, and the area S of the confined layer; obviously $V = d \times S$. Actually, exactly as found for bulk crystal, Eq. (6.28) establishes the relationship between the sheet concentration, n_{2D}, and the Fermi energy, E_F.

The sheet concentration of the electrons can be explicitly calculated in the limit of a degenerate electron gas. Let us suppose that the temperature is low and that only the lowest subband is populated by the electrons. Then, in the sum over l_3 in Eq. (6.28) we should keep only one term with $l_3 = 1$:

$$n_{2D} = \frac{2}{S} \sum_{\mathbf{k}_{||}} \frac{1}{1 + e^{(E_1(\mathbf{k}_{||}) - E_F)/(k_B T)}}. \tag{6.29}$$

Calculation of the right-hand side of this equation can be simplified by replacing the summation over $\mathbf{k}_{||}$ by integration, similarly to that of Eq. (6.25):

$$\sum_{\mathbf{k}_{||}} (\ldots) = \frac{S}{(2\pi)^2} \int \int dk_x \, dk_y (\ldots), \tag{6.30}$$

where k_x and k_y are the components of the two-dimensional vector $\mathbf{k}_{||}$. In the limit of $T \to 0$, the sheet concentration n_{2D} can be calculated by defining a two-dimensional Fermi wavevector $\mathbf{k}_{||,F}$ via the relationship

$$E_F = \frac{\hbar^2 k_{||,F}^2}{2m^*}. \tag{6.31}$$

The Fermi "surface" for two-dimensional carriers is a "disk" of radius $k_{||,F}$ in $\mathbf{k}_{||}$ space, as shown in Fig. 6.5(b). On performing the integration of Eq. (6.30) over the "disk," we find

$$k_{||,F} = (2\pi n_{2D})^{1/2} \quad \text{and} \quad E_F = \frac{\pi \hbar^2}{m^*} n_{2D}, \quad \text{at } T \to 0. \tag{6.32}$$

That is, the Fermi energy, E_F, increases as the first power of the electron sheet concentration, n_{2D}.

For a nanostructure in which the electron motion is restricted in two directions, i.e., for the quantum wire, the energy spectrum is given by Eq. (3.50):

$$E_{l_2,l_3}(k_x) = \varepsilon_{l_2,l_3} + \frac{\hbar^2 k_x^2}{2m^*}, \tag{6.33}$$

where it is assumed that the x direction is the only direction of free motion. By applying a procedure similar to that used above, we may see that for low temperatures the Fermi "surface" shrinks to *two points* in the one-dimensional k space: $k_x = \pm k_F$ with $k_F = \frac{1}{2}\pi n_{1D}$. Here, n_{1D} is the *linear electron concentration*, which can be defined by the formula $n_{1D} = N/L$, with N being the total number of electrons in the wire and L being the wire length. Subsequent evolution of the Fermi surface of the electrons as the dimensionality is lowered is illustrated in Figs. 6.5(a)–(c). The Fermi energy of one-dimensional electrons is given by

$$E_F = \frac{\pi^2 \hbar^2}{8m^*} n_{1D}^2, \quad \text{for } T \to 0. \tag{6.34}$$

On comparing the Fermi energies obtained for different dimensionalities of the electron gas, we can conclude that lowering the dimensionality of the gas gives rise to a more rapid increase of E_F with increasing concentration for low-dimensional systems.

The concept of Fermi statistics is one of the fundamental ideas of modern solid-state physics and extremely important for nanoelectronics. The Fermi statistics is applied widely in the description of nanoelectronic devices.

6.4   The density of states of electrons in nanostructures

To complete the analysis of the occupation of the energy levels by electrons in nanostructures, we shall study another physical quantity characterizing the occupation, known as the *density of states*. The reason for introducing this quantity is the following. From the results presented in the previous chapters, we see that the electron energy spectra in nanostructures with conduction electrons are complex and consist of a series of subbands. The distances between subbands are determined by the profile of the confining potential, while inside each subband the spectrum is continuous and these continuous spectra overlap. To characterize these complex spectra, it is convenient to introduce a special function known as the density of states, $\varrho(E)$, which gives the number of quantum states $d\mathcal{N}(E)$ in a small interval dE around energy E:

$$d\mathcal{N} = \varrho(E)dE. \tag{6.35}$$

If the set of quantum numbers corresponding to a certain quantum state is designated as ν, the general expression for the density of states is defined by

$$\varrho(E) = \sum_{\nu} \delta(E - E_\nu), \tag{6.36}$$

where E_ν is the energy associated with the quantum state ν. At this point, it is useful to introduce Dirac's famous δ-function:

$$\delta(x) = \begin{cases} 0, & \text{for } x \neq 0, \\ \infty, & \text{for } x \to 0, \end{cases} \tag{6.37}$$

though

$$\int_{-\infty}^{+\infty} \delta(x)dx = 1. \tag{6.38}$$

Dirac's δ-function is used when performing an integration. The main rule of such an integration is the following:

$$\int_{a}^{b} dx\, \delta(x - x_0)\Phi(x) = \Phi(x_0), \quad \text{if } a < x_0 < b, \tag{6.39}$$

where $\Phi(x)$ is an arbitrary well-behaved function.

As a simple example of a calculation of the density of states by using the definition of Eq. (6.36), let us calculate this quantity for electrons in a bulk crystal. As discussed previously, for such a case the set of quantum numbers is $\{\mathbf{k}, s\}$. Assuming the energy to be independent of the spin, $E(\mathbf{k}) = \hbar^2 k^2/(2m^*)$. Replacing the summation by an integration as in Eq. (6.25), we obtain

$$\varrho_{3D}(E) = \frac{2V}{(2\pi)^3} \int \int \int d\mathbf{k}\, \delta(E - E(\mathbf{k})). \tag{6.40}$$

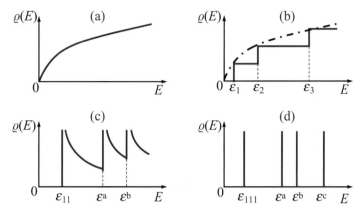

Figure 6.6 The density of states of electrons, $\varrho(E)$, in systems of different dimensionalities: (a) a bulk crystal, (b) a quantum well, (c) a quantum wire, and (d) a quantum dot. Here ε_1, ε_{11}, and ε_{111} are the ground states in a quantum well, quantum wire, and quantum dot, respectively; ε_2 and ε_3 are higher states in a quantum well and ε^a, ε^b, and ε^c are higher states in a quantum wire and in a quantum dot.

Because $E(\mathbf{k})$ actually depends on the modulus $\mathbf{k}$, spherical coordinates may be used. Integration over two angles gives 4π and the triple integral is reduced to the single integral

$$\varrho_{3D}(E) = \frac{2V}{(2\pi)^3} \times 4\pi \int_{-\infty}^{\infty} dk\, k^2 \delta(E - E(\mathbf{k})). \qquad (6.41)$$

Now, we replace the integration over k by an integration over $E(\mathbf{k}) = \mathcal{E}$. Then, taking into account that $k^2 = (2m^*/\hbar^2)\mathcal{E}$ and $dk = \sqrt{m^*/(2\hbar^2\mathcal{E})}\,d\mathcal{E}$,

$$\varrho_{3D}(E) = \frac{V}{2\pi^2}\left(\frac{2m^*}{\hbar^2}\right)^{3/2} \int_0^{\infty} d\mathcal{E}\, \sqrt{\mathcal{E}}\delta(E - \mathcal{E}). \qquad (6.42)$$

Finally, by using the rule of Eq. (6.39), we obtain

$$\varrho_{3D}(E) = \left(\frac{m^*}{\hbar^2}\right)^{3/2} \frac{V}{\pi^2}\sqrt{2E}. \qquad (6.43)$$

The obtained density of states for three-dimensional electrons is depicted in Fig. 6.6(a). For electrons in a quantum well with the energy spectrum given by Eq. (6.27), the set of quantum numbers includes a spin quantum number, s, a quantum number, l_3, characterizing the transverse quantization of the electron states, and a continuous two-dimensional vector, $\mathbf{k}_{||}$. Hence, $\nu \equiv \{s, l_3, \mathbf{k}_{||}\}$. There is a two-fold spin degeneracy of each state, $(s = \pm\frac{1}{2})$, so that

$$\varrho_{2D}(E) = 2 \sum_{l_3, k_x, k_y} \delta\left(E - \varepsilon_{l_3} - \frac{\hbar^2\left(k_x^2 + k_y^2\right)}{2m^*}\right). \qquad (6.44)$$

In order to calculate the sum over k_x and k_y, we can apply the replacement, Eq. (6.30), with S being the area of the surface of the quantum well, $S = L_x \times L_y$, where L_x and

L_y are the sizes of the quantum well in the x and y directions, respectively. Calculations of the integrals give us

$$\varrho_{2D}(E) = \frac{m^*S}{\pi\hbar^2}\sum_{l_3}\int_0^\infty d\mathcal{E}\, \delta(E - \varepsilon_{l_3} - \mathcal{E}) = \frac{m^*S}{\pi\hbar^2}\sum_{l_3}\Theta(E - \varepsilon_{l_3}), \quad (6.45)$$

where $\Theta(x)$ is the Heaviside step-function:

$$\Theta(x) = \begin{cases} 1, & \text{for } x > 0, \\ 0, & \text{for } x < 0. \end{cases} \quad (6.46)$$

Very often the density of states per unit area, ϱ_{2D}/S, is used to eliminate the size of a sample. Each term in the sum of Eq. (6.45) corresponds to the contribution from one subband. The contributions of all subbands are equal and independent of energy. As a result, the density of states of two-dimensional electrons exhibits a staircase-shaped energy dependence with each step being associated with one of the energy states, ε_{l_3}. The height of each step is universal and depends only on the effective electron mass. Figure 6.6(b) depicts the two-dimensional density of states. By comparing the densities of states for the electrons in bulk crystals per unit volume and those of quantum wells per unit surface we may see that the differences between the two- and three-dimensional cases are most pronounced in the energy regions of the lowest subbands. For large l_3 the staircase function lies very close to the bulk curve $\varrho_{3D}(E)$ and coincides with it asymptotically.

Similarly, we can find the density of states of a one-dimensional electron gas with the energy spectrum given by Eq. (6.33). The result of calculations is

$$\varrho_{1D}(E) = \sum_{l_2,l_3}\varrho_{l_2,l_3}(E),$$

where

$$\varrho_{l_2,l_3}(E) = \frac{L}{\pi}\sqrt{\frac{2m^*}{\hbar^2}}\frac{1}{\sqrt{E - \varepsilon_{l_2,l_3}}}\Theta(E - \varepsilon_{l_2,l_3}). \quad (6.47)$$

Here, L is the length of the wire. Schematically, $\varrho_{1D}(E)$ for one-dimensional electrons is shown in Fig. 6.6(c). The characteristic feature of the one-dimensional density of states is its divergence near the bottom of each of the one-dimensional subbands. The density of states then decreases as the kinetic energy increases. This behavior is very remarkable because it leads to a whole class of new electrical and optical effects peculiar to quantum wires.

Now, we can consider the ultimate case of the density of states for zero-dimensional electrons, i.e., for electrons in quantum dots. According to the definition of Eq. (6.36), in the case of quantum dots or boxes the spectra are discrete. Thus, the density of states is simply a set of δ-shaped peaks, as depicted in Fig. 6.6(d). For an idealized system, the peaks are very narrow and infinitely high, as illustrated in Fig. 6.6(d). In fact, interactions between electrons and impurities as well as collisions with lattice vibrations bring about a broadening of the discrete levels and, as a result, the peaks for physically realizable systems have finite amplitudes and widths. Nevertheless, the major trend of sharpening of the spectral density dependences as a result of lowering the dimensionality of the system is a dominant effect for nearly perfect structures at low temperatures.

The dramatic changes in the electron density of states caused by dimensionally confining crystals manifest themselves in a variety of major modifications in conductivity, optical properties, etc. Indeed, as we will see, these modifications in the density of states also lead to new physical phenomena.

6.5　Electron transport in nanostructures

As indicated in Section 6.2, there are distinct regimes of electron motion in nanostructures. In this section, we will study several instructive examples of electron transport induced by an electric bias. We start with the simplest low-field dissipative classical electron transport, which is realized at large intercontact distances, according to Eqs. (6.10) and (6.12). This transport regime can occur in bulk samples, quantum wells, and quantum-wire structures.

Classical dissipative transport

In a solid the electrons are in constant motion. However, this motion is chaotic as a result of random scattering by imperfections, lattice vibrations, interface roughness, etc. As a result there is no preferred direction for electron motion. Thus, the net electron flux and electric current are equal to zero. If an electric field $\mathbf{F}$ is applied to the solid, an electric force, $-e\mathbf{F}$, acts on each of the electrons (here we suppose that the electron charge is equal to $-e$). Though the chaotic character of electron motion can remain, a directional net drift of electrons is induced by the electric force. Figure 6.7 demonstrates chaotic electron motion without and with the electric field applied. To describe this motion, we can use the Newton equation (2.8) for the average velocity of the electrons, $\mathbf{v}$. To take into account the scattering processes, which lead to the loss in the directed component of the velocity, we introduce an additional term that contains a friction force:

$$m^* \frac{\mathrm{d}\mathbf{v}}{\mathrm{d}t} = -\frac{m^*}{\tau_\mathrm{e}}\mathbf{v} - e\mathbf{F}, \tag{6.48}$$

where τ_e can be interpreted as the momentum relaxation, or free-flight time, which was discussed in Section 6.2. Obviously, the electron motion described by the above equation is dissipative in nature.

For the stationary state, $\mathrm{d}\mathbf{v}/\mathrm{d}t = 0$, we obtain

$$\mathbf{v} = -\frac{e\tau_\mathrm{e}}{m^*}\mathbf{F} = -\mu\mathbf{F}, \tag{6.49}$$

where we have introduced the parameter μ, called the *electron mobility*. The mobility is one of the basic characteristics of electron transport in the case of low electric fields. The definition of μ,

$$\mu = \frac{e\tau_\mathrm{e}}{m^*}, \tag{6.50}$$

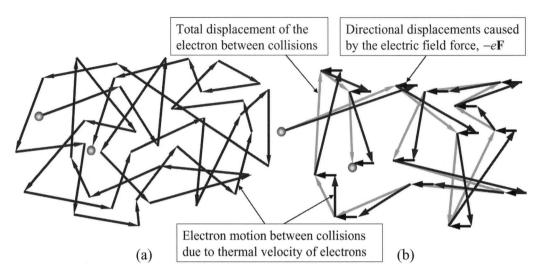

| Total displacement of the electron between collisions | Directional displacements caused by the electric field force, $-e\mathbf{F}$ |

| Electron motion between collisions due to thermal velocity of electrons |

(a)   (b)

Figure 6.7 Chaotic diffusion motion of an electron at equilibrium (a) and when an electric field $\mathbf{F}$ is applied (b).

shows that the mobility is larger for materials with small effective mass m^* and with suppressed scattering, i.e., for large τ_e. The negative sign in Eq. (6.49) reflects the fact that the electrons move in the direction opposite to the electric field, because of their negative charge.

Having the electron average (drift) velocity, $\mathbf{v}$, and the electron concentration, n, we can calculate the electric current density as

$$\mathbf{J} = -e\mathbf{v}n = e\mu n\mathbf{F} = \sigma\mathbf{F}. \qquad (6.51)$$

Here,

$$\sigma = e\mu n \qquad (6.52)$$

is the so-called *specific conductivity*, and, as we can see, it depends on both the electron concentration, n, and the electron mobility, μ. The result given by Eq. (6.51) is known as *Ohm's law*. If the specific conductivity and the geometrical dimensions of a sample are known, we can readily calculate the total electric current:

$$\mathbf{I} = \mathbf{J} \cdot S = \sigma S\mathbf{F}, \quad |\mathbf{I}| = \frac{\Phi_0}{R}, \qquad (6.53)$$

where we introduce the cross-section of the sample S, the voltage drop across the sample $\Phi_0 = L_x \cdot F$, and the sample length L_x. The resistance R is

$$R = \frac{L_x}{\sigma S}. \qquad (6.54)$$

Equations (6.49)–(6.51) are valid for systems of any dimensionality, where dissipative transport occurs. These include three-dimensional, two-dimensional, and one-dimensional systems. For all of them, the electron mobility is a representative characteristic of the classical transport regime.

To define measurement units of the mobility, we must first discuss briefly the units of measurement of several entities introduced in the preceding discussion. (1) The charge is measured in units of coulombs, C. The elementary charge of the electron, e, is equal to 1.6×10^{-19} C. (2) The electron concentration is measured in m^{-3}, m^{-2}, and m^{-1} for three-dimensional, two-dimensional, and one-dimensional systems, respectively. (3) The electric field is measured in units of volts per meter, $V\,m^{-1}$. (4) Accordingly, the units of mobility are $m^2\,V^{-1}\,s^{-1}$. (5) The current is measured in amperes, A, i.e., coulombs per second, $C\,s^{-1}$. (6) For the current density and conductivity of Eq. (6.51), we obtain different units for electron systems with different dimensionalities. For example, in a bulk crystal the current density is measured in amperes per unit cross-section, $A\,m^{-2}$, and the units of conductivity are $1\,\Omega^{-1}\,m^{-1}$ with Ω being the ohm – the unit of electrical resistance.

Returning to the discussion of mobility, we shall stress that the magnitude of the mobility depends on the particular material and scattering mechanisms in this material. Typically, the scattering mechanisms in semiconductors include scattering by impurities, imperfections, and lattice vibrations. If the first two scattering mechanisms can be avoided in pure and high-quality samples, lattice or, as one frequently says, phonon scattering can not be avoided in principle. The total scattering rate is a sum of rates of scattering due to particular mechanisms. Each of the rates is proportional to a scattering probability and inversely proportional to the mean free time. This leads to the conclusion that also the mobilities due to two or more scattering mechanisms should be added inversely:

$$\frac{1}{\mu} = \frac{1}{\mu_{\text{ph}}} + \frac{1}{\mu_{\text{im}}} + \cdots, \tag{6.55}$$

where μ_{ph}, μ_{im}, ... are partial mobilities determined by phonon scattering, impurity scattering, etc. Phonon scattering depends in an obvious way on temperature: at low temperatures, when thermal lattice vibrations are suppressed, the rate of this type of scattering is small, but it increases with the temperature. This results in a finite value of mobility even in pure crystals. For example, at room temperature the electron mobilities of Si and GaAs bulk crystals are limited to $1350\,cm^2\,V^{-1}\,s^{-1}$ and $8500\,cm^2\,V^{-1}\,s^{-1}$, respectively. The same order of magnitude for μ is realized in low-dimensional systems at room temperature. When the temperature decreases, mobility increases and the scattering by imperfections and impurities becomes the limiting factor, as shown in Fig. 6.8. These scattering mechanisms can be avoided in low-dimensional systems, where mobility can reach values above 10^5–$10^6\,cm^2\,V^{-1}\,s^{-1}$.

The electron current in the form of Eq. (6.51) is valid for *uniform* conductors. If the electron concentration, n, is spatially dependent, the electrons, naturally, diffuse from a region of high concentration to one of a low concentration. This produces an electron flux opposite to the gradient of the electron concentration: $\sim -dn/d\mathbf{r}$. The diffusive contribution to the current can be written as

$$\mathbf{J}_{\text{D}} = eD\,\frac{dn}{d\mathbf{r}} \equiv eD\,\nabla_{\mathbf{r}} n, \tag{6.56}$$

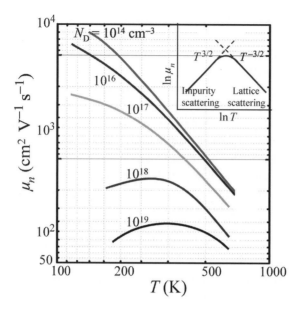

Figure 6.8 The temperature dependence of the electron mobility in Si for a system with two mechanisms of scattering, lattice scattering and impurity scattering. The concentration of impurities, N_D, is indicated near each curve. After W. F. Beadle, J. C. C. Tsai *et al.* (Eds.), *Quick Reference Manual for Semiconductor Engineers* (New York, Wiley, 1985).

where D is the diffusion coefficient, that was discussed in Section 6.2. Thus, in a nonuniform conductor, the total current is composed of both drift and diffusive contributions:

$$\mathbf{J} = e\mu \mathbf{F} n + e D \nabla_\mathbf{r} n. \tag{6.57}$$

It is easy to establish the fundamental relationship between two kinetic coefficients, the mobility, μ, and the diffusivity, D. Indeed, let us apply Eq. (6.57) to equilibrium conditions. Under equilibrium, the electric current is absent, i.e., $\mathbf{J} = 0$. The electron concentration, $n(\mathbf{r})$, can be expressed via the electrostatic potential, $\Phi(\mathbf{r})$, according to Boltzmann's distribution in the form of Eq. (6.14): $n(\mathbf{r}) = n_0 \exp[e\Phi(\mathbf{r})/(k_B T)]$. On putting this dependence and the electric field, $\mathbf{F}(\mathbf{r}) = -\mathrm{d}\Phi/\mathrm{d}\mathbf{r}$, into Eq. (6.57) and equating the current to zero, we obtain the so-called Einstein relationship:

$$\frac{D}{\mu} = \frac{k_B T}{e}. \tag{6.58}$$

Thus, having the electron mobility μ, one can easily calculate the diffusion coefficient, D.

The results just discussed are relevant to steady-state electron transport. However, the Newton equation (6.48) can describe the behavior of electrons in an arbitrary time-dependent electric field, $\mathbf{F}(t)$. Since any dependence, $\mathbf{F}(t)$, can be represented by using a Fourier transform, without loss of generality we can analyze the case of a harmonic dependence of the field:

$$\mathbf{F}(t) = \mathbf{F}_\omega \cos(\omega t), \tag{6.59}$$

where $\mathbf{F}_\omega$ is the field magnitude and ω is the frequency of field oscillations. For further analysis, we use the well-known formula $\cos(\omega t) = \mathcal{R}e[\mathrm{e}^{-\mathrm{i}\omega t}]$. Here $\mathcal{R}e[\ldots]$ means calculation of the real part of an expression in the brackets. Thus, the physical field is $\mathbf{F}(t) = \mathcal{R}e[\mathbf{F}_\omega \mathrm{e}^{-\mathrm{i}\omega t}]$.

It is worth noting that calculations with exponential functions like $\mathrm{e}^{-\mathrm{i}\omega t}$ are always simpler than those with sine and cosine functions. For this reason, the following approach is commonly used. Instead of the field in the form of Eq. (6.59) in the Newton equation, one uses the complex representation of the field:

$$\tilde{\mathbf{F}}(t) = \mathbf{F}_\omega \mathrm{e}^{-\mathrm{i}\omega t}.$$

When a solution to Eq. (6.48) with the complex field is found, the real part of this solution, that has a physical meaning, can be calculated easily.

To apply this approach, we look for the solution of Eq. (6.48) with a complex field of the exponential form $\mathbf{v}_\omega \mathrm{e}^{-\mathrm{i}\omega t}$. On substituting this form into Eq. (6.48), we immediately obtain

$$\mathbf{v}_\omega = -\frac{e}{m^*}\frac{\tau_\mathrm{e}}{1 - \mathrm{i}\omega\tau_\mathrm{e}}\mathbf{F}_\omega. \tag{6.60}$$

In this case, the electron velocity should be calculated as $\mathbf{v}(t) = \mathcal{R}e[\mathbf{v}_\omega \mathrm{e}^{-\mathrm{i}\omega t}]$; i.e.,

$$\mathbf{v}(t) = -\frac{e\tau_\mathrm{e}}{m^*}\mathbf{F}_\omega\left(\frac{\cos(\omega t)}{1 + \omega^2\tau_\mathrm{e}^2} + \frac{\omega\tau_\mathrm{e}\,\sin(\omega t)}{1 + \omega^2\tau_\mathrm{e}^2}\right). \tag{6.61}$$

Thus, the cosine electric field of Eq. (6.59) induces electron motion, with both cosine and sine contributions. It is instructive to rewrite Eq. (6.61) as

$$\mathbf{v}(t) = -\frac{e\tau_\mathrm{e}}{m^*}\mathbf{F}_\omega\frac{\cos(\omega t - \varphi)}{\sqrt{1 + \omega^2\tau_\mathrm{e}^2}}, \tag{6.62}$$

where we introduce the *phase shift* φ of the electron velocity with respect to the phase of the field (6.59). This phase shift can be found from the equation $\tan\varphi = \omega\tau_\mathrm{e}$. Here, $\varphi > 0$, which implies a delay of the electrons with respect to the field changes. Obviously, the delay exists only for non-zero frequencies ω and arises because of the friction force in the Newton equation. The second conclusion, which follows from Eq. (6.62), is that the magnitude of velocity oscillations decreases when the field frequency increases. In the limit of very high frequencies $\omega\tau_\mathrm{e} \gg 1$, the alternating electron velocity vanishes.

Having the electron velocity, $\mathbf{v}(t)$, and concentration, n, we can calculate the time-dependent electric current similarly to Eqs. (6.51) and (6.52). However, more common is the use of the complex representation of the current density:

$$\tilde{\mathbf{J}}(t) = \sigma(\omega)\mathbf{F}_\omega \mathrm{e}^{-\mathrm{i}\omega t}, \tag{6.63}$$

where we introduce the *complex conductivity*:

$$\sigma(\omega) = \frac{e^2\tau_\mathrm{e}n}{m^*}\left(\frac{1}{1 + \omega^2\tau_\mathrm{e}^2} + \mathrm{i}\frac{\omega\tau_\mathrm{e}}{1 + \omega^2\tau_\mathrm{e}^2}\right). \tag{6.64}$$

This conductivity reduces to its steady-state value in the limit $\omega\tau_\mathrm{e} \to 0$, and, in this limit, $\mathcal{I}m[\sigma] \to 0$. At finite frequencies, both contributions, $\mathcal{R}e[\sigma]$ and $\mathcal{I}m[\sigma]$, are

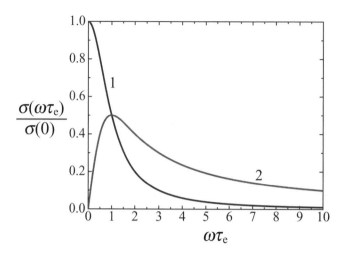

Figure 6.9 Dependences of the real (curve 1) and imaginary (curve 2) parts of the complex conductivity of Eq. (6.64) on frequency ω; $\sigma(0) = \sigma(\omega = 0)$.

important for an alternating current; in particular, at $\omega = 1/\tau_e$ the two contributions are equal numerically. The latter property is used typically to determine τ_e by varying the frequency, ω. Importantly, the electron scattering time, τ_e, determines the high-frequency properties of a material. In particular, at $\omega\tau_e \gg 1$ the conductivity vanishes and the electron subsystem of the material does not react to a high-frequency field. The relationship given by Eq. (6.64) is known as the *Drude* formula. In Fig. 6.9 the dependences of $\mathcal{R}e[\sigma(\omega)]$ and $\mathcal{I}m[\sigma(\omega)]$ versus ω are shown.

The conductivity $\sigma(\omega)$ is a specific characteristic of a material. If the material is homogeneous and its geometrical parameters are known, we can use σ to relate the total alternating current $\tilde{\mathbf{I}}$ and the alternating voltage $\tilde{\Phi} = \Phi_\omega e^{-i\omega t}$, both in complex representation:

$$\tilde{\mathbf{I}}(t) = \mathbf{I}_\omega e^{-i\omega t}, \quad I_\omega = \frac{\Phi_\omega}{Z(\omega)}, \quad Z(\omega) = \frac{L_x}{\sigma(\omega)S}. \tag{6.65}$$

Here, in the second relation, we omit the vector designation; L_x and S are the intercontact distance and the cross-section of the sample, respectively. By comparing these results with those of the steady-state case given by Eqs. (6.53) and (6.54), one may see that, instead of the resistance, R, a frequency-dependent parameter, $Z(\omega)$, is introduced. It is called the *impedance*. The impedance is a complex function that characterizes the electrical properties of the whole sample. The impedance can be introduced for any nonhomogeneous sample and device.

Dissipative transport in short structures

The mobility is a specific characteristic of a uniform conducting material. In a sample of extended dimension L_x, the electric field is almost uniform. It can be estimated as $F = \Phi_0/L_x$, with Φ_0 being the applied voltage, and the mobility determines basically

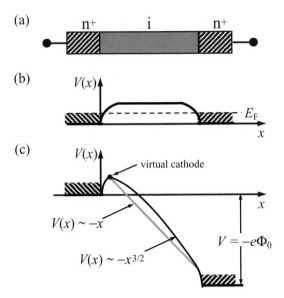

Figure 6.10 (a) The structure schematic of an n^+-i-n^+ diode. Potential energy profiles, $V(x)$, in a diode with charge-limited dissipative transport: (b) for an unbiased diode and (c) for a biased diode.

the electric resistance and the current, according to Eq. (6.53). In short samples, another electrical effect can contribute essentially toward the electron transport and the current. The effect arises due to nonuniform redistribution of the electrons and, thus, of the electric charge across the short sample. The charge induced by the current affects the potential distribution along the sample, making it strongly nonuniform and dependent on the current. As a result, the dependence of the total current on the applied voltage is no longer linear. The electron transport in this regime is called *space-charge-limited transport*. Since in this text we are interested particularly in very short structures, it is instructive to review briefly the space-charge-limited transport under the condition of Eq. (6.2), when we can introduce mobility.

Let us consider a short sample with two contacts as depicted in Fig. 6.2(b). The two-contact device is called the *diode* and the body of the sample between the contacts is called the *base* of the diode. Assume for simplicity that the base is not doped, i.e., there are no conducting electrons in the base. The contacts can be fabricated by heavy n-type doping of the contact regions; they are called n^+ regions. In such a case, this device is an n^+-i (insulator)$-n^+$ diode as shown in Fig. 6.10(a). Let Φ designate the electrostatic potential. Then, the potential energy of the electrons is $V = -e\Phi$. In Fig. 6.10(b), a sketch of the potential energy of an unbiased diode is presented: the electrons in the contact regions on the left and right are separated by a high potential barrier in the base. If an electric voltage Φ_0 is applied, the potential relief changes, as shown in Fig. 6.10(c). The potential energy, $V(x) = -e\Phi(x)$, decreases and some of the electrons can overcome the potential barrier, penetrate from the emitter electrode (the *cathode*) into the diode base, and contribute to the current. This bias-induced effect is referred to as electron *injection* from the electrode to the base. Obviously, with increasing bias, the maximum of the potential shifts toward the cathode and becomes lower, and the injection current increases. This physical picture can be described by the following simple model, which is valid at large electric biases. The current density, J, can be written via the mobility characteristic for the undoped base, μ, the injected electron concentration, $n(x)$, and the electrostatic field, $F(x) = -d\Phi/dx$, as in Eq. (6.51):

$$J = e\mu n(x)F(x). \tag{6.66}$$

Importantly, for the chosen configuration, see Fig. 6.10(a), the cathode is on the left and the anode is on the right. The potential, Φ, increases with x, and the electric field, F, and current density, J, are both negative. As a result of the continuity condition, in the stationary state the current density through the diode is conserved: $J = -J_0 = $ constant, where J_0 is the absolute value of the current density. From the latter relationship, we can determine the concentration of the injected electrons, $n(x)$:

$$n(x) = -J_0/(e\mu F(x)). \tag{6.67}$$

For the electrostatic field, we can use Poisson's equation,

$$\frac{d^2\Phi}{dx^2} = -\frac{dF}{dx} = \frac{en(x)}{\epsilon_0\epsilon}, \tag{6.68}$$

where ϵ is the dielectric constant of the base material and ϵ_0 is the permittivity of free space. After substitution of $n(x)$ from Eq. (6.67), Eq. (6.68) can be rewritten in terms of the field, F:

$$F\frac{dF}{dx} = \frac{J_0}{\epsilon_0\epsilon\mu} \quad \text{or} \quad F\,dF = \frac{J_0}{\epsilon_0\epsilon\mu}\,dx. \tag{6.69}$$

This equation should be supplemented by a boundary condition for the field. We can use the fact that at the maximum of the potential barrier:

$$-\frac{d\Phi}{dx} = F = 0. \tag{6.70}$$

At a large electric bias, this maximum is shifted close to the cathode and we can set $F(x = 0) \approx 0$. This approach neglects the processes which occur in a very narrow region between the real cathode and the potential maximum. It is called the *virtual-cathode approximation* and it is used widely for such a simplified analysis. Integration of Eq. (6.69) leads to the solution

$$\frac{1}{2}F^2(x) = \frac{J_0}{\epsilon_0\epsilon\mu}x \quad \text{or} \quad F(x) = -\left(\frac{2J_0}{\epsilon_0\epsilon\mu}\right)^{1/2}x^{1/2}, \tag{6.71}$$

where the negative value of the square root is chosen for the solution in accordance with the above-discussed sign of the electric field. Thus, the distribution of the potential, $\Phi(x)$, is given by

$$\Phi(x) = -\int_0^x dx\, F(x) = \left(\frac{8J_0}{9\epsilon_0\epsilon\mu}\right)^{1/2}x^{3/2}. \tag{6.72}$$

From the total voltage drop $\Phi(L_x) = \Phi_0$, we can find the relationship between the current density, J_0, and the bias, Φ_0, i.e., the current–voltage characteristic:

$$J_0 \equiv J_{0,\mathrm{d}} = \frac{9\epsilon_0\epsilon\mu}{8} \frac{\Phi_0^2}{L_x^3}. \tag{6.73}$$

This is the so-called *Mott–Gurney law* for the diode with dissipative electron transport. We have found that the current–voltage characteristic becomes strongly nonlinear because of the space-charge effect. The space-charge effect can be characterized, particularly, by the average concentration of the injected electrons, $\overline{n}$:

$$\overline{n} = \frac{1}{L_x} \int_0^{L_x} n(x)\mathrm{d}x = \frac{3}{2} \frac{\epsilon_0\epsilon\Phi_0}{eL_x^2}, \tag{6.74}$$

which increases with the applied voltage.

Equation (6.49) and $F(x)$ facilitate the calculation of the average time of electron transit through the diode, $t_{\mathrm{tr,d}}$:

$$t_{\mathrm{tr,d}} = \int_0^{L_x} \mathrm{d}x \, \frac{1}{v(x)} = \int_0^{L_x} \mathrm{d}x \, \frac{1}{\mu F(x)} = \frac{4}{3} t_{0,\mathrm{d}}, \tag{6.75}$$

where

$$t_{0,\mathrm{d}} = \frac{L_x^2}{\mu\Phi_0}. \tag{6.76}$$

Here, $t_{0,\mathrm{d}}$ is the transit time of the electrons drifting in the average electric field, Φ_0/L_x, i.e., when the space-charge effects are neglected. As follows from Eq. (6.75), these effects increase the transit time by the factor $4/3$.

In general, non-stationary electrical properties of a device are characterized by the current induced in response to a time-dependent external voltage bias. If the bias is of a frequency ω, the current response is given by the impedance according to Eq. (6.65). For a Mott–Gurney diode biased with a steady-state voltage, Φ_0, the complex impedance can be calculated exactly:

$$Z(\omega) = \frac{6R_\mathrm{d}}{\Omega^3} \left[(\Omega - \sin\Omega) + \mathrm{i}\left(\frac{\Omega^2}{2} - 1 + \cos\Omega \right) \right], \tag{6.77}$$

where

$$R_\mathrm{d} = \frac{\mathrm{d}\Phi_0}{\mathrm{d}J_0} = \frac{4L_x^3}{9\epsilon_0\epsilon\mu S\Phi_0}$$

is the differential resistance of the diode at steady state calculated from Eq. (6.73), $I_0 = J_0 S$, $S = L_y \times L_z$ is the cross-section of the sample, and $\Omega = \omega t_{\mathrm{tr,d}}$; i.e., it is determined via the average time, $t_{\mathrm{tr,d}}$, of electron transit through the diode; see Eq. (6.75). The dependences of $\mathcal{R}e[Z(\Omega)]$ and $\mathcal{I}m[Z(\Omega)]$ are presented in Fig. 6.11. The impedance of Eq. (6.77) differs considerably from the impedance of a material with the Drude-like high-frequency behavior. Importantly, as the voltage bias increases, the impedance decreases and the current response increases $\propto \Phi_0$; moreover, the spectral width of the response becomes larger ($\propto 1/t_{\mathrm{tr,d}} \propto \Phi_0$). Both conclusions provide evidence that a short

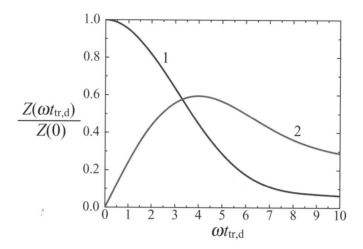

Figure 6.11 Dependences of the real (curve 1) and imaginary (curve 2) parts of the impedance of Eq. (6.77) on frequency ω; $Z(0) = Z(\omega = 0)$.

diode with space-charge-limited dissipative transport under a high bias can operate at higher frequencies than can a doped diode with Drude conductivity.

To conclude this analysis of dissipative electron transport in short samples, we want to point out that the main results obtained are valid at large biases, when the diffusive contribution to the current can be neglected:

$$\mu |\mathbf{F}(x)| \gg D \left| \frac{dn(x)}{dx} \right|. \tag{6.78}$$

(A related analysis is proposed in Problem 7.) For such conditions, the transport is determined to a large extent by the space-charge effects, the electric field is highly nonuniform, and the injection current increases quadratically with the bias, as Eq. (6.73) shows. Under high bias, the transit time through the diode decreases and the device remains electrically active in a frequency range enlarged proportionally with the bias.

Hot electrons

Now, we return to large uniform crystals to study the effect of high electric fields on the drift velocity and the current–voltage characteristics. Equations (6.49)–(6.51) were obtained under the assumption that the relaxation time, τ_e, is constant; i.e., it does not depend on the magnitude of the electric field. When the field increases, the electron gas is far from equilibrium. In particular, the average electron energy increases. This may be understood from the following qualitative considerations. The electron energy in an electric field may be described through the relationship

$$\frac{dE}{dt} = e(\mathbf{vF}) - \frac{E - E_{eq}}{\tau_E}. \tag{6.79}$$

The first term on the right-hand side corresponds to the power gained by the electron from the electric field; the second term represents the rate of electron-energy losses.

The losses are proportional to the deviation of electron energy, E, from its equilibrium value, E_{eq}, and τ_E is the energy relaxation time introduced in Section 6.2. The electron energy increases in the electric field until the overall energy balance is maintained. In the stationary case, when $dE/dt = 0$, Eq. (6.79) yields an electron energy, E, given by

$$E = E_{eq} + e(\mathbf{vF})\tau_E. \tag{6.80}$$

According to Eq. (6.49), the velocity, $\mathbf{v}$, is linearly proportional to the electric field, $\mathbf{F}$.

Thus, the mean electron energy increases as the square of the electric field and exceeds the equilibrium value, E_{eq}. It is convenient – and conventional in semiconductor electronics – to consider the electron *effective temperature*, T_e, instead of the mean electron energy. The relationship between the temperature and the mean energy can be found at thermal equilibrium and is $E = \alpha k_B T_e/2$, where the factor α is the dimensionality of the structure. Obviously, under thermal equilibrium the electron temperature, T_e, coincides with the lattice temperature, T. Under nonequilibrium conditions the two temperatures, T, and, T_e, may differ. The effective electron temperature expressed through the mean electron energy serves as a gauge of the nonequilibrium state. If T_e exceeds T only slightly, and the electron transport still obeys Ohm's law, we have *warm electrons*. The case with $T_e \gg T$ corresponds to the situation in which the electrons are far from equilibrium. For such a situation, the electrons frequently are called *hot electrons*. The electron temperature can reach magnitudes of about several thousand degrees Kelvin, while the lattice can remain cold. For simple estimates it is accepted that transition from the warm-electron regime to the hot-electron regime occurs at the electric field $F = F_{he}$, when $e(\mathbf{vF})\tau_E$ is equal to $E_{eq} = \alpha k_B T$. It is easy to estimate the *heating* electric field, F_{he}. Indeed, in Eq. (6.80) the drift velocity $\mathbf{v}$ can be estimated from $\mathbf{v} = -\mu\mathbf{F}$; thus, the critical heating field is

$$F_{he} = \sqrt{\frac{\alpha k_B T}{e\mu\tau_E}}. \tag{6.81}$$

For hot electrons, the scattering processes themselves become dependent on the field. The linear relationship of Ohm's law of Eq. (6.51) is no longer valid and the current–voltage characteristics, $J = J(F)$, and the drift-velocity–field dependence, $v = v(F)$, can exhibit strongly nonlinear behavior, which depends on the electron bandstructure and specific scattering mechanisms. In particular, these dependences are of different shapes in the two most important materials. In group IV semiconductor materials, the $J(F)$ and $v(F)$ dependences lead to the saturation effect at large fields, as shown in Fig. 6.12(a), whereas in III–V compounds, after a nonlinear portion with a rise, the current and the velocity undergo a decrease in some interval of the fields, as shown in Fig. 6.12(b). As follows from the analysis of different transport regimes given in Section 6.2, for high-speed devices the important material parameter is the maximum of the drift velocity, which can be achieved in high electric fields. A comparison of the saturated and maximum velocities for relevent materials is presented in Table 6.2.

One can see that some of the III–V compounds (including GaAs, InP, and InSb) have velocities several times larger than those of Si, GaP, and AlAs. For Si the drift velocity is restricted to a magnitude of about 10^7 cm s^{-1} that can be reached for electric fields above

Table 6.2 Values of the saturated (for Si, SiC, SiO$_2$, AlAs, and GaP) and maximum (for GaAs, InP, InAs, and InSb) drift velocities for semiconductor materials at room temperature

Material	Velocity (10^7 cm s^{-1})
Si	1
SiC	2
SiO$_2$	1.9
AlAs	0.65
GaP	1.1
GaAs	2
InP	2.5
InAs	4.4
InSb	6.5

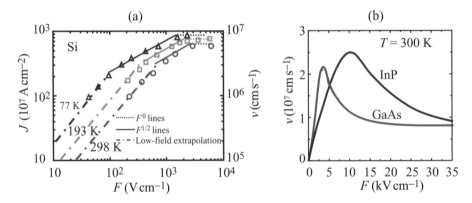

Figure 6.12 Current–voltage characteristics at large electric fields: (a) Si and (b) GaAs and InP. Reprinted with permission, (a), from E. J. Ryder, "Mobility of holes and electrons in high electric fields," *Phys. Rev.*, **90**, 766 (1953). © 1953 by the American Physical Society.

several kV cm^{-1}. For GaAs, the velocity maximum of 2×10^7 cm s^{-1} can be achieved at fields ≈ 3.5 kV cm^{-1}.

Transient overshoot effects

Limitations in the drift-velocity characteristic for a particular material can be overcome by utilizing another hot-electron effect known as *velocity overshoot*. In order to explain this phenomenon, let us mention that our previous conclusions have been made for the steady-state case, in which the electron distribution is stationary. In other words, the previous analysis has referred to the electron properties averaged over times much greater than the characteristic times of the system such as the mean-free-flight time,

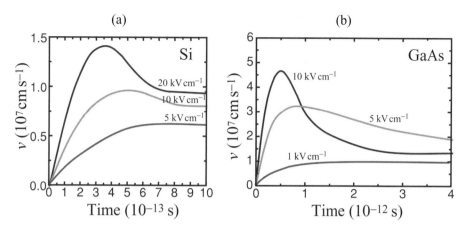

Figure 6.13 The overshoot effect – transient response of the electron drift velocity to step-like pulses of the electric field at room temperature. The field magnitudes are indicated for each plot: (a) Si and (b) GaAs. Reprinted with permission, from V. Mitin, V. Kochelap, and M. Stroscio, *Quantum Heterostructures*, Fig. 7.13 (Cambridge, Cambridge University Press, 1999).

momentum and energy relaxation times, etc. Now, we will consider processes that occur in the electron system immediately following its deviation from the equilibrium state. In this particular case, we will focus on the response to the electric field in the form of a step-function.

In general, the momentum relaxation time, τ_e, entering Eq. (6.49) is shorter than the energy relaxation time, τ_E, which determines the electron energy; see Eq. (6.79). Therefore, the velocity response to the electric field step is faster than the energy response for the case described by Eq. (6.79). Typically, if the scattering rate increases with increasing electron energy, the electron velocity may exceed the stationary velocity during a time interval of the order of τ_E. In other words, the transient velocity is not just a function of the electric field but also a function of the electron energy. Indeed, the velocity "adjusts" itself quickly to the quite slowly changing energy, and it "follows" that energy until the energy reaches the steady state. Initially, when the electron energy has not reached the stationary value, the electron velocity corresponding to the transient energy is higher than the velocity corresponding to the stationary energy.

The overshoot effect is shown in Fig. 6.13 for Si and GaAs. The drift velocity is presented as a function of time for several electric fields. The electric field is assumed to be switched on at time $t = 0$. The overshoot effect is pronounced in high electric fields. The maximum of the transient velocity can exceed the stationary saturated velocity by as much as two to four times.

From a physical explanation of the overshoot effect, one can understand how it is possible to utilize this effect. Let us imagine that cold electrons enter an active region of a semiconductor device through a contact. If there is a high electric field in the active region, the electrons will be accelerated. At some distance from the injecting contact the

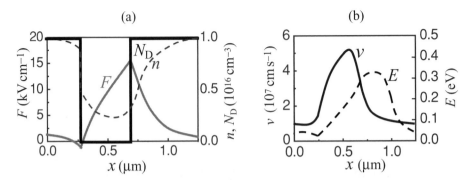

Figure 6.14 Space-charge-limited transport with the velocity overshoot effect in an n^+-i-n^+ GaAs diode at $L_x = 0.37\,\mu$m, $T = 77$ K, and $\Phi_0 = 0.5$ V. (a) The doped regions, the electric fields, F, and the injected electron concentration, n. (b) The drift velocity, v, and the average energy of the injected electrons, E. From A. Ghis, E. Constant, and B. Boittiaux, "Ballistic and overshoot electron transport in bulk semiconductors and in submicronic devices," *J. Appl. Phys.*, **54**, 214–221 (1982). Reprinted wth permission from A Ghis, E Constant, and B. Boittiaux, *Journal of Applied Physics*, **54**, 214 (1393). © 1983 American Institute of Physics.

electrons will reach the maximum overshoot velocity and after that their velocity will gradually decrease to the stationary value. If the active region of the device is short and comparable to the distance over which the overshoot effect takes place, electron transit through this active region will occur at a velocity higher than the stationary velocity and the overall transit time will be shorter. Consequently, the device will be able to operate at higher speed and frequency.

As follows from Fig. 6.13, the characteristic time of the overshoot effect is about $\tau_{tr} = 0.5 \times 10^{-12}$ s for GaAs. Estimating the average velocity under the overshoot as $v_m = (2$–$4) \times 10^7$ cm s^{-1}, we obtain an estimate for the device length necessary to realize the discussed effect: $L_x \leq v_m \tau_{tr} = 0.1$–$0.2\,\mu$m. That is, the overshoot effect and ultra-high-speed electrons are achievable in short (submicron) samples.

At high biases and currents, the space-charge effects should be taken into account together with the velocity overshoot. In Fig. 6.14, we present the calculations of space-charge-limited electron transport in a short n^+-i-n^+ GaAs diode, where n^+ are heavily doped contacts and i is the undoped base of the diode. The step-like doping of the contact regions is marked by N_D. The field, the concentration of the injected electrons and their average drift velocity, and the average energy are shown. The electric field and the injected electron concentration are extremely nonuniform, which is similar to the previously analyzed case of dissipative electron transport. However, for a short diode, the velocity overshoot in the diode base is clearly observed. The average energy increases in the base, and drops only in the receiving contact because of strong scattering in the heavily doped region. Interestingly, the drift velocity reaches a value of above 4×10^7 cm s^{-1} and the electron energy exceeds 3500 K at a crystal temperature of 77 K.

For the low-field, hot-electron, and overshoot transport regimes studied, the processes of dissipation play a dominant role in controlling the electron drift velocity. By scaling down the sample dimensions, one can achieve an ultimately fast collisionless electron transport.

Classical ballistic transport

According to Eq. (6.11), the classical ballistic (collisionless) regime occurs in very short samples, where during their flight the electrons do not undergo any scattering. The finite electric current and electric resistance in such a case occur exclusively due to the space-charge effects. We can account for these effects in a ballistic diode with the geometry depicted in Fig. 6.2(b). Similarly to the case of the above-presented short diodes with dissipative and overshoot transport, the distribution of the electrostatic potential, $\Phi(x)$, the concentration, $n(x)$, and the velocity, $v(x)$, have to be determined under collisionless electron flight. We will exploit Poisson's equation (6.68) and use the current density in the form

$$\mathbf{J} = -en(x)\mathbf{v}(x). \tag{6.82}$$

The velocity, $\mathbf{v}$, can be found from the Newton equation (2.8); however, we will use the energy conservation law, which reads as

$$\frac{m^* v^2(x)}{2} - e\Phi(x) = \frac{m v_c^2}{2} - e\Phi_c = \text{constant.} \tag{6.83}$$

Here, v_c and Φ_c are parameters corresponding to the injecting electrode (cathode). In the case of a large electric bias, we can simplify the latter equation by considering that the electrons are injected over the barrier with small velocity $v_c \to 0$. Then, we exploit the virtual-cathode approximation discussed previously. It is convenient to set $\Phi_c = 0$ and obtain

$$v(x) = \sqrt{\frac{2e\Phi(x)}{m^*}} \tag{6.84}$$

and

$$n(x) = \frac{J_0}{ev(x)}, \tag{6.85}$$

where J_0 is the absolute value of the current density through the diode. As we noted in the section devoted to dissipative transport, J is negative and it is convenient to introduce an absolute value of current density, J_0, which is constant throughout the entire sample. On combining the latter formula with Poisson's equation (Eq. (6.68)) we obtain the equation for $\Phi(x)$:

$$\frac{d^2\Phi(x)}{dx^2} = \frac{J_0}{\epsilon_0\epsilon}\sqrt{\frac{m^*}{2e\Phi(x)}}. \tag{6.86}$$

On multiplying both sides of the last equation by $d\Phi(x)/dx$, we obtain an equation that can be integrated. The result of integration is

$$\frac{1}{2}\left(\frac{d\Phi}{dx}\right)^2 = \frac{2J_0}{\epsilon_0\epsilon}\sqrt{\frac{m^*\Phi}{2e}} + C. \tag{6.87}$$

The integration constant, C, can be set to zero, since within our approach as $x \to 0$ we have $d\Phi/dx \to 0$ and $\Phi \to 0$. Now, we can rewrite the result as

$$\frac{1}{\Phi^{1/4}} \frac{d\Phi}{dx} = 2\sqrt{\frac{J_0}{\epsilon_0 \epsilon} \sqrt{\frac{m^*}{2e}}}. \tag{6.88}$$

Further integration gives us the final result for the distribution of the electrostatic potential:

$$\Phi(x) = \frac{3}{2} \left(\frac{J_0}{\epsilon_0 \epsilon} \right)^{2/3} \left(\frac{m^*}{2e} \right)^{1/3} x^{4/3}. \tag{6.89}$$

Taking into account that $\Phi(L_x) = \Phi_0$, we can find the current–voltage characteristic of the ballistic diode:

$$J_{0,\mathrm{b}} = \left(\frac{2}{3} \right)^{3/2} \frac{\epsilon_0 \epsilon}{L_x^2} \sqrt{\frac{2e}{m^*}} \Phi_0^{3/2}. \tag{6.90}$$

This equation is known as *Child's law*. It was found and verified first for vacuum diodes, where collisions of electrons were totally absent. Child's law differs considerably from the current–voltage characteristic of the dissipative diode given by Eq. (6.73).

Using Eq. (6.89), we can calculate other parameters of the ballistic diode. For example, the transit time for the electrons is

$$t_{\mathrm{tr,b}} = \frac{3}{2} t_{0,\mathrm{b}}, \quad t_{0,\mathrm{b}} = L_x \sqrt{\frac{2m^*}{e\Phi_0}}, \tag{6.91}$$

with $t_{0,\mathrm{b}}$ being the transit time of the ballistic electrons in the absence of space-charge effects. The space-charge effects increase the transit time, $t_{\mathrm{tr,b}}$, by the factor $3/2$.

It is instructive to compare the results obtained for space-charge-limited transport in the dissipative and ballistic diodes. To make this comparison we assume that the lengths of the diodes, L_x, are equal and that the two diodes are biased equally. Then, from Eqs. (6.75) and (6.91), for the ratio of the transit times we obtain

$$\frac{t_{\mathrm{tr,b}}}{t_{\mathrm{tr,d}}} = \frac{9\sqrt{m^*\Phi_0}\mu}{4\sqrt{2e}L_x} = \frac{9\tau_{\mathrm{e}}\sqrt{e\Phi_0}}{4\sqrt{2m^*}L_x} = \frac{9l_{\mathrm{e}}}{8L_x} \ll 1, \tag{6.92}$$

where we used the relationship $\mu = e\tau_{\mathrm{e}}/m^*$, and introduced the maximum mean-free path, $l_{\mathrm{e}} = \tau_{\mathrm{e}} v_{\mathrm{m}}$, and the maximum electron velocity in the diode, $v_{\mathrm{m}} = \sqrt{2e\Phi_0/m^*}$. The ratio obtained is small, in accord with the condition of dissipative transport of Eq. (6.12). That is, the ballistic diodes can provide much faster regimes of operation. Similarly, it can be shown that, for a given electric bias, the currents in the ballistic diode are much greater than those of the dissipative diode: $J_{0,\mathrm{b}}/J_{0,\mathrm{d}} = \frac{8}{9}(\frac{2}{3})^{3/2}L_x/l_{\mathrm{e}} \gg 1$. Hence, the ultimate collisionless transport has a number of advantages over the dissipative transport due to the higher velocity of electrons.

In fact, in a real situation some collisions always occur. Then, this "intermediate" case can be treated numerically by using a much more complex model. In Figs. 6.15(a) and 6.15(b) the results of such numerical modeling for a GaAs diode of length $L_x = 0.4 \ \mu\mathrm{m}$ at temperature $T = 300$ K are presented. A typical value of mobility for high-quality GaAs material is $\mu = 7500 \ \mathrm{cm}^2 \ \mathrm{V}^{-1} \ \mathrm{s}^{-1}$. For this mobility, the time of free flight

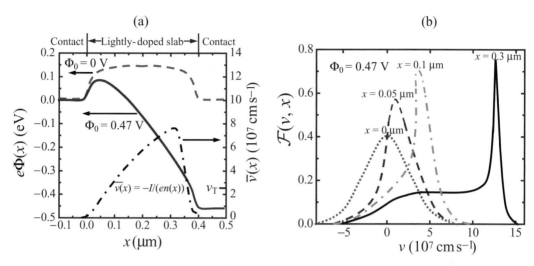

Figure 6.15 The results of numerical modeling of a GaAs diode. (a) The distribution of electron potential energy, $e\Phi(x)$, and average velocity, $\bar{v}(x)$, across the diode. (b) The distribution function, $\mathcal{F}(v, x)$, at $x = 0, 0.05, 0.1$, and 0.3 μm. The applied voltage, Φ_0, is equal to 0.47 V; the thermal velocity, v_T, is equal to 2.6×10^7 cm s^{-1}. Reprinted with permission from H. U. Baranger and J. W. Wilkins, "Ballistic electrons in an inhomogeneous submicron structure: thermal and contact effects," *Phys. Rev.* B, **30**, 7349 (1984). © 1984 by the American Physical Society.

is $\tau_e = 2.9 \times 10^{-13}$ s. At the given temperature, the thermal velocity of the electrons is $v_T = \sqrt{k_B T / m^*} \approx 2.6 \times 10^7$ cm s^{-1}. The average mean-free path of an electron is $l_e = 0.075$ μm and $L_x/l_e \approx 5$. Thus, on average the electrons with velocity v_T would undergo as many as five collisions during the flight across a uniform sample of the chosen length, L_x. For such a sample, the transport would be almost dissipative. For a strongly biased diode, we reach quite different conclusions. In Fig. 6.15(a), the potential-energy profile and the average electron velocity are presented for an applied voltage $\Phi_0 = 0.47$ V. For comparison, the thermal velocity v_T is also indicated. One can see that the maximum average velocity is above 7×10^7 cm s^{-1}. The actual mean free path is estimated to be larger, 0.07 μm $\leq l_e \leq 0.2$ μm. Thus, this example corresponds to an intermediate case between pure ballistic and dissipative transport. For such a case, different electrons have different velocities and it is possible to characterize these electrons by a distribution function, $\mathcal{F}$, over the velocity, v. The distribution depends also on the distance along the diode, x: $\mathcal{F} = \mathcal{F}(v, x)$. In Fig. 6.15(b), the distribution function is presented for various distances, x. The thermal distribution of the electrons in the cathode is shown by the dotted line, and it is obviously the Maxwellian distribution of Eq. (6.16) with the temperature $T = 300$ K. Actually, this distribution determines the velocities of the injected electrons. The initial "spreading" of the electrons over velocities in the cathode is one of the main differences from the simple model analyzed previously, in which the electrons are injected with a near-zero velocity. Inside the diode, the distribution becomes strongly anisotropic: electrons with negative velocities are almost absent. As the distance increases, the distribution becomes more and more anisotropic, with a

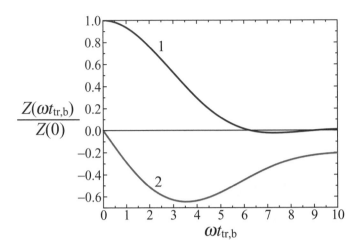

Figure 6.16 Dependences of the real (curve 1) and imaginary (curve 2) parts of the impedance of Eqs. (6.93) and (6.94) on frequency ω; $Z(0) = Z(\omega = 0)$.

well-pronounced maximum. For larger distances, this maximum corresponds approximately to values given by the previously discussed model of pure ballistic transport. This example highlights the real nature of an almost ballistic diode: some of the electrons participate in transport and there occurs some spreading over the velocity. However, the average properties of the diode are very close to those obtained in the simple model based on the classical Newton and Poisson equations.

The previously discussed simple model allows one to investigate high-frequency properties of the ballistic diode and calculate its impedance. The real and imaginary parts of the impedance are

$$\mathcal{R}e[Z(\Omega)] = \frac{12 R_d}{\Omega^4}[2(1 - \cos \Omega) - \Omega \sin \Omega], \tag{6.93}$$

$$\mathcal{I}m[Z(\Omega)] = \frac{12 R_d}{\Omega^4}[2 \sin \Omega - \Omega(1 + \cos \Omega) - \Omega^3/6]. \tag{6.94}$$

Here, $R_d = d\Phi_0/dJ_0$ is the differential resistance calculated with the use of the steady-state current–voltage characteristic of Eq. (6.90); $\Omega = \omega t_{tr,b}$, i.e., the frequency measured in units of the inverse transient time defined by Eq. (6.91). Both dependences, $\mathcal{R}e[Z(\Omega)]$ and $\mathcal{I}m[Z(\Omega)]$, are presented in Fig. 6.16. According to Eqs. (6.93) and (6.94) and the definition of R_d, the magnitudes of $\mathcal{R}e[Z(\Omega)]$ and $\mathcal{I}m[Z(\Omega)]$ decrease as the voltage bias, Φ_0, increases:

$$R_d = \frac{d\Phi_0}{dJ_0} = \frac{1}{2}\sqrt{\frac{3m^*}{e\Phi_0} \frac{L_x^2}{\epsilon_0 \epsilon}}.$$

That is, the current response becomes larger at larger biases ($R_d \propto 1/\sqrt{\Phi_0}$). Simultaneously, the frequency region within which the diode is still active increases with the bias ($\propto \sqrt{\Phi_0}$). These features are quite similar to those obtained for the dissipative Mott–Gurney diode; see Eq. (6.77). A principally new feature is the observable oscillations of $\mathcal{R}e[Z(\omega)]$ and $\mathcal{I}m[Z(\omega)]$ with ω and the "frequency windows" arising with negative

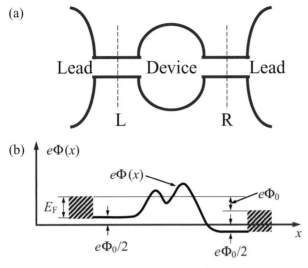

Figure 6.17 (a) The structure schematic of a nanostructure mesoscopic device and (b) the potential energy profile, $e\Phi(x)$, in the biased mesoscopic device.

values of $\mathcal{R}e[Z(\omega)]$. The first "window" with $\mathcal{R}e[Z(\omega)] < 0$ occurs for ω between $6.3/t_{\mathrm{tr,b}}$ and $9/t_{\mathrm{tr,b}}$. It is known that a negative $\mathcal{R}e[Z(\omega)]$ corresponds to an electrical instability at frequency ω. Such an instability can be used for generation of high-frequency electromagnetic oscillations, and we will discuss this phenomena in detail in Section 8.2.

In conclusion, ballistic nanoscale devices are the fastest devices based on classical electron transport. Both a finite electric current and a finite electric resistance in the absence of electron scattering arise because of the electrostatic effects induced by redistribution of charged electrons in the device. This is the reason why this case is associated with *charge-limited transport*. The ballistic diodes are characterized also by "frequency windows" within which the dynamic resistance of the device is negative, which signifies an electrical instability and the possibility of generating ultra-high-frequency electromagnetic radiation.

Quantum ballistic transport: the Landauer formula

The general theoretical description of the various quantum transport regimes is too complex a problem to be presented in this book. We consider the simplest limiting case of time-independent and low-temperature transport, when inelastic processes are negligible. Thus, according to the classification given in Table 6.1 we consider now the mesoscopic regime of electron transport. As has been pointed out previously, transport through nanostructure mesoscopic devices depends on both the geometry of the nanostructure and the "leads" (electrodes, contacts, wires, interconnects, etc.) which connect the device to an external electric circuit. Hence, we should consider the whole system: the device and the leads. Figure 6.17 illustrates the essential features of the simplified model. The system consists of two leads, to which a bias is applied, two "electron waveguides," L and R (they can be thought of as quantum wires), and the device itself.

In order to avoid the need for a detailed description of the leads, we assume that the leads are reservoirs of electrons, where energy and momentum relaxation processes are so effective that the electrons remain in equilibrium even under a given applied voltage bias. Hence, the boundary conditions at the interface between the leads and the device are assumed to be determined by the equilibrium Fermi distribution function of Eq. (6.18). The electron concentration in the leads is so high that the electrostatic potential in each lead is taken to be constant, as for the case of a metal. Let E_F be the Fermi energy of the electrons in the leads in the absence of a bias. On applying a voltage bias, Φ_0, the Fermi level in one of the leads becomes $E_F - \frac{1}{2}e\Phi_0$, while that of the second lead becomes $E_F + \frac{1}{2}e\Phi_0$, as depicted in Fig. 6.17(b). Thus, the electron distribution functions are

$$\mathcal{F}_F\left(E + \frac{1}{2}e\Phi_0 - E_F\right) \qquad \text{and} \qquad \mathcal{F}_F\left(E - \frac{1}{2}e\Phi_0 - E_F\right), \qquad (6.95)$$

at the left and right leads, respectively. Here E is the kinetic energy of the electrons.

In studying quantum wires in Section 3.3, we found that the electron spectrum consists of a series of one-dimensional subbands with the energies

$$E_{k_x,n,m} = \varepsilon_{n,m} + \frac{\hbar^2 k_x^2}{2m^*}. \qquad (6.96)$$

Here, n and m are integers and k_x is a one-dimensional wavevector directed along the axis of the quantum wire. The electron wavefunction can be written as

$$\psi(x, y, z) = \psi_\perp(y, z) \times \psi_{||}(x). \qquad (6.97)$$

The component of the wavefunction $\psi_\perp(y, z)$ describes the transverse distribution of the electrons in the quantum wire, while $\psi_{||}(x)$ describes the translational motion of electrons along the connecting quantum wires. It is generally composed of plane waves $e^{\pm ik_x x}$. Since electron transport occurs along the wires, we must analyze the wavefunction $\psi_{||}(x)$. We can write the wavefunction of the electrons, $\psi_{||}(x)$, at the interfaces between the leads and the device. These interfaces are indicated in Fig. 6.17 by the L and R cross-sections. The wavefunction, $\psi_{||,l}(x)$, of the electrons coming from the left lead is

$$\psi_{||,l}(x) = \begin{cases} e^{ik_l(x-x_l)} + r_l e^{-ik_l(x-x_l)}, & x \sim x_l, \\ t_r e^{ik_r(x-x_r)}, & x \sim x_r. \end{cases} \qquad (6.98)$$

Here, x_l and x_r denote the coordinates of the cross-sections L and R, respectively. That is, at the cross-section L the wavefunction consists of incident and reflected waves, while at the cross-section R there is only the wave that has passed through the device. The coefficients t_r and r_l are the amplitudes of these transmitted and reflected waves, respectively. These coefficients depend on the *particular potential profile* in the device, its geometry, etc. In a similar manner, one can write the wavefunction, $\psi_{||,r}(x)$, of the electrons coming from the right lead:

$$\psi_{||,r}(x) = \begin{cases} t_l e^{-ik_l(x-x_l)}, & x \sim x_l, \\ e^{-ik_r(x-x_r)} + r_r e^{ik_r(x-x_r)}, & x \sim x_r. \end{cases} \qquad (6.99)$$

In Eqs. (6.98) and (6.99), k_l and k_r are the wavevectors in the left and right cross-sections, respectively.

The complex coefficients t_r, r_l, t_l, and r_r are connected by several fundamental relationships, which do not depend on the specific design of the device. The requirement of current continuity for both wavefunctions, $\psi_{||,l}(x)$ and $\psi_{||,r}(x)$, gives

$$k_l(1 - |r_l|^2) = k_r|t_r|^2 \qquad \text{and} \qquad k_r(1 - |r_r|^2) = k_l|t_l|^2. \tag{6.100}$$

Other important relationships are

$$|r_l|^2 = |r_r|^2 \qquad \text{and} \qquad t_r^* t_l = t_r t_l^*. \tag{6.101}$$

Then, using Eq. (6.100), we get

$$k_r^2 |t_r|^2 = k_l^2 |t_l|^2. \tag{6.102}$$

By substituting the wavefunctions of Eqs. (6.98) and (6.99) into Eq. (3.10), which defines the particle flux, we calculate the incoming, i_{in}, and outgoing, i_{out}, electron flows:

$$i_{in} = v_l \qquad \text{and} \qquad i_{out} = v_r |t_r|^2, \tag{6.103}$$

where $v_l = \hbar k_l/m^*$ and $v_r = \hbar k_r/m^*$ are the velocities at the L and R cross-sections, respectively. The ratio of these quantities defines the *transmission coefficient* for electrons moving through the device from the left to the right:

$$T_{l \to r}(E) = \frac{i_{out}}{i_{in}} = \frac{k_r}{k_l} |t_r|^2. \tag{6.104}$$

The transmission coefficient corresponding to the electrons moving through the device from the right to the left equals

$$T_{r \to l}(E) = \frac{k_l}{k_r} |t_l|^2. \tag{6.105}$$

From Eqs. (6.102), (6.104), and (6.105), one can find that

$$T_{l \to r}(E) = T_{r \to l}(E) = T(E_{||}), \tag{6.106}$$

where

$$E_{||} = \frac{\hbar^2 k_x^2}{2m^*} \tag{6.107}$$

is the kinetic energy corresponding to the longitudinal component of the electron's momentum. Thus, the transmission coefficients are the same for both directions of incoming electrons. The ratio of reflected and incoming electron fluxes defines the reflection coefficient:

$$R(E) = \frac{i_r}{i_{in}} = |r_l|^2 = |r_r|^2. \tag{6.108}$$

It is obvious that

$$T(E) + R(E) = 1. \tag{6.109}$$

Now, we can take into account contributions to the total current from all electrons entering the device from both leads. Consider a state of the electrons, say in the left lead, with quantum numbers k_x, n, and m. The number of electrons in this state is given by $2\mathcal{F}_{\mathrm{F}}(E(k_x, n, m) + \frac{1}{2}e\Phi_0 - E_{\mathrm{F}})$, where the factor 2 comes from spin degeneracy. If the connection length is L_{c}, the number of electrons per unit length of the connection is $2\mathcal{F}_{\mathrm{F}}/L_{\mathrm{c}}$. The total contribution to the electric current from the electrons entering from the left is

$$I_1 = -\frac{2e}{L_{\mathrm{c}}} \sum_{n,m} \sum_{k_x > 0} v_{\parallel} T(E_{\parallel}) \mathcal{F}_{\mathrm{F}}\left(E(k_x, n, m) + \frac{1}{2}e\Phi_0 - E_{\mathrm{F}}\right). \tag{6.110}$$

Similarly, for the electrons from the right lead one gets

$$I_{\mathrm{r}} = \frac{2e}{L_{\mathrm{c}}} \sum_{n,m} \sum_{k_x < 0} v_{\parallel} T(E_{\parallel}) \mathcal{F}_{\mathrm{F}}\left(E(k_x, n, m) - \frac{1}{2}e\Phi_0 - E_{\mathrm{F}}\right). \tag{6.111}$$

Therefore, the total current through the device is

$$\begin{aligned}
I &= I_1 - I_{\mathrm{r}} \\
&= -\frac{2e}{L_{\mathrm{c}}} \sum_{n,m} \sum_{k_x > 0} v_{\parallel} T(E_{\parallel}) \Bigg[\mathcal{F}_{\mathrm{F}}\left(E(k_x, n, m) + \frac{1}{2}e\Phi_0 - E_{\mathrm{F}}\right) \\
&\quad - \mathcal{F}_{\mathrm{F}}\left(E(k_x, n, m) - \frac{1}{2}e\Phi_0 - E_{\mathrm{F}}\right) \Bigg].
\end{aligned} \tag{6.112}$$

Since, for our model, the electron velocity, $v_{\parallel}$, and the transmission coefficient, T, are independent of the transverse quantum numbers, n and m, we can calculate the sum over n and m. Taking into account the explicit form of the Fermi function, $\mathcal{F}_{\mathrm{F}}$, it is convenient to introduce the distribution function dependent only on the kinetic energy $E_{\parallel}$:

$$\mathcal{F}_{\mathrm{F}}(E_{\parallel}) = 2 \sum_{n,m} \frac{1}{1 + \exp\left(\dfrac{E_{\parallel} + \varepsilon_{n,m} - E_{\mathrm{F}}}{k_{\mathrm{B}}T}\right)}. \tag{6.113}$$

Next, as in Eq. (6.30) the summation over k_x in Eq. (6.112) may be replaced by an integration:

$$\sum_{k_x} \{\ldots\} \to L_{\mathrm{c}} \int \frac{dk_x}{2\pi} \{\ldots\} = L_{\mathrm{c}} \int \frac{dE_{\parallel}}{2\pi\hbar v_{\parallel}} \{\ldots\}. \tag{6.114}$$

Finally, we obtain the following expression for the total current:

$$I = -e \int \frac{dE_{\parallel}}{2\pi\hbar} T(E_{\parallel}) \left[\mathcal{F}_{\mathrm{F}}\left(E_{\parallel} + \frac{1}{2}e\Phi_0 - E_{\mathrm{F}}\right) - \mathcal{F}_{\mathrm{F}}\left(E_{\parallel} - \frac{1}{2}e\Phi_0 - E_{\mathrm{F}}\right) \right], \tag{6.115}$$

where the integration range runs over the kinetic energy of the longitudinal motion $E_{\parallel}$. Note that *the electron velocity does not appear in the final expression* for the total current, I. This general result may be applied to a variety of different cases. Let us consider two of them.

A device macroscopically large in the transverse directions

In this case the transverse quantum numbers are the wavevectors $n = k_y$ and $m = k_z$, and

$$\varepsilon_{n,m} = \frac{\hbar^2}{2m^*}\left(k_y^2 + k_z^2\right).$$

The function of Eq. (6.113) can be calculated explicitly,

$$\mathcal{F}_F(E_\parallel) = S \frac{m^* k_B T}{\pi \hbar^2} \ln\left[1 + \exp\left(\frac{E_F - E_\parallel}{k_B T}\right)\right], \qquad (6.116)$$

where S is the cross-sectional area. The term $\mathcal{F}_F(E_\parallel)$ has the meaning of the number of electrons with energy $E_\parallel$. Then the total current density, $J = I/S$, is

$$J = -e \frac{m^* k_B T}{\pi \hbar^2} \int \frac{dE_\parallel}{2\pi \hbar} T(E_\parallel) \ln\left[\frac{1 + \exp\left(\dfrac{E_F - E_\parallel + \frac{1}{2} e \Phi_0}{k_B T}\right)}{1 + \exp\left(\dfrac{E_F - E_\parallel - \frac{1}{2} e \Phi_0}{k_B T}\right)}\right]. \qquad (6.117)$$

This is indeed a useful result since Eq. (6.117) allows one to calculate the current–voltage characteristic of a nanostructure device and its dependence on the electron concentration, temperature, etc. Interestingly, we found a *finite current* in this biased device, where *no scattering occurs*. That is, the device possesses an electric resistance. The resistance is explained by two factors: (1) quantum-mechanical reflection of electron waves inside the device and (2) the finite number of electrons which can be injected into the device. Note that no electrical charge redistribution was considered, i.e., relatively small currents were supposed.

Device conductance at low temperatures. The Landauer formula

Let us turn to the general result of Eq. (6.112). It can be simplified significantly for near-equilibrium transport at low temperatures. We know that in the limit of zero temperature the Fermi distribution function, $\mathcal{F}_F$, becomes a step function:

$$\lim_{T \to 0} \mathcal{F}_F(E - E_F) = \Theta(E_F - E). \qquad (6.118)$$

If the applied voltage, Φ_0, is small, the difference between the distribution functions in the sum of Eq. (6.112) is equal to

$$\mathcal{F}_F\left(E(k_x, n, m) + \frac{1}{2} e \Phi_0 - E_F\right) - \mathcal{F}_F\left(E(k_x, n, m) - \frac{1}{2} e \Phi_0 - E_F\right) = e \Phi_0 \delta(E_F - E), \qquad (6.119)$$

where we have taken into account the fact that the derivative of the step-like Fermi distribution function is a δ-function. Hence, the current is proportional to the voltage bias, Φ_0.

One can introduce the *conductance* of a nanostructure device as the ratio

$$G = \frac{I}{\Phi_0}. \qquad (6.120)$$

From Eqs. (6.112), (6.114), (6.119), and (6.120) we obtain the conductance at low temperatures in the form

$$G = \frac{e^2}{h} \sum_{n,m,s} T(E_F, n, m) = 2 \frac{e^2}{h} \sum_{n,m} T(E_F, n, m), \qquad (6.121)$$

where the sum extends only over the electron states (n, m) with energy $E < E_{\mathrm{F}}$. The coefficient in front of the sum of Eq. (6.121),

$$G_0 = \frac{e^2}{h},\tag{6.122}$$

is called the *quantum of conductance*; here h is Planck's constant; $h = 2\pi\hbar$, where $\hbar$ is the reduced Planck constant introduced previously in Chapter 2. The quantum of conductance is equal to $G_0 = 39.6\ \mu\mathrm{S}$ (1 siemens $= 1\ \mathrm{S} = 1\ \mathrm{A\ V^{-1}}$) and its inverse value is $1/G_0 = 25.2\ \mathrm{k\Omega}$. Equation (6.121) is often called the *Landauer formula*.

Sometimes it is convenient to consider electron states corresponding to different quantum numbers (n, m) in terms of separate electron conduction channels. In this latter formulation, Eq. (6.121), may be rewritten in the form

$$G = 2G_0 \sum_{n,m} T_{n,m} \quad \text{with } T_{n,m} = T(E_{\mathrm{F}}, n, m).\tag{6.123}$$

Here each channel (n, m) contributes $G_0 T_{n,m}$ to the conductance, G. If the channel corresponding to the (n, m) state is transparent to electrons, $T_{n,m} = 1$, and the contribution of this channel is equal to the quantum of conductance of Eq. (6.122).

This regime is also referred to as *quantum ballistics*. Thus, a nanoscale device with quantum ballistic transport exhibits *finite conductance* (*finite resistance*). Generally, the conductance depends both on the transmission coefficient of the device and on the Fermi distribution functions in the leads. In particular, even if the device is entirely transparent, $T = 1$, the conductance remains finite and is equal to the quantum of conductance, G_0. An increase in the occupation of the upper low-dimensional subbands gives rise to an unusual behavior of the conductance. Indeed, at low temperature, actually at $k_{\mathrm{B}}T \ll E_{\mathrm{F}}$, only those subbands for which the energy of the subband bottoms $\varepsilon_{n,m} < E_{\mathrm{F}}$ are occupied. If, on changing the Fermi energy, E_{F}, a new subband starts to populate, the conductance of the device increases in a step-like manner, independently of how many electrons occupy this new subband. In conclusion, the Landauer formula describes quantum transport in mesoscopic devices. It is valid at low temperature and small voltage bias.

The simplest device with quantum ballistic transport is the so-called *quantum point contact*. It can be fabricated in a number of fashions. The essence of the quantum point contact is that two electron reservoirs are connected through a conducting region with transverse dimensions, L_z and L_y, comparable to the de Broglie wavelength of the electrons. In Fig. 6.18(a), a sketch of a point contact on the $\{x, y\}$-plane is depicted. The electron transport and current occur along the x direction. For simplicity, we can assume that the channel has a constant dimension, L_x, in the x direction. If the variation of the channel dimension in the y direction, $L_y(x)$, is adiabatically smooth, we can use Eq. (3.50) for the one-dimensional energy subbands:

$$E_{k_x,n,m} = \frac{\hbar^2 \pi^2 n^2}{2m^* L_z^2} + \frac{\hbar^2 \pi^2 m^2}{2m^* [L_y(x)]^2} + \frac{\hbar^2 k_x^2}{2m^*},\tag{6.124}$$

where we used approximate quantization of electron motion in the y direction. The resulting potential profiles are shown in Fig. 6.18(b), for a few subbands in the structure

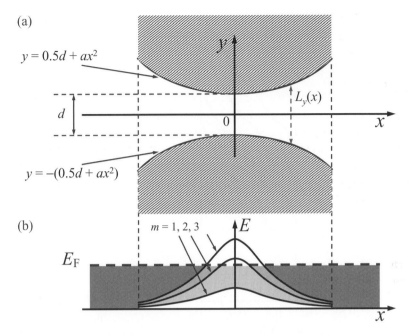

Figure 6.18 Geometrical representations of point contact (a) and quantized energy $\varepsilon_{1,m}$ in the region of point contact ($L_z \ll L_{y,\text{min}}$) (b). It is shown schematically that electrons can be in all three subbands in the region with the densest hatching. Electrons can be only in the subband $\varepsilon_{1,1}$ in the region shown by the least-dense hatching.

with $L_z \ll L_{y,\text{min}}$. For the indicated Fermi energy, we have only one channel open for electron motion and the second channel has a small barrier. As soon as the Fermi energy reaches the bottom of the next subband, a new channel opens for conducting electrons and the conductance of this point contact undergoes a step-like increase. Typically, such narrow constrictions clearly demonstrate the quantization of conductance. Since the potential barriers in the constriction are quite wide, tunneling processes have small probability. As a result, each of the channels is either almost open (the transmission coefficient $T = 1$) or almost closed ($T = 0$); i.e., one will expect steps in the conductance close to their universal value $2G_0$.

Consider as an example the narrow constriction defined by shallow etching on the AlGaAs/GaAs structure with two-dimensional electrons. In order to perform the shallow etching with well-controlled edges, a 13-nm-thick Al mask was formed by electron-beam lithography and lift-off, using a field-emission scanning electron microscope at an acceleration voltage of 2.5 kV. The sample was then shallow etched, as described in Chapter 4. After the etching mask was removed, an Al gate electrode was deposited, covering the constriction. By applying a voltage to the gate, one can control the electron density and, thus, the Fermi energy. A scanning electron micrograph of a shallow etched constriction is shown in the inset of Fig. 6.19. The shape of the etched constriction is characterized by two parameters: (1) the width of the constriction, d, and (2) the curvature, a, of the parabolas describing the shallow etched walls of the constriction

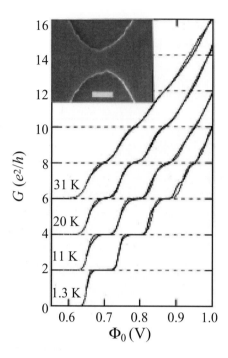

Figure 6.19 The quantization of conductance, G, at temperatures $T = 1.3, 11, 20,$ and 31 K. Inset: a shallow etched constriction. From A. Kristensen, J. B. Jensen *et al.*, "Conductance quantization above 30 K in GaAlAs shallow-etched quantum point contacts smoothly joined to the background 2DEG," *J. Appl. Phys.*, **83**, 607 (1998). Reprinted with permission from A. Kristensen, J. B. Jensen, M. Zaffalon, C. B. Sørensen, S. M. Reimann, P. E. Lindelof, M. Michel, and A. Forchel, *Journal of Applied Physics*, **83**, 607 (1998). © 1998 American Institute of Physics.

seen in the picture. If the $\{x, y\}$-coordinate system is placed in the center of symmetry of the constriction, the etched walls are described by

$$y = \pm(0.5d + ax^2). \tag{6.125}$$

The quantum contact presented in Fig. 6.19 has the parameters $d = 50$ nm and $a = 0.001\ 25$ nm^{-1}. Using the second term in Eq. (6.124), we can estimate the energy distances between the lowest subbands in the narrow cross-section of the constriction; they are 20 meV, 16 meV, 11 meV, etc. The sample exhibits well-defined conductance quantization at low temperatures, as depicted in Fig. 6.19 for a series of measurements at various temperatures. The rise of the conductance for each step is close to the universal value $2G_0$.

Single-electron transport

In the previously discussed cases of electron transport, we assumed that the number of electrons participating in this transport is so large that the discrete nature of the electrons does not matter. For small devices operating with weak currents, this assumption is no longer valid. Instead, a new type of charge-dependent electron transport, *single-electron transport*, develops. Below we present the basic concepts of this transport regime.

In general, in bulk-like materials and macroscopic devices the electron discreteness is not manifested in average characteristics such as the local electron density, total electric current, etc. However, it is well known that this discreteness manifests itself by contributing to current noise (fluctuations) even for macroscopic samples. For example, the so-called *shot noise* is entirely due to the discreteness of the electron charge; it arises because of the random process of electrons entering the device.

When the dimensions of the devices are scaled down, the role of the discreteness of charge increases. In the case of ultra-small devices, the discreteness of the electron charge gives rise to principally new effects in electron transport. This transport becomes *correlated*; i.e., a transfer of one electron through the device depends upon the transfer of others. The correlation appears because of the Coulomb interaction of individual electrons. A new class of devices, referred to as *single-electron devices*, is based on such processes.

Single-electron effects rely on a charging process that occurs when electrons enter a tiny conducting sample. When the tiny conducting sample, often called the *metallic island*, is extremely small, the electrostatic potential of the island significantly increases even when only one electron enters it. In general, the charging energy of a sample is

$$E_C = \frac{Q^2}{2C},$$
(6.126)

where Q is the charge and C is the capacitance of the sample. For a spherical island of radius r, the capacitance, C, can be estimated as

$$C = 4\pi\epsilon_0\epsilon r.$$
(6.127)

For example, an island with radius $r = 10$ nm has a capacitance of the order of 10 aF = 10^{-17} F, where F is the farad – the unit of capacitance, C (1 aF = 1 attofarad = 10^{-18} F and 1 F = 1 C V^{-1}). Then, with an increase in the voltage, $\Delta\Phi$ ($\Delta\Phi = \Delta Q/C$), which is equal to e/C with the electron charge $e = 1.602 \times 10^{-19}$ C, the increase in the energy, $\Delta E = e\,\Delta\Phi$, reaches 16 meV. This is comparable to the "thermal noise energy" at room temperature, $k_B T \approx 26$ meV. If one electron is transferred to the island, the Coulomb repulsion prevents additional electrons from entering the island unless the island potential is intentionally lowered by an external bias. If the island potential is lowered gradually, the other electrons can enter the island only one by one, with negligibly small power dissipation.

We can define a *single-electron device* as a nanostructure with so small a capacitance that a single electron added to the device generates a measured voltage change on the device. The basic properties of single-electron devices can be described by using the example of a customary system: an insulated tunnel junction, I, between two conducting electrodes, M (heavily doped semiconductor regions or metals), as shown in Fig. 6.20 for a metal–insulator–metal (M–I–M) system. Let this junction be characterized by the

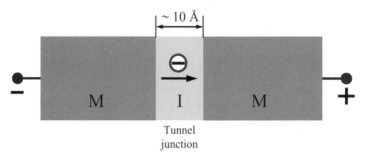

Figure 6.20 The simplest M–I–M (metal–insulator–metal) system of single electronics: a tunnel junction with a small capacitance, C. Single-electron transfer is shown schematically.

capacitance, C, and the conductance, G. We can suppose that the capacitance is roughly proportional to the cross-section, S, of this junction. Thus, a small S implies a small capacitance. The conductance of the junction, G, is small enough for one to consider the system as a leaking capacitor. Let the system have a charge, Q. Then, when a single electron leaks through the insulator, this event changes the initial electrostatic charge just by the elementary charge, e: $Q \rightarrow (Q - e)$. Hence, according to Eq. (6.126), the electrostatic energy of the junction changes by

$$\Delta E_C = \frac{(Q - e)^2}{2C} - \frac{Q^2}{2C} = -\frac{e(2Q - e)}{2C}. \tag{6.128}$$

The potential difference between the plates changes from

$$\Phi_1 = Q/C \tag{6.129}$$

to

$$\Phi_2 = (Q - e)/C. \tag{6.130}$$

Then, the single-electron transfer leads to a voltage fluctuation across the junction equal to

$$\Delta\Phi = \Phi_1 - \Phi_2 = e/C, \tag{6.131}$$

and a corresponding fluctuation of the leakage current,

$$\Delta I = G\,\Delta\Phi. \tag{6.132}$$

The current related to the single-electron transfer can be estimated from the following qualitative considerations. The uncertainty relation of Eq. (3.8) between the energy and time allows us to estimate the extreme limit of the tunneling time, τ_t, for such a junction:

$$\tau_t \geq \frac{h}{\Delta E_C}. \tag{6.133}$$

In turn, the electric current

$$I_t = \frac{\Delta Q}{\Delta t} \tag{6.134}$$

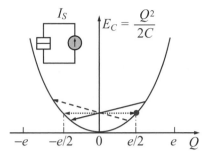

Figure 6.21 The charging energy, E_C, for a tunnel junction with the capacitance, C, and charge, Q, of one of the plates. The solid arrow indicates energetically favorable electron-tunneling events, the dashed arrow indicates an energetically unfavorable event, and the dotted double arrow indicates tunneling events without a change in energy of the capacitor, E_C. Inset: a circuit with charging of a tunnel junction by a current source.

associated with the tunneling of a single electron ($\Delta Q = e$, $\Delta t = \tau_t$, and $\Delta E_C = -e^2/(2C)$) is

$$I_t \approx \frac{e}{\tau_t} \geq \frac{e^3}{2hC}. \tag{6.135}$$

If this current, I_t, exceeds the current fluctuation due to the voltage fluctuation, $\Delta\Phi$, namely

$$I_t \geq \Delta I = G\,\Delta\Phi = G\frac{e}{C}, \tag{6.136}$$

the single-electron transfer processes will control the electric current through the junction. The latter inequality leads to the criterion

$$\frac{e^2}{h} \gg G. \tag{6.137}$$

Note that the value on the left-hand side of this criterion just coincides with the quantum of conductance, $G_0 \equiv e^2/h$, introduced by the Landauer formula of Eq. (6.121).

As a result of this analysis, we can state that, if the Coulomb energy of charging of the device is greater than the thermal energy, $k_B T$, and the current, I, associated with the single-electron transport is greater than fluctuations of the leakage current, then electron transport is correlated and single-electron effects are important. The criteria for such single-electron transport can be formulated as follows:

$$C \ll \frac{e^2}{k_B T} \quad \text{and} \quad G \ll \frac{e^2}{h}; \tag{6.138}$$

i.e., both the capacitance and the leakage conductance of the device should be small.

Now we can discuss the physics of single-electron transfer by using the electric circuit shown in Fig. 6.21. We assume a situation in which the tunnel junction in the circuit is charged by a current source, I_S. Starting with no net charge on the capacitor plates at time $t = 0$, the current source slowly begins to charge the junction. For a given charging rate and short time scales, the excess charge on the capacitor plates $Q < e$. Importantly,

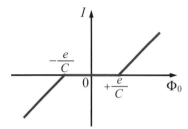

Figure 6.22 The current–voltage characteristic under Coulomb blockade. Within the voltage-bias range from $-e/C$ to $+e/C$ the current is suppressed by Coulomb correlations.

macroscopic plates can be charged with a charge less than the charge of a single electron just by shifting the conduction electrons in the plates with respect to the positive ions of the lattice (the so-called *polarization charge*). When the current source has charged one of the plates of the capacitor to a charge equal to $+e/2$ and another plate to $-e/2$, the plates have the charge difference of $1e$ between them. The electrostatic energy of the junction reaches the value marked by the dot in Fig. 6.21. Now $1e$ charge can tunnel through the junction as shown by the dotted arrow in Fig. 6.21. Thus, before the tunneling the initial charges on the plates were $(+e/2, -e/2)$, whereas after the tunneling the charges are $(-e/2, +e/2)$. We see from Fig. 6.21 that the charge transfer does not change the energy of the junction. The two charge configurations are, actually, identical. Because of this, nothing can stop the electron from tunneling back to its initial state and so on. However, the current source continues to charge the plates and the tunneling shown by the solid arrow becomes energetically favorable (ΔE_C in Eq. (6.128) becomes negative), and it becomes energetically unfavorable for the electron to tunnel back to its initial capacitor plate. Thus, in reality, we obtain tunneling of a single electron and the system starts to accumulate the charge to provide for the tunneling of the next electron, which will happen after some time has elapsed. The succession of electron-tunneling events means that the electron flux is highly correlated. In Fig. 6.21 all discussed tunneling events are shown schematically. From the considerations based on Eq. (6.128) one can see that any electron transfer is *prohibited* for a small initial capacitor charge: $-e/2 < Q < e/2$ (ΔE_C for such Q is positive). This physical effect is called *Coulomb blockade*. If the conditions of Eq. (6.138) are met, the charging energy plays the dominant role in the system, the tunneling of an electron is energetically unfavorable, and at low temperature tunneling is not possible at all (if it is blocked). This results in the specific current–voltage characteristic presented in Fig. 6.22. The main feature of such a characteristic is the total suppression of the current in some finite interval of external voltage biases, $-e/C < \Phi_0 < +e/C$.

Importantly, the manifestation of the correlated transport strongly depends on the external circuit to which the single-electron tunnel junction is attached. Let the external circuit impedance be $Z(\omega)$. If the impedance is as low as $|Z(\omega)| \ll G_0^{-1}$ for the most important frequencies, $\omega \sim 1/\tau_t \sim e^2/(hC)$, then charge fluctuations in the circuit are greater than the elementary charge, e, and all correlation effects are suppressed.

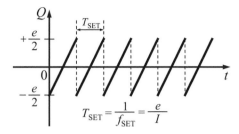

Figure 6.23 Bloch oscillations, showing the dependence of junction charging on time.

If the external circuit impedance is in the intermediate interval $G_0^{-1} \ll |Z(\omega)| < G^{-1}$, the junction exhibits Coulomb blockade in the bias range from $-e/C$ to $+e/C$, as shown in Fig. 6.22, but outside this range there is no correlation between tunneling events.

Finally, an interesting current regime occurs if $|Z(\omega)| \gg G^{-1} \gg G_0^{-1}$, when the external circuit can be considered as a source of a fixed direct current, I. This current causes a recharging of the junction inside the range of Coulomb blockade without electron tunneling; it corresponds to a linear change in the charge with time: $dQ/dt = I \approx$ constant. When the edge of the blockade range is reached, an electron tunnels through the junction. The system finds itself again in the blockade range – near the opposite edge – and the process repeats; see Fig. 6.23. Thus, one gets temporal oscillations of the charging with a frequency, f_{SET}, determined by the current:

$$f_{SET} = \frac{I}{e}. \tag{6.139}$$

This is the so-called *frequency of single-electron tunneling (or Bloch) oscillations*.

The previously discussed nonlinear current–voltage characteristics, oscillations, and other single-electron effects give rise to a principally new approach to low-energy electronics. This field of single-electron devices is developing rapidly and has many potential applications. Devices such as the single-electron transistor, the turnstile, and the single-electron pump, have been proposed and realized on the basis of these effects. Though these results were achieved at low temperatures, modern technology portends their extension to liquid-nitrogen temperature and even room temperature.

6.6   Closing remarks

In this chapter, we have shown that a number of different electron-transport regimes can occur in semiconductors and their nanostructures. These regimes are characterized by different values of the electron velocity and magnitude of the current, very different current–voltage dependences, etc. The electrons can behave as semiclassical particles, or as quantum ones. If the device dimension along the current is much larger than the electron de Broglie wavelength, electron motion, typically, is classical. It can occur in a dissipative manner when electrons undergo multiple collisions caused by crystal defects and lattice vibrations. The rate of these collisions determines the average electron velocity achievable at a given electric field.

If the field is small, the average velocity is a linear function of the field. The coefficient in the linear relationship between the average velocity and the field is the electron (hole) mobility. In perfect materials and structures the mobility is high and limited by scattering by lattice vibrations. For such systems, at low temperatures, at which these vibrations are reduced, the mobility reaches a maximum magnitude that is limited by scattering by impurities. The mobility is one of the most important characteristics for electronic applications. The electron concentration, mobility, and geometry of a sample determine its electrical conductance or electrical resistance. We have analyzed modification of the transport in oscillating electric fields and found that the current response is defined by a complex and frequency-dependent characteristic – the impedance.

Then, we considered the behavior of the electrons in high electric fields, explaining hot-electron effects and transit-time effects. Among the latter, the velocity-overshoot phenomenon, i.e., achievement of very high speed for a short period of time, is the most important for ultra-high-speed devices.

Practically, the transient time effects, including velocity overshoot, can be realized in short devices. In short devices, another type of physical effects should also be considered. Specifically, the formation of space charge affects the current. The electron space charge increases the electric resistance; this case is known as *space-charge-limited transport*. We have considered and compared several short diodes of different lengths, from the dissipative diode to the ballistic device. We have found that their electrical properties are markedly different from those of bulk-like samples; in particular, the current–voltage characteristics are essentially nonlinear. Calculations of the impedance, which determines the high-frequency properties of these diodes, showed that the range of frequencies within which the device response is high is directly related to the time of electron transit through the device. Thus, ultra-high frequencies can be achieved only for very short (nanoscale) devices.

We have analyzed the simplest types of quantum transport and have come to a very unusual conclusion – the quantum device possesses a finite resistance (conductance) even if scattering is entirely absent. The physical reason for such a finite resistance in the absence of scattering is quantum-mechanical reflection of electron waves from the interfaces between terminals and the quantum device. We have discussed the quantization of the conductance in a quantum device with essentially one-dimensional electron motion. The conductance quantization is related to electron energy quantization and manifests itself as a multiple-step-like behavior of the electric current.

Finally, we have considered electric currents in devices so small that the electrons can be transferred only one by one, because of Coulomb repulsion. The associated single-electron-transport regime has very unusual properties, which can be exploited in ultra-small electronic devices.

Additional information on the various regimes of electric current can be found in

K. Hess, *Advanced Theory of Semiconductor Devices* (New York, Wiley–IEEE Press, 1999).

D. K. Ferry, *Semiconductors* (New York, Macmillan Publishing Company, 1991).

I. I. Ipatova and V. Mitin, *Introduction to Solid State Electronics* (New York, Addison-Wesley Publishing Company, 1996).

V. V. Mitin, V. A. Kochelap, and M. A. Stroscio, *Quantum Heterostructures* (New York, Cambridge University Press, 1999).

For much more information on quantum transport and its applications to device physics the reader can consult the following references:

R. Landauer, "Electrical resistance of disordered one-dimensional lattices," *Philos. Mag.*, **21**, 863 (1970).

H. Grubin, D. K. Ferry, and C. Jacoboni (Eds.), *The Physics of Submicron Semiconductor Devices* (New York, Plenum Press, 1989).

C. Jacoboni, L. Reggiani, and D. K. Ferry (Eds.), *Quantum Transport in Semiconductors* (New York, Plenum Press, 1992).

Discussions of charge-limited transport and single-electron transport are presented, for example, in the following references:

M. A. Lampert and P. Mark, *Current Injection in Solids* (New York, Academic Press, 1970).

K. K. Likharev, "Correlated discrete transfer of single electrons in ultrasmall tunnel junctions," *IBM J. Res. Develop.*, **12**, 144 (1988).

H. Koch and H. Lubbig (Eds.), *Single-Electron Tunneling and Mesoscopic Devices* (Berlin, Springer-Verlag, 1992).

6.7 Problems

1. Consider the bulk crystal of GaAs with an electron effective mass $m^* = 0.067 m_0$, where m_0 is the free-electron mass, and mobility $\mu = 10^5$ cm^2 V^{-1} s^{-1} at $T = 77$ K. Using the relationships introduced in Sections 6.2 and 6.5, and Eq. (6.50) ($\mu = -e\tau_e/m^*$), calculate

the de Broglie wavelength, λ;
the scattering time, τ_e;
the thermal electron velocity, v_T;
the mean free path, l_e; and
the diffusion coefficient, D ($D = v_T^2 \tau_e/\alpha$, $\alpha = 3$ for bulk material).

Determine the transport regimes for devices with feature sizes $L_x = 0.05, 0.5$, and 5 µm.

2. Consider a bulk crystal, a quantum well of thickness 10 nm, and a quantum wire of cross-section 10 nm × 10 nm. For such samples, the specific electron concentrations of

10^{18} cm^{-3}, 10^{12} cm^{-2}, and 10^6 cm^{-1} correspond to the same three-dimensional electron density. For these three cases, by using the results of Section 6.3, calculate and compare values of

the Fermi wavevector, k_F, and corresponding de Broglie wavelength ($\lambda = 2\pi/k_F$);
the Fermi energy, E_F; and
the Fermi velocity, $v_F = \sqrt{2E_F/m^*}$.

3. Assuming an ambient temperature $T = 4$ K, prove that the electron gas considered in the previous problem is degenerate for all three cases. For a nondegenerate gas the thermal energy is $E_T = (\alpha/2)k_B T$, where $\alpha = 3, 2$, and 1 for bulk, quantum well, and quantum wire, respectively, and the thermal velocity is $v_T = \sqrt{2E_T/m^*}$. Calculate these parameters for further comparison. Using the Pauli exclusion principle, explain why the Fermi energy and the Fermi velocity of the electrons are higher than the corresponding characteristics E_T and v_T for a nondegenerate gas of particles with the same mass m^*.

4. As studied in Section 4.4 for many semiconductor materials, the energy spectra of the holes are degenerate and consist of two branches: the light holes, $E_{lh} = \hbar^2 k^2/(2m_{lh}^*)$, and the heavy holes, $E_{hh} = \hbar^2 k^2/(2m_{hh}^*)$. Let both hole subbands be populated, with concentrations n_{lh} and n_{hh}, respectively. At low temperature, the holes in both subbands have the same Fermi energy. Find the relationship between the concentrations of the heavy and light holes.

5. The density of states as a function of the energy has a staircase character for quantum wells. Using Eq. (6.45), estimate the height of the stairs for the density per unit area in the case of a GaAs quantum well.

6. Use the energy spectra of light and heavy holes presented in Problem 4 and compare the densities of states for the light and heavy holes in bulk samples. Which of these two hole subbands is more populated under equilibrium conditions?

7. Consider a short diode of length L_x with space-charge-limited dissipative transport. Let Φ_0 be the voltage applied to the diode. To make estimates for this device one should specify the criteria of validity of the concept of the dissipative diode. For this, one should combine Eq. (6.12) and the condition of the absence of electron heating in the diode, i.e., $F(x) < F_{he}$ with F_{he} defined by Eq. (6.81).

Using the estimate for the field $F \leq \Phi_0/L_x$, prove that the diode concept works for the voltage $k_B T/e < \Phi_0 < L_x\sqrt{k_B T/(e\mu\tau_E)}$. Show that these criteria can be met under the condition $L_x > \sqrt{\mu\tau_E k_B T/e}$.

Assuming room temperature and the material parameters $\mu = 10^3$ cm^2 V^{-1} s^{-1} and $\tau_E = 10^{-10}$ s, calculate the critical length, L_x. For $L_x = 5$ μm and $\Phi_0 = 0.1$ V, determine the electric current density, J, and the electron transit time, τ_{tr}.

8. Consider a GaAs diode, in which the electrons have the effective mass $m^* = 0.067m_0$. Assume liquid-nitrogen temperature, $T = 77$ K. For such a temperature in high-quality GaAs material, the typical mobility is $\mu = 10^5$ cm^2 V^{-1} s^{-1}.

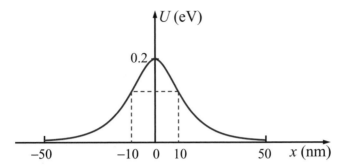

Figure 6.24 A potential barrier $U(x) = U_0/\cosh(x/d)$ with $U_0 = 0.2$ eV and $d = 10$ nm.

Estimate the mean free path, $l_e = \tau_e \times v_T$. Set the diode length $L_x = l_e$; i.e., assume that the diode is a ballistic device.

Calculate the current density, J, and the transit time, τ_{tr}, for the ballistic diode at $\Phi_0 = 0.1$ V and compare your result with parameters obtained for Problem 7.

9. The Landauer formula of Eq. (6.121) includes the energy-dependent transmission coefficient, $T(E)$. It is instructive to discuss this formula by using an example for which the exact solution of the quantum-mechanical Schrödinger equation is known. One such case is the electron motion in the potential of a special dependence: $U(x) = U_0/\cosh(x/d)$ (Fig. 6.24). One can see that this potential barrier has height equal to U_0 and the characteristic spatial scale d. It was found that the transmission coefficient has the form

$$T(E) = \frac{\sinh^2(\pi k_x d)}{\sinh^2(\pi k_x d) + \cosh^2\left(\frac{\pi}{2}\sqrt{\frac{8m^*U_0d^2}{\hbar^2} - 1}\right)}, \quad \text{for} \quad \frac{8m^*U_0d^2}{\hbar^2} > 1,$$

where $k_x = \sqrt{2m^*E}/\hbar$ is the wavevector of the incident electron. Assume that only the one-dimensional subband with $n = m = 1$ is populated by electrons. Set the following parameters: $U_0 = 0.2$ eV, $d = 10$ nm, and $m^* = 0.067m_0$, as for GaAs.

Using the dependence $T(E)$, calculate the conductance, G, as a function of the Fermi energy.

7 Electrons in traditional low-dimensional structures

7.1   Introduction

Now, we begin our analysis of novel developments in electronics that have resulted from the use of nanostructures in modern electronic devices. Importantly, the attributes of nanotechnology make it possible to pursue both devices with smaller dimensional scales and novel types of device. Though the ongoing trend of miniaturization in electronics is extremely important, the unique properties of electrons in nanostructures give rise to novel electrical and optical effects, and open the way to new device concepts. The electric current and voltage in a device are determined by two major factors: the concentration and the transport properties of the charge carriers. In nanostructures, these factors can be controlled over wide ranges. In this and the next chapter we will study nanostructures for which these basic factors that are important for the electronics are engineered, which are being exploited intensively both in research laboratories and in practical nanoelectronics.

To distinguish the nanostructures already having applications from the newly emerging systems, we refer to the former as *traditional low-dimensional structures*.

7.2   Electrons in quantum wells

In this section, we consider a few particular examples of nanostructures with two-dimensional electrons.

As a basis for the further analysis, we will recall and develop several of the previously introduced definitions and properties of an electron gas. In what follows, the effect of the band offset arising at a junction of two semiconducting materials, which was defined in Section 4.5 via electron affinities, is critically important. As discussed previously, the electron affinity is the energy required to remove an electron from the bottom of the conduction band of a specific material to vacuum. The difference in affinities of two adjacent materials is equal to the conduction-band offset at their heterojunction. Another useful parameter of a material is the *work function*, which is defined as an energy required to remove an electron from the Fermi level of the material to vacuum. Finally, we will use also the fundamental property of electron statistics which implies that in equilibrium the Fermi level is constant throughout the whole system.

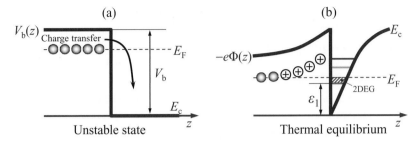

Figure 7.1 The charge-transfer effect in a selectively doped single heterostructure. (a) The step-like profile is unstable. (b) Charge transfer results in ionization of a layer of the doped material, band bending, lining up the Fermi levels across the structure, and formation of an electron channel at the interface.

Single modulation-doped heterojunctions

First, let us analyze qualitatively a single semiconductor heterojunction. We consider a junction of two semiconductors with affinities such that they result in a discontinuity of the conduction band as indicated by Fig. 7.1(a). Then, we assume that the barrier semiconductor material is n-doped while the narrow-bandgap material on the right-hand side of the structure is undoped. In real situations, the latter material is, usually, lightly doped by acceptors; i.e., it is a p-type material.

The doped regions of the system fix the positions of the Fermi level, E_F. At equilibrium, the energy levels below E_F are occupied by electrons. Accordingly, the scheme sketched in Fig. 7.1(a) is unstable. The electrons will move toward the undoped crystal until an electrostatic field, brought about by the redistribution of electrical charge, bends the band edges so that the Fermi level becomes constant across the materials; i.e., we obtain the same Fermi level in the n-doped region and in the undoped region. Instead of having an energy step as in Fig. 7.1(a), one obtains the situation shown in Fig. 7.1(b). That is, the band edges are bent, there is ionization of impurities in some region of the doped part of the system, and there are free electrons inside the potential well. This well is formed by both (1) the bandgap discontinuity and (2) the electrostatic potential. The situation just described is frequently referred to as *formation of an electron gas at the interface*.

Three important conclusions can be derived immediately from these considerations.

(i) Though both the initial materials were insulators (at least at low temperatures), now at the interface, near the junction, one obtains an electron channel and electron concentration that are finite even at $T = 0$ K.

(ii) Charge carriers in the potential well are separated spatially from their parent impurities from the barrier side of the structure. Charged impurities, which usually lead to large scattering rates and low mobilities, serve only as sources of carriers, and scattering of electrons from the potential well by the impurities in the barrier is suppressed as a result of the spatial separation of electrons from impurities.

(iii) The bending of the energy band creates a confining potential for carriers in one direction, say along the z-axis. Hence there is a quantization of electrons in the z direction and the establishment of a two-dimensional electron gas is quite possible.

To complete the discussion of this simple model, let us note that in the real situation the side of the heterostructure with the lower conduction-band edge is usually doped lightly with acceptors. In this case, the right barrier of the well is higher than half of the bandgap of the narrow-bandgap material. Of course, the residual acceptors reduce the electron channel mobility but their effect is weak.

Here, we considered a heterostructure with the Fermi level defined only by the modulation doping. These structures are called *ungated heterostructures*. For *gated heterostructures* the physical picture is slightly different. Such gated heterostructures will be studied subsequently.

Basic equations describing the physics of the electrons at an interface

It is important to have an idea of the basic equations describing the electrons at an interface. We began by considering a step-like discontinuity of the energy band and found that the potential well is formed by charge transfer in space. Thus, the shape of the potential is determined by the charges of all electrons on the interface and ionized impurities. On the other hand, this many-body potential determines the motion of each electron and the total number of ionized impurities. Thus, we face the so-called *self-consistent problem*: the potential is defined by the concentration of electrons and ionized impurities, and it, in turn, affects their redistribution. The simplest approach to this self-consistent problem is to treat the electron quantization in the scheme wherein the potential is described by a self-consistent electrostatic field. The corresponding electrostatic potential, $\Phi(z)$, can be thought of as a function dependent on only the coordinate z perpendicular to the interface and is a solution of Poisson's equation:

$$\frac{d^2\Phi(z)}{dz^2} = \frac{e}{\epsilon_0 \epsilon}\left[\sum_\nu |\psi_\nu(\mathbf{r})|^2 \mathcal{F}(E_\nu) - N_D(z) + N_A(z)\right]. \tag{7.1}$$

Here, e is the elementary electrical charge, ϵ is the dielectric constant, and ϵ_0 is the permittivity of free space. The total charge consists of the charges of the electrons and impurities. Let the wavefunctions of the electrons be $\psi_\nu(\mathbf{r})$, with ν being the set of electron quantum numbers. Then, $|\psi_\nu(\mathbf{r})|^2$ represents the probability of finding the electron of the state labeled by ν at the point $\mathbf{r}$. In order to calculate the contribution of the electrons to the space charge we introduce the energy-dependent electron distribution function, $\mathcal{F}(E_\nu)$, which determines the probability of the electron occupying the energy level E_ν. Thus, the electron charge density is $-e\sum_\nu |\psi_\nu(\mathbf{r})|^2 \mathcal{F}(E_\nu)$. The charge of the nonuniformly distributed positive donors and negative acceptors is determined by their densities, $eN_D(z)$, and, $-eN_A(z)$, respectively.

We assumed that the potential, $\Phi(z)$, does not depend on the x- and y-coordinates. Thus, $\psi_\nu(\mathbf{r})$ can be factorized as discussed in Section 3.2:

$$\psi_\nu(\mathbf{r}) = \frac{1}{\sqrt{S}}e^{i(k_x x + k_y y)}\chi_j(z), \quad \nu \equiv \{j, \mathbf{k}_{||}\}, \tag{7.2}$$

where S is the area of the junction and $\mathbf{k}_{||} = \{k_x, k_y\}$. The factorization (7.2) leads to

$$\psi_\nu(\mathbf{r})\psi_\nu^*(\mathbf{r}) = |\psi_\nu(\mathbf{r})|^2 = \frac{1}{S}|\chi_j(z)|^2, \tag{7.3}$$

which makes Eq. (7.1) dependent only on the z-coordinate. We also obtain the one-dimensional Schrödinger equation for the z direction:

$$\left(-\frac{\hbar^2}{2m^*}\frac{d^2}{dz^2} + V_b(z) - e\Phi(z)\right)\chi_j(z) = \varepsilon_j\chi_j(z).$$ (7.4)

Here, the total potential energy consists of two contributions: the built-in potential of the heterostructure, $V_b(z)$, and the self-consistent potential, $-e\Phi(z)$. In our case, $V_b(z)$ corresponds to the energy-band discontinuity at the junction – see Fig. 7.1(a) – which depends only on the z-coordinate. We set

$$V_b(z) = -V_b\Theta(z),$$

where $\Theta(z)$ is the Heaviside step-function. In Eq. (7.4), ε_j are the energies of two-dimensional subbands. Thus, the total electron energy is given by Eq. (6.27):

$$E_j(k_{||}) \equiv \varepsilon_j + \frac{\hbar^2 k_{||}^2}{2m^*}.$$

Now, we can calculate the electron concentration:

$$\sum_\nu |\psi_\nu|^2 \mathcal{F}(E_\nu) = \frac{1}{S}\sum_{s,j,\mathbf{k}_{||}} |\chi_j(z)|^2 \mathcal{F}\left(\varepsilon_j + \frac{\hbar^2 k_{||}^2}{2m^*}\right) = \sum_j |\chi_j(z)|^2 n_j.$$ (7.5)

Here, we also introduce the sheet density of electrons for the level j:

$$n_{s,j}(E_F) \equiv \frac{1}{S}\sum_{s,k_x k_y} \mathcal{F}(s, \mathbf{k}_{||}, j) = \frac{m^* k_B T}{\pi\hbar^2} \ln\left[1 + \exp\left(\frac{E_F - \varepsilon_j}{k_B T}\right)\right].$$ (7.6)

Actually, this result repeats Eq. (6.116). The quantity $n_{s,j}$ is a function of the temperature, T, and takes the simplest form as $T \to 0$:

$$n_{s,j}(T = 0) = \frac{m^*}{\pi\hbar^2}(E_F - \varepsilon_j)\Theta(E_F - \varepsilon_j).$$ (7.7)

Equation (7.7) indicates that the level j is occupied if the Fermi level exceeds the corresponding energy of quantization of transverse electron propagation, ε_j.

It is necessary to formulate boundary conditions for the coupled differential equations represented by Eqs. (7.4) and (7.1). For localized electron states, we may set $\chi_j(z) \to 0$ for $z \to \pm\infty$. For the electrostatic potential we suppose that $d\Phi/dz \to 0$ for $z \to \pm\infty$.

In conclusion, Eqs. (7.1), (7.4), and (7.5) together with the boundary conditions complete the formulation of the self-consistent problem that describes the formation of a two-dimensional electron gas at the interface as well as the electron quantization. By integrating Eq. (7.1) over an infinite range of z and using the boundary conditions, one can find a neutrality equation in the form

$$\sum_j n_{s,j} + \int_{-\infty}^{+\infty} dz[N_A(z) - N_D(z)] = 0.$$ (7.8)

This result implies that, despite the charge transfer, the entire system remains electrically neutral.

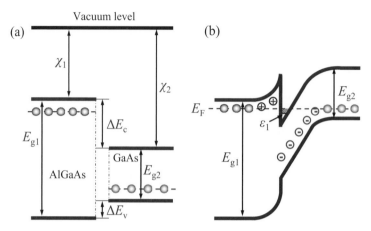

Figure 7.2 Schematic energy band diagrams of a selectively doped AlGaAs/GaAs heterostructure before (a) and after (b) charge transfers have occurred. Relative positions of the valence and conduction bands are given for both materials. The electron affinities are shown conditionally (left). The Fermi level in the AlGaAs material is supposed to be pinned on the donor level. The narrow-bandgap GaAs is slightly p-doped. After C. Weisbuch and B. Vinter, *Quantum Semiconductor Structures* (San Diego, Academic Press, 1991).

Numerical analysis of a single heterojunction

Accurate analysis of the problem of electron-channel formation at a heterojunction can be accomplished through numerical solution of the system of Eqs. (7.1)–(7.5). Frequently, to solve the Schrödinger equation one exploits the so-called *variational method*. Briefly, the method is based on the use of so-called *trial wavefunctions* containing a few well-chosen parameters, which then are determined from the condition that the total energy of the system should be a minimum.

If we assume that only one lowest subband with $j = 1$ is occupied by the electrons, a simple trial wavefunction can be chosen in the form

$$\chi_1(z) = 0 \quad \text{for } z \leq 0 \qquad \text{and} \qquad \chi_1(z) = \sqrt{\frac{b^3}{2}} z e^{-bz/2} \quad \text{for} \quad z \geq 0, \qquad (7.9)$$

which implies that the wavefunction is localized near the interface and is equal to zero on the wide-bandgap side of the heterojunction.

Before discussing the numerical results, let us return briefly to the picture of the electron energy at the heterojunction in order to present both the conduction- and the valence-band profiles for specific heterojunctions. The left-hand side of Fig. 7.2 shows the band edges of separated AlGaAs and GaAs; the distance to the vacuum level is shown provisionally. It is assumed that AlGaAs is heavily n-doped and GaAs is doped lightly by acceptors (p-doped). This doping brings about the pinning of the Fermi level at the donor level on the AlGaAs side of the structure. On the right-hand side of the picture, one can see the energy structure of the AlGaAs/GaAs junction with two depletion regions: one, positively charged, on the AlGaAs side and another, negatively charged, on the GaAs

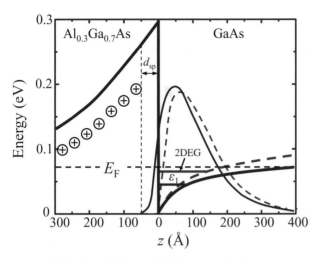

Figure 7.3 Calculated self-consistent potentials, energy levels, and wavefunctions of an $Al_{0.3}Ga_{0.7}As/GaAs$ selectively doped heterostructure. The junction is situated at $z = 0$. The spacer thickness is 50 Å and the donor binding energy of AlGaAs is chosen as $E_D = 50$ meV. After T. Ando, "Self-consistent results for a $GaAs/Al_xGa_{1-x}As$ heterojunction. I. Subband structure and light-scattering spectra," *J. Phys. Soc. Japan*, **51**, 3893 (1982).

side. For such a doping the quantum well occurs only for electrons and does not occur for holes, since the AlGaAs/GaAs heterojunction is of type I (according to the classification of Chapter 4.)

In Fig. 7.3, the results of numerical solution of Eqs. (7.1)–(7.5) are presented for a selectively doped $Al_{0.3}Ga_{0.7}As/GaAs$ heterojunction. Heavily n-doped $Al_{0.3}Ga_{0.7}As$ and lightly p-doped GaAs materials are separated by a thin undoped $Al_{0.3}Ga_{0.7}As$ layer with thickness d_{sp}. Such an undoped layer is called the *spacer layer* or simply a *spacer*. In calculations, the sheet electron concentration, n_s, in the channel at the junction is equal to the number of ionized donors per unit area, $N_s = 5 \times 10^{11}$ cm^{-2}. The sheet concentration of ionized acceptors, N_{depl}, is supposed to be much smaller: $N_{depl} = 5 \times 10^{10}$ cm^{-2}. The self-consistent potentials, energy levels, and wavefunctions are shown for two different cases. For the first case, shown by the dashed curve in Fig. 7.3, the trial electron wavefunction is defined by Eqs. (7.9). The second case, shown by the solid curve in Fig. 7.3, corresponds to a more sophisticated trial wavefunction, which can penetrate under the barrier into the wide-bandgap part of the structure. In the region of classically allowed motion, the wavefunctions for all of these approximations are similar to each other. However, they are drastically different in the barriers. Generally, the trial function of Eqs. (7.9) gives a reasonable value of the energy ε_1, as well as the behavior of the potential at large distances. However, the second trial function yields more accurate results. In particular, it gives a slightly lower energy level. Results obtained for various concentrations, n_s, of electrons at the interface are summarized in Table 7.1.

In this table, the lowest energy level, ε_1, the spatial scale of the electron confinement, $\langle z \rangle$, and the probability of finding electrons in the barrier, P_b, are presented for three

Table 7.1 Results of calculations of electron parameters for
two wavefunction approximations: I corresponds to the
function of Eqs. (7.9), and II to a more accurate trial function

		n_s, $(10^{11}$ cm$^{-2})$		
		2	4	8
I	ε_1 (meV)	38.6	56.6	86.1
	$\langle z \rangle$ (Å)	116	99	82
	P_b (%)	0	0	0
II	ε_1 (meV)	32.00	45.6	66.3
	$\langle z \rangle$ (Å)	100	82.6	65.6
	P_b (%)	0.7	1.11	1.95

values of the electron concentration, n_s. The value of $\langle z \rangle$ was calculated as the quantum-mechanical average of the z-coordinate of an electron, and $P_b = \int_{-\infty}^{0} dz |\chi_1(z)|^2$. One can see that the relative height of the first energy level, ε_1, increases as the concentration increases. The value of $\langle z \rangle$ decreases with increasing n_s as well as with the confining electrostatic potential. Thus, we can see that the width of the electron channel is in the range 60–100 Å. The probability of finding electrons in the wide-bandgap barrier is small, but increases with increasing electron confinement. Obviously, P_b is equal to zero for the wavefunction of Eqs. (7.9).

The examples given in our discussions illustrate the following major features of a selectively doped heterojunction: (1) the formation of electron-conducting channels with concentrations in the range $n_s = 10^{11}$–10^{12} cm^{-2} at any temperature, including $T = 0$ k; (2) the spatial separation of the electrons from their parent donors with very low probability of electron penetration into the barrier – less than or about 1% – as well as spatial isolation of the electrons from the p-doped narrow-bandgap material; (3) the formation of a potential well for electrons with the potential profile self-consistently dependent on the electron concentration; and (4) the quantization of electrons inside the potential well with the resultant two-dimensional character of the electron spectrum and with the electrons confined in the two-dimensional channel with a width of less than 100 Å.

Control of charge transfer

We have considered a heterojunction formed by two semi-infinite semiconductors with fixed concentrations of donors and acceptors as illustrated in Fig. 7.2. This results in a conducting channel at the interface with a constant electron concentration that is determined by the doping profile. The control of the conductivity, or more exactly, the control of the resistance – or its inverse, the conductance – of the structure is necessary in order to realize useful devices.

Let us consider the possibility of changing the conductance of a heterojunction by controlling the electron concentration. For that purpose, we study the so-called

(a)   (b)

                                  Gated heterostructure

Ungated heterostructure

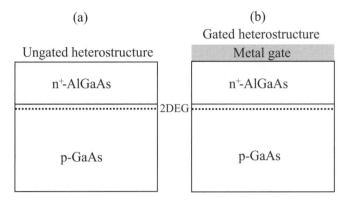

Figure 7.4 Schematic diagrams of (a) ungated and (b) gated heterostructures with two-dimensional electron gas (2DEG).

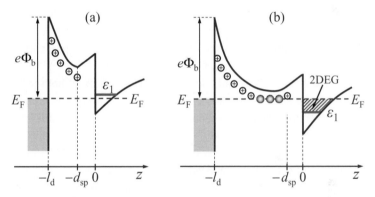

Figure 7.5 Conduction-band diagrams for M/AlGaAs/GaAs heterostructures. The built-in Schottky voltage controls the depletion region under the metallic gate. It results in (a) a normally-off device for a narrow barrier and (b) a normally-on device for a wide barrier with 2DEG in the potential well (60 nm or wider).

gated heterojunction which is presented schematically in Fig. 7.4(b). For comparison, Fig. 7.4(a) depicts the ungated heterojunction considered previously. The only difference between them is the metal (M) contact placed on the top of the n^+ layer of the AlGaAs barrier material in the gated structure. This metal semiconductor system (MES) is called a MES structure. For GaAs-like materials, MES structures are of the most importance for device applications, because these materials do not have a stable natural oxide, unlike in the case of SiO_2 on silicon. Therefore, most of the electronic devices based on GaAs use MES structures. One also refers to these structures as *Schottky-gate structures*.

Typically, under a metallic gate in GaAs-like materials there are extended depletion regions, which occur because of a high built-in *Schottky voltage*, Φ_b, of about 0.8 V. Such a depletion region is known as a *Schottky depletion region*. The conduction-band energy diagram for an M/AlGaAs/GaAs heterostructure is presented in Fig. 7.5 for two thicknesses of the AlGaAs layer. The n-doped region is separated from the junction by an undoped spacer. The structure is shown in Fig. 7.5(a) for a relatively thin AlGaAs layer, and Fig. 7.5(b) depicts a structure with a thicker AlGaAs layer. For both cases, there exists an extended depletion region, which affects the electron channel formed at the AlGaAs/GaAs interface and provides two possibilities of controlling the structure. The *normally-off* structure corresponds to Fig. 7.5(a). The depletion region extends through both a thin AlGaAs layer and the junction. The bottom of the quantum well shifts up. The Fermi level lies under the lowest energy subband. Thus, there are no electrons inside the channel and the conductivity along the heterostructure is almost zero. The donors in the AlGaAs doped region are ionized, and the electrons have left the semiconductor part of the structure which is charged positively. In order to turn on the conductivity of the device, it is necessary to apply a *positive voltage* to the metal gate. normally-off structures can be fabricated by using a *thin* AlGaAs barrier.

The *normally-on* structure is illustrated in Fig. 7.5(b). In this case, the built-in voltage drops across a thick AlGaAs layer so that the Fermi level lies above the lowest subband and electrons populate the channel without an external voltage bias. This channel has a finite conductivity under normal conditions. This case can be realized for sufficiently thick AlGaAs layers. Thus, in normally-on devices, one can control the conductance of the channel by applying a *negative voltage* to the metal. A large voltage leads to a depopulation of the channel and can switch the device off.

The calculated electron potential energy and quantized levels both for normally-off and for normally-on devices are shown in Fig. 7.6 for various gate voltages, Φ_G. The Fermi level is taken to be at the zero energy. The values for the normally-off device were calculated for an AlGaAs layer thickness of about 400 Å; see Fig. 7.6(a). The quantum well formed on the interface contains up to four quantized levels. A positive voltage lowers the bottom of the conduction band of GaAs at the interface. The bottom touches the Fermi level at $\Phi_G = +0.3$ V and the device is turned on at a threshold voltage of about $+0.8$ V when the first quantized level touches the Fermi level. It is clearly seen that in the AlGaAs barrier layer a potential minimum occurs and tends to be lowered with increasing gate voltage. This phenomenon can result in a negative effect: the formation of a second channel in this layer, which will collect electrons, screen the gate voltage, and result in loss of control of the concentration of the two-dimensional electron gas at the interface. The normally-on device is shown in Fig. 7.6(b). It has an AlGaAs layer thickness greater than 600 Å. The device can be switched off when a negative voltage of about -0.5 V is applied to the gate.

The results just presented were obtained by numerical calculations. Let us consider some experimental data related to the problem of modulation-doped heterostructures where carrier concentrations and their mobilities have been measured simultaneously. In Fig. 7.7 the sheet electron concentration controlled by the gate voltage is shown for Al/AlGaAs/GaAs systems fabricated for high-electron-mobility transistors (HEMTs). The curves correspond to various spacer thicknesses d_{sp}. One can see that the electron concentration can be varied over one order of magnitude. Saturation of the sheet concentration at high positive voltage is caused by transitions of electrons to the potential well which is formed in the middle depleted barrier region, as discussed previously. Figure 7.7

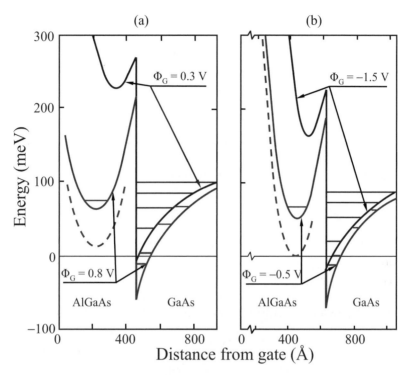

Figure 7.6 Calculated self-consistent potentials for conduction electrons in two M/AlGaAs/GaAs heterostructures (a) corresponds to the normally-off device; (b) corresponds to the normally-on device at room temperature. The Fermi level is at $E = 0$. Horizontal lines indicate the bottom energy of the lowest four subbands. Dashed lines show donor levels. After B. Vinter, "Subbands and charge control in a two-dimensional electron gas field-effect transistor," *Appl. Phys. Lett.*, **44**, 307 (1984). Reused with permission from B. Vinter, *Applied Physics Letters*, **44**, 307 (1984). © 1984 American Institute of Physics.

shows that considerable changes in the properties of the structures occur with changes in spacer thickness. Among the various structures presented in Fig. 7.7, the particular structure with $d_{sp} = 0$ is most likely to be useful for fabricating a normally-off device. Indeed, by applying a positive voltage to the gate one can increase the electron concentration in the channel. Structures with a thick spacer are well suited for normally-on devices; in these structures, positive voltage does not change the concentration, whereas a negative voltage reduces it sharply.

The spacer is an important element of modulation-doped structures because it partially prevents the scattering of channel electrons by the heavily doped side of the heterostructure and increases the electron mobility. On the other hand, there is one negative effect of a spacer. Increasing the spacer thickness leads to an increase of the potential drop on the spacer and consequently to a lowering of the electrostatic potential that confines electrons near the interface. Hence, a thick spacer causes a decrease in the electron concentration. This trade-off between the mobility and the carrier concentration requires an optimization of the structural design for each particular device application.

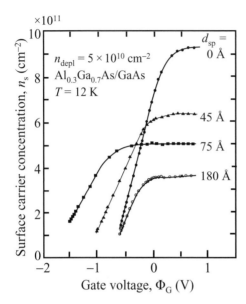

Figure 7.7 Measured gate-voltage dependences of the channel density of two-dimensional electrons in $Al_{0.3}Ga_{0.7}As/GaAs$ structures at $T = 12$ K and various spacer thicknesses d_{sp}. All samples are Si-doped with $N_D = 4.6 \times 10^{17}$ cm^{-3}, except for the sample with 180-Å spacer, which has $N_D = 9.2 \times 10^{17}$ cm^{-3}. The thickness of the doped $Al_{0.3}Ga_{0.7}As$ layer is 700–1200 Å. From K. Hirakawa, H. Sakaki, and J. Yoshino, "Concentration of electrons in selectively doped GaAlAs/GaAs heterojunction and its dependence on spacer-layer thickness and gate electric field," *Appl. Phys. Lett.*, **45**, 253 (1984). Reprinted with permission from K. Hirakawa, H. Sakaki, and J. Yoshino, *Applied Physics Letters*, **45**, 253 (1984). © 1984 American Institute of Physics.

Here, we have considered systems with a single heterojunction. These systems can be fabricated by a relatively simple technology and they have numerous applications, but they suffer from greatly limited carrier concentration in the conduction channel. As can be seen from Fig. 7.7, the typical surface concentrations are less than 10^{12} cm^{-2} for single-heterojunction devices. Higher concentrations of carriers in the conducting channel can be obtained in a double-junction system. This system is illustrated by Fig. 7.8, where possible double-junction heterostructures are presented schematically. For heterostructures of type I, quantum wells are formed both for electrons (two quantized levels are shown) and for holes (one quantized level is shown), whereas for a type-II heterostructure a quantum well arises only for the electrons.

7.3   Electrons in quantum wires

A quantum wire is a conductive structure, wherein electron transport is constrained primarily to be along a single direction. Let this direction be along x. For the two other directions – the y and z directions – the quantum-mechanical confinement of the electrons is imposed by the heterointerface potentials, or by suitable externally applied electrostatic potentials. Such a system is also called a *one-dimensional electron system*. Electron

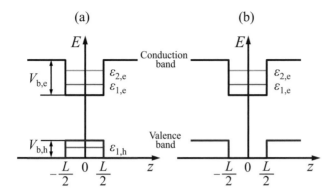

Figure 7.8 Two types of band diagrams of double heterostructures: (a) type I and (b) type II.

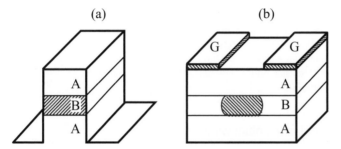

Figure 7.9 Quantum wire formed by etching (a) and split-gating (b) of the two-dimensional electron gas.

motion along the free x direction is characterized by a one-dimensional wavevector k_x. The wavefunction has the form

$$\Psi(x, y, z) = e^{ik_x x} \psi_{n_1, n_2}(y, z),$$

where $\psi_{n_1, n_2}(y, z)$ corresponds to confined transverse motion of the electron. The bound states corresponding to transverse motion are enumerated by integers n_1 and n_2 and usually are called one-dimensional "subbands." In Section 3.3, we considered electron confinement in two directions and found subband energies of the form of Eq. (3.50). If distances between the lowest subband and excited subbands are small in comparison with both the thermal energy, $k_B T$, and the Fermi energy of electrons, E_F, scattering of electrons from the lowest subband into higher subbands is relatively weak. The electrons occupy primarily the lowest subband and behave almost as one-dimensional particles.

Currently, there are several methods for the fabrication of quantum wires, including the direct growth of the wires. The most obvious method is to start with a two-dimensional structure and to impose an additional confinement to two-dimensional electron gas, as shown schematically in Fig. 7.9. In the case of Fig. 7.9(a), the etching process is used to realize geometrical restrictions for electron motion. In the case of Fig. 7.9(b), additional confinement is induced by an electrostatic potential that is applied to a metallic split gate placed on the top of the heterostructure.

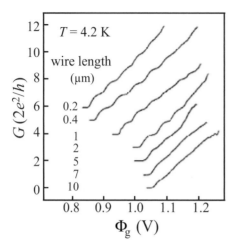

Figure 7.10 The dependence of the conductance, G, of a quantum wire versus gate voltage, Φ_g, for seven different lengths of wires as indicated near each curve. After L. Worschech, F. Beuscher, and A. Forchel, "Quantized conductance in up to 20 μm long shallow etched GaAs/AlGaAs quantum wires," *Appl. Phys. Lett.*, **75**, 578 (1999). Reused with permission from L. Worschech, F. Beuscher, and A. Forchel, *Applied Physics Letters*, **75**, 578 (1999). © 1999 American Institute of Physics.

Electron transport in quantum wires

The principal distinct features of electron transport in quantum wires are related to Landauer quantization of the conductance. This quantization can be observable for perfect wire structures at low temperatures. As an example, we discuss measurements of the conductance for AlGaAs/GaAs quantum wires obtained by the etching method, as in Fig. 7.9(a). The quantum wires were fabricated on modulation-doped AlGaAs/GaAs heterostructures grown by molecular-beam epitaxy. High-resolution electron-beam lithography was used to define the masks of the quantum wires. With the help of wet chemical etching, the doped layer and the spacer – both fabricated with AlGaAs – as well as 70 nm of GaAs were removed. From scanning electron microscopy, the geometrical width of the quantum wires was determined to be 135 nm. The micrographs indicated a very small sidewall roughness of the wires. Before etching, the electrons at the selectively doped AlGaAs/GaAs heterointeface were characterized by a density of about 3×10^{11} cm^{-2} and high mobility in the range $(1-2) \times 10^6$ cm^2 V^{-1}s^{-1}. At low temperature (4 K), the electron gas is degenerate and the mean free path of electrons having the Fermi energy is estimated to be in the range 10–20 μm. Thus, for wires fabricated with lengths less than 10 μm, we can expect Landauer quantization. To control one-dimensional subbands in the wires, an aluminum top gate was evaporated onto the top of the structure. A voltage applied to this gate changed the confining potential and subband energies. The results are presented in Fig. 7.10 for various wire lengths. The conductance quantization is seen clearly for the wires shorter than 5 μm. Up to ten "steps" in the conductance are seen; i.e., about ten one-dimensional subbands can be observed by changing the gate voltage. Experiments with magnetic field and temperature variation also reveal the quantization effect and allow one to determine the concentration of one-dimensional electrons. The typical concentration and subband spacing were found to be $n_{1D} \approx 6 \times 10^6$ cm^{-1} and $\Delta\varepsilon_{1,2} \approx 12.5$ meV, respectively. The relatively large subband spacing – equivalent to $\approx$140 K – implies that the one-dimensional character of electron motion should be unchanged even at temperatures of several tens of degrees Kelvin.

For higher temperatures, at which electron motion becomes semiclassical, the transport properties of quantum wires based on semiconductor heterostructures are similar to those of quantum wells and pure bulk-like materials. Thus, selectively doped quantum wires should exhibit high electron mobilities under low electric fields and high drift velocities characteristic of hot electrons under high fields.

The situation with electron transport in carbon nanotubes is quite different. In high fields the electron drift velocities can reach magnitudes in the range $(2–4) \times 10^7$ cm s^{-1}, which are larger than those in perfect III–V semiconductor compounds. However, the electrical properties are affected significantly by the surrounding environment, which influences the removal of heat from current-carrying nanotubes. For example, suspended nanotubes display drastically different electron transport from that of those on substrates. Figure 7.11 illustrates this difference. In the upper panel of this figure, scanning electron microscope images of two single-walled carbon nanotubes with Pt contacts are shown. The left nanotube segment is not suspended and is in contact with a nitride-based substrate. The right nanotube segment is suspended over a 0.5-μm-deep trench. The diameters of the nanotubes were measured by the atomic-force microscope to be in the range 2–3 nm. In the lower panel, the results of measurements of the current–voltage characteristics for the 3-μm length of suspended and non-suspended segments of the nanotube are given. The measurements were conducted at room temperature in vacuum. One can see that the non-suspended nanotube portion displays a monotonic increase in the current, approaching 20 μA under increasing voltage, V, while the current in the suspended tube reaches a peak of $\approx$5 μA followed by a pronounced current drop. This strikingly different behavior is due to significant self-heating effects of the wires carrying ultra-high current densities. Indeed, in the case of the suspended tube the Joule heating can not be removed effectively, because thermal fluxes are possible only through nanotube contacts. As a result, the temperature of the tube increases, especially in its central region.

7.4 Electrons in quantum dots

A quantum dot can be defined as a material system in which electrons are confined in all three directions. Some particular examples of such systems were described in previous chapters. They include the self-forming nanosized semiconductor islands, clusters, and nanocrystals studied in Chapter 5, etc. Quantum dots can be fabricated by various methods, including a direct way, starting from a structure with two-dimensional electrons and using an etching process to impose geometrical restriction on the electrons in two additional directions. A small metallic gate fabricated upon a two-dimensional heterostructure also can provide the necessary electron confinement under a negative

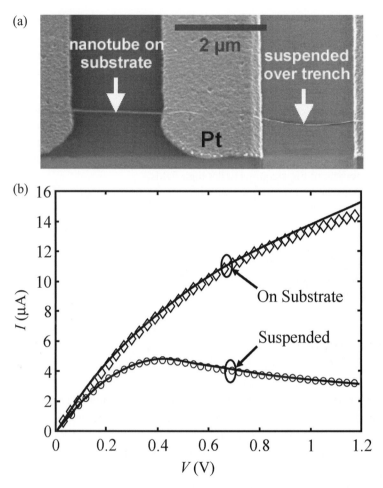

Figure 7.11 (a) Carbon nanotubes on a substrate and suspended over a trench; (b) their current–voltage characteristics. Reprinted with permission from E. Pop, D. Mann *et al.*, "Negative differential conductance and hot phonons in suspended nanotube molecular wires," *Phys. Rev. Lett.*, **95**, 155–505 (2005). © 2005 by the American Physical Society.

voltage applied to the gate. In Fig. 7.12, three types of quantum dots are shown: a nanosized island self-forming under a special growth regime, colloidal nanocrystals, and a "quantum box" fabricated in a controlled way from a two-dimensional heterostructure. An example of a quantum dot formed by using the technique of fabricating a gate upon a two-dimensional heterostructure will be discussed in the next chapter; see Section 8.4.

The most notable feature of quantum dots is that all existing degrees of freedom of electron propagation are quantized, if a confining potential is deep and the dimensions of the structure are comparable to the de Broglie wavelength of the electron, as defined in Eq. (6.5). According to the analysis given in Section 6.2, such a system can be called a *zero-dimensional system*. The latter terminology stresses the dramatic changes of the electronic properties in quantum dots. Indeed, in a crystalline solid, as was discussed in Chapter 4, electrons possess continuous energy bands with the occupation, width, and

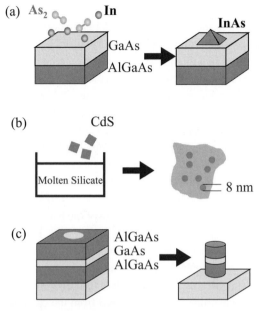

Figure 7.12 Schematic representations of three different approaches for the fabrication of quantum dots: (a) self-organized growth of nanosized islands, (b) nanocrystals in a colloid, and (c) artificial patterning and etching of a heterostructure with two-dimensional electron gas.

separation of these bands determining the fundamental electrical and optical properties of the solid. At the other end of the length scale, for individual atoms the electronic state density is discrete, resulting in, for example, the absence of simple electron transport and intrinsically sharp spectral optical lines. In some respects, the electronic structure of quantum dots might be said to fall somewhere between these two extremes. Thus, quantum-dot structures are like large artificial atoms, "macroatoms."

For the analysis of the electronic properties of a quantum dot, a model of a confining potential may be used. In Section 3.3, we considered two examples of these models: the quantum-box model given by Eq. (3.51) and the spherical-dot model of Eq. (3.53). The results obtained with these models for the energy spectra and wavefunctions can be applied to semiconductor quantum dots, if the masses are taken to be the effective masses of the material being analyzed. The height of the confining potential is determined by the band offsets, see Fig. 4.12.

Consider semiconductor quantum dots fabricated from a two-dimensional heterostructure by using an etching process, or by colloidal synthesis, as illustrated in Figs. 7.12(b) and 7.12(c). In these cases, depending on the heterostructure type, it is possible to realize different types of carrier confinement. The classification of heterostructure types was presented in Section 4.5. If the initial heterostructure is of type I, both electrons and holes can be confined in the same quantum-dot structure. If the initial heterostructure is of type II, only one type of carriers, either electrons or holes, can be confined.

If electron motion is quantized in all three possible directions, we obtain a new physical object, a macroatom. Questions concerning the usefulness of such objects for applications naturally arise from the point of view of their electronic applications. A fundamental question is the following: what is the current through a macroatom? A valid answer is that there exists the possibility of passing an electric current through an artificial atom due to *tunneling* of electrons through quantum levels of the macroatom.

Importantly, that transport of charge occurs *always* in quanta of the elementary charge of the electron. This discreteness has no practical consequences for current flowing in bulk materials, in two-dimensional structures, and even in quantum wires, because the numbers of electrons transported are large. Charge transport in quantum-dot systems is completely different. Indeed, it occurs as electron tunneling from a cathode through the quantum dot to an anode. That is, the clearance space between the cathode and the anode should be considered as a potential barrier in which a quantum dot is embedded. In Section 3.3, while studying the tunneling effect, we found that the probability of tunneling drastically depends on the thickness and height of the barrier: for large enough thickness and barrier height, tunneling is essentially suppressed. However, tunneling through a discrete state localized inside the barrier has much higher probability. The whole tunneling process can be thought of as a sequential process: the capture of an electron from the cathode to the localized state and then its emission to the anode. As a result, the easiest way for the electrons to be transferred from the cathode to the anode is for them to tunnel through the quantum dot. Then, an electron being captured to the quantum dot blocks tunneling of other electrons. Electron transfer occurs in highly correlated manner, one by one, at least at low temperatures. Examples of such "single-electron" transport were discussed in Section 6.5. Thus, the electric current through a quantum dot occurs in the regime of single-electron transport. Devices that use such transport will be discussed in the next chapter.

In fact, electrical methods applied to quantum dots to realize useful devices are not the only methods possible. The control of the electric current through the dots can also be realized by means of light, sound waves, etc. Consider here optical control of the dots and optoelectronic functions of zero-dimensional devices. The main peculiarities of the optical properties of quantum dots arise due to electron and hole quantization. In quantum dots fabricated using a type-I heterostructure, the carrier energies have the form

$$E_e^{QD} = E_g^{QD} + \varepsilon_n(n_1, n_2, n_3), \qquad E_h^{QD} = -\varepsilon_p(n_1', n_2', n_3').$$

Here, E_g^{QD} is the fundamental bandgap of the material of the quantum dots. For a type-I heterostructure, E_g^{QD} is less than the bandgap E_g of the surrounding material into which the dots are embedded; ε_n and ε_p depend on sets of three discrete quantum numbers, $\{n_1, n_2, n_3\}$ and $\{n_1', n_2', n_3'\}$, for the electrons and holes, respectively. Specific dependences of E_e^{QD} and E_h^{QD} on the quantum numbers are determined by the materials used for quantum dots and their environment, the geometry of dots, etc. The model dependences of Eqs. (3.51) and (3.53) can be used for estimation of the energy levels. Owing to these discrete energy spectra, quantum dots interact primarily with photons of discrete energies:

$$\hbar\omega \equiv \frac{2\pi\hbar c}{\lambda} = E_g^{QD} + \varepsilon_n(n_1, n_2, n_3) + \varepsilon_p(n_1', n_2', n_3'), \qquad (7.10)$$

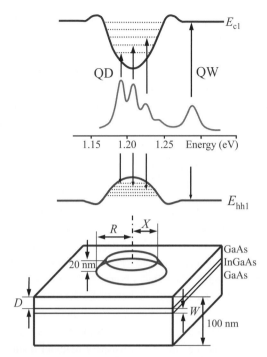

Figure 7.13 A sketch of quantum dot confining potentials for electrons and holes. Peaks in optical spectra correspond to phototransitions between quantized states of the electrons and holes in the quantum dot. Reprinted with permission from J. Tulkki and A. Heinamaki, "Confinement effect in a quantum well dot induced by an InP stressor," *Phys. Rev.* B, **52** (11), 8239 (1995). © 1995 by the American Physical Society.

where c is the velocity of light and λ is the wavelength of the light. The different combinations of quantum numbers $\{n_1, n_2, n_3\}$ and $\{n'_1, n'_2, n'_3\}$ give a series of optical spectral lines, for which interaction between the dots and light is efficient. Importantly, the fact that $E_g^{QD} < E_g$ implies that the light interacting with the dots is not absorbed by the surrounding material. These considerations are illustrated in Fig. 7.13. The potential profiles of the conduction band, E_c, and the upper (heavy-hole) valence band, E_{hh}, are depicted. The potential wells for electrons and holes represent the confinement potentials of the quantum dot. Quantized levels for electrons and holes are shown; arrows indicate the possible phototransitions between different quantized states. For each such transition, a characteristic line of the optical spectrum arises. For the example of an InGaAs quantum dot grown inside a GaAs quantum well, the quantum-dot spectral lines arise in the region of photon energies from 1.1 to 1.3 eV, as presented in Fig. 7.13. The width of observed spectral lines depends primarily on two factors. First, the spectral lines are broadened because of relaxation processes in the system. Second, it is typical that in experiments light interacts with numerous dots and there is some dispersion of their sizes. This produces a dispersion in energy-level positions and additional line broadening.

The optical control of the electric current flowing through a quantum dot can be explained with the help of a device that can be called a single-quantum-dot photodiode.

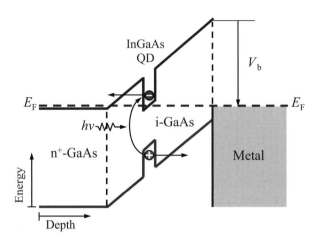

Figure 7.14 A schematic energy band diagram of a single-quantum-dot photodiode for photocurrent experiments, with an n^+-contact on the left and a metal contact (Schottky contact) on the right. After Figs. 1 and 13(b), H. Kenner, S. Sufler *et al.*, "Recent advances in exciton-based quantum information processing in quantum dot nanostructures," *New J. Phys.*, **7**, 184 (2005). © IOP Publishing Limited.

The diode base is an i region (undoped, i.e., intrinsic region), into which a single quantum dot is embedded. A schematic band diagram of such a photodiode under equilibrium, when the Fermi level, E_F, is the same throughout the whole structure, is shown in Fig. 7.14. At the semiconductor/metal heterojunction a potential barrier V_b (the Schottky barrier) arises. An electric bias is to be applied to two terminals; in the case under consideration they are the n^+ region (on the left) and a metal contact (the Schottky contact) on the right. Under inverse biasing, an electric current is possible, in principle, due to the interband tunneling mechanism. The probability of such tunneling is extremely small and the current is negligible even if a quantum dot is embedded in the base. Illumination of the diode by light resonant with phototransitions between discrete electron and hole levels in the quantum dot leads to the excitation of an electron–hole pair inside the dot. Now, the electron and hole can tunnel more readily from the dot into the i region and contribute to the electric current, the "photocurrent." In Fig. 7.15, we present the experimental results obtained from excitation of the ground state ($n_1 = n_2 = n_3 = n'_1 = n'_2 = n'_3 = 1$) of a single self-assembled $In_{0.5}Ga_{0.5}As$ quantum dot embedded into a 360-nm-thick intrinsic GaAs layer. Since InGaAs/GaAs heterostructures are of type I, the only *optically active part* is the single $In_{0.5}Ga_{0.5}As$ quantum dot. The experiments were carried out at 4.2 K. In Fig. 7.15, the photocurrent is plotted as a function of the electric bias for various wavelengths of illuminating light. The wavelengths are indicated on the photocurrent curves. One can see that photocurrent–voltage dependences have, in fact, a pronounced resonant character. This is explained by the fact that the quantized electron and hole energies are shifted under an applied electric field, as expected from the so-called *Stark effect* observable for atoms and molecules. When these energies are such that the wavelength given by Eq. (7.10) corresponds to the illuminating light, the light excites

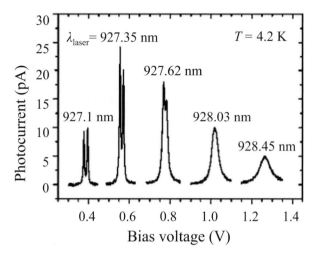

Figure 7.15 Photocurrent resonances for several different excitation wavelengths versus bias voltage. Reprinted with permission, from H. Kenner, S. Sufler *et al.*, "Recent advances in exciton-based quantum information processing in quantum dot nanostructures," *New J. Phys.*, **7**, 184 (2005). © IOP Publishing Limited.

electrons and holes inside the quantum well, which produces the measured photocurrent. As the applied bias increases, the energies are shifted to smaller values and the resonance wavelength increases. In Fig. 7.15, spectra for excitation of the same ground state of the dot for different biases are shown. The observed photocurrent spectra are very narrow because a single dot is involved. Spectral broadening becomes visible at high biases when the electron and hole energy levels decay as a result of the increased rate of tunneling from the dot.

A photocurrent excited resonantly with an electrically tunable optical resonance can find numerous applications. Specifically, such a quantum-dot photodiode facilitates optical manipulation of individual quantum states tunable with an electric bias. Such a capability is necessary for devices required as components of systems used in quantum information technology.

7.5 Closing remarks

In this chapter, we focussed on traditional quantum structures that are already being exploited in nanoelectronics and optoelectronics. These include quantum wells, wires, and dots. These structures provide electron confinement in one, two, and three dimensions, respectively. We found that confinement of the electrons in any direction forbidding free propagation in this direction leads to quantization of the electron energy.

We showed that the confinement effect can be very strong. In particular, for one of the most used and practical cases – when electrons are confined at a heterointerface – we found that the electrons are localized within a thin spatial layer of thickness ranging from

5 nm to 10 nm depending on the electron concentration. This confinement results in high electron densities. For example, a surface concentration of confined two-dimensional electrons of about 10^{12} cm^{-2} corresponds to a bulk concentration of from 10^{18} cm^{-3} to 2×10^{18} cm^{-3}, which are typical values for heavily doped semiconductor bulk materials.

In quantum wells and wires, the carriers can be spatially separated from layers with dopants, which generate free carriers. Such techniques lead to quantum structures exhibiting large electron mobility relative to the case of bulk structures. Together with high and controllable electron density, this enhanced mobility results in high electric current densities and high speeds of operation.

Quantum dots possess totally quantized spectra for electrons and holes. Electron transport via the dots can occur in the tunneling regime. Accordingly, the optical properties of the dots differ considerably from those of bulk samples, as well as from those of quantum wells and wires. Quantum-dot spectra consist of a series of separate lines. The positions of these spectral lines can be controlled by dot dimensions and geometry. Quantum dots find applications in optoelectronics.

For those who want to study traditional quantum structures in detail, we recommend the following additional reading.

Results on electron quantization in SiO$_2$/Si structures are presented in the review

> T. Ando, A. B. Fowler, and F. Stern, "Electronic properties of two-dimensional systems," *Rev. Mod. Phys.*, **54**, 437 (1982).

Discussions of particular examples of the quantum wells and wires based on III–V compounds can be found in the following references:

> G. Bastard, *Wave Mechanics Applied to Semiconductor Heterostructures* (New York, Halsted Press, 1988).
> C. Weisbuch and B. Vinter, *Quantum Semiconductor Structures* (New York, Academic Press, 1991).
> V. V. Mitin, V. A. Kochelap, and M. A. Stroscio, *Quantum Heterostructures* (New York, Cambridge University Press, 1999).

Fabrication methods and properties of quantum dots are analyzed in the book

> D. Bimberg, M. Grundmann, and N. N. Ledentsov, *Quantum Dot Heterostructures* (Chichester, John Wiley & Sons, 1999).

7.6   Problems

1. Consider a quantum-well structure with the electron energy given by Eq. (6.27). Assume that for the lowest two-dimensional subbands the intersubband distance $\Delta\varepsilon_{21} = \varepsilon_2 - \varepsilon_1$ is given. For $T \to 0$, find the formula for the critical concentration n_c of two-dimensional electrons at which the second subband starts to be populated. Using ε_1, ε_2, and $\Delta\varepsilon_{21}$ for an infinitely deep quantum well, estimate the critical concentration, n_c, for a quantum well of width $L = 10$ nm.

2. To obtain characteristics of the electrons confined at an interface, one can use the electron wavefunction of the lowest quantized state in the form of Eq. (7.9), which depends solely on the parameter b. Apply the definition of the expectation value of a physical quantity presented in Eq. (3.12) and find an expression for the average distance of the electrons from the interface $\langle z \rangle$. Calculate the mean square deviation of the location of electrons from its average position; i.e., calculate $\langle (z - \langle z \rangle)^2 \rangle$.

3. Near the semiconductor heterojunction the electrons can be confined in the direction normal to the heterojunction. In Figs. 7.1 and 7.2, sketches of confining potentials are presented. If the potential has high barriers, the lower energy levels may be studied with some accuracy by applying the following "triangular" approximation:

$$V(x) = \begin{cases} eFx, & \text{for } x > 0, \\ \infty, & \text{for } x \leq 0. \end{cases}$$

Here, $x = 0$ corresponds to the heterojunction position, and F can be interpreted as the confining electrostatic field. For this triangular model of the quantum well, the Schrödinger equation for the transverse component of the electronic wavefunction can be solved exactly. The quantized electron energies are

$$\varepsilon_n = \left(\frac{e^2 \hbar^2 F^2}{2m^*} \right)^{1/3} p_n,$$

where p_n are parameters defined by some algebraic equation. The lowest parameters are known to be $p_1 \approx 2.35$ and $p_2 \approx 4.09$. Using the effective mass of GaAs, $m^* = 0.067 m_0$, evaluate the positions of the lowest energy levels for the AlGaAs/GaAs heterojunction as a function of the field F.

The electrostatic field F can be related easily to the electron concentration n_s inside the triangular well. Indeed, ionized (positive) impurities in the wide-bandgap part of the heterojunction ($x < 0$) determine the confining field F. The number of these impurities approximately equals the number of the electrons. Thus, in the region $x > 0$, Gauss' law gives the field as $F = 4\pi e n_s / \epsilon$, where ϵ is the dielectric constant. Assuming a concentration of $n_s = 10^{12}$ cm^{-2} and taking $\epsilon = 13$, estimate the field F and the energies ε_1 and ε_2. Compare the separation between the first two levels with the thermal energy $k_B T$ at room temperature. Discuss the population of these levels by the electrons at $T = 300$ K.

4. In Section 4.4 on analyzing different types of crystals, we found that in silicon there are six energy-equivalent minima (valleys). The electrons are mainly accumulated near these minima. The minima are at the following crystalline directions: [100], [$\bar{1}$00], [010], [0$\bar{1}$0], [001], and [00$\bar{1}$] (see Table 4.4). Within these minima, the electron energy spectrum is given by Eq. (4.17) with effective mass tensors of the type of Eq. (4.19). Let us consider a Si quantum-well layer oriented perpendicular to the [001] direction. For free electrons collected in the pair of valleys along directions [001] and [00$\bar{1}$], the electron Hamiltonian (kinetic energy) is

$$-\frac{\hbar^2}{2m_{\mathrm{t}}}\left(\frac{\partial^2}{\partial x^2}+\frac{\partial^2}{\partial y^2}\right)-\frac{\hbar^2}{2m_{\mathrm{l}}}\frac{\partial^2}{\partial z^2},$$

while for the other two pairs of valleys the Hamiltonians are

$$-\frac{\hbar^2}{2m_{\mathrm{l}}}\frac{\partial^2}{\partial x^2}-\frac{\hbar^2}{2m_{\mathrm{t}}}\frac{\partial^2}{\partial y^2}-\frac{\hbar^2}{2m_{\mathrm{t}}}\frac{\partial^2}{\partial z^2},$$

and

$$-\frac{\hbar^2}{2m_{\mathrm{t}}}\frac{\partial^2}{\partial x^2}-\frac{\hbar^2}{2m_{\mathrm{l}}}\frac{\partial^2}{\partial y^2}-\frac{\hbar^2}{2m_{\mathrm{t}}}\frac{\partial^2}{\partial z^2}.$$

These Hamiltonians imply that the electron motion along the quantum-well layer and that in the perpendicular direction are completely independent. The wavefunctions and quantized energies should have the form of Eq. (7.2).

Assume that the potential barriers of the quantum well are infinitely high. Employing the results of Eqs. (3.27) and (3.28) for an infinitely deep potential, find energy subbands for the different electron valleys. Discuss the energy splitting of the valleys in the quantum well. Using the data for m_{l} and m_{t} presented in Table 4.4, determine which pair of valleys will be of the lowest energy. In bulk Si crystals, the numbers of electrons in different equivalent minima are equal. Can we expect an electron redistribution to be caused by quantization in the well? Which valleys will be overpopulated and which will be depleted?

5. A nanostructure consists of two "semi-infinite" two-dimensional electron gases connected through a quantum wire. The entire structure is fabricated from a single "infinite" quantum-well layer. Does a potential barrier arise for electron motion from one two-dimensional gas to the another? Explain why the barrier arises.

Using the adiabatic approach described by Eq. (6.124), discuss the barrier heights for the electrons from different two-dimensional subbands.

Assume that the thickness of the quantum-well layer fabricated from GaAs is 5 nm and that the quantum-wire width is 5 nm. Estimate the height of this barrier for the electrons populating the lowest two-dimensional subband.

6. In III–V compound semiconductors the energies of heavy and light holes coincide at zero wavevector, as shown in Fig. 4.10. Energy levels exhibiting such a coincidence in energy are known as degenerate energy levels. Quantization of the two types of holes lifts this degeneracy. Using models with infinite-potential walls for quantum wells, wires, and dots, calculate the splitting between the lowest states of heavy and light holes in these low-dimensional structures.

Obtain numerical estimates of this splitting for GaAs structures with the following geometrical parameters: a quantum well of thickness 5 nm, a quantum wire of cross-section 5 nm $\times$ 5 nm, and a quantum dot of volume 5 nm $\times$ 5 nm $\times$ 5 nm. Parameters of heavy and light holes are given in Table 4.5.

7. It is known that GaAs/AlGaAs heterostructures are of type I. Thus, both electrons and holes are confined in low-dimensional structures fabricated from these materials. Consider GaAs quantum dots embedded in an AlGaAs matrix. Use the parameters of electrons and of heavy and light holes presented in Table 4.5. Explain what kind of holes forms the hole ground state in a GaAs quantum dot.

To describe the quantum dots, apply the model of a potential box of Eq. (3.51). Using Eq. (7.10), calculate the positions of spectral lines for quantum dots with the following dimensions: (a) $L_x = L_y = L_z = 5$ nm and (b) $L_x = 3$ nm, $L_y = 5$ nm, and $L_z = 7$ nm. The bandgap of GaAs is given in Table 4.5.

8 Nanostructure devices

8.1 Introduction

In previous chapters, fundamental physical processes on the nanoscale, analysis of nano-materials, and nanofabrication methods were all discussed extensively. The knowledge gained in these previous discussions makes it possible to consider and analyze a variety of different nanostructure devices. In this chapter, we consider electronic, optical, and electromechanical devices. Some of these devices mimic well-known microelectronic devices but with small dimensional scales. This approach facilitates applications to devices with shorter response times and higher operational frequencies that operate at lower working currents, dissipate less power, and exhibit other useful properties and enhanced characteristics. Such examples include the field-effect transistors and bipolar transistors considered in Sections 8.3 and 8.5.

On the other hand, new generations of the devices are based on new physical principles, which can not be realized in microscale devices. Among these novel devices are the resonant-tunneling devices analyzed in Section 8.2, the hot-electron (ballistic) transistors of Section 8.5, single-electron-transfer devices (Section 8.4), nanoelectromechanical devices (Section 8.7), and quantum-dot cellular automata (Section 8.8).

As a whole, the ideas presented in this chapter provide an understanding of the future development of nanoelectronic and optoelectronic devices that may be realized through the wide use of nanotechnology.

8.2 Resonant-tunneling diodes

Diodes or, in other words, two-terminal electrical devices, are the simplest active elements of electronic circuits. Some applications of diodes are based on their nonlinear current–voltage characteristics. Another important capability required of diodes is their operational speed. Such high-speed operation implies that the feature sizes of diodes should be as small as possible.

In previous sections, we studied two types of short $n^+ - i - n^+$ diode, both with space-charge-limited classical electron transport. These diodes used homostructures, i.e., it was supposed that they are made from the same material with nonuniform doping. Among the cases studied, the highest speed is achieved for the ballistic diodes. According to the classification given in Section 6.2 for the extreme scaling down of the diode size,

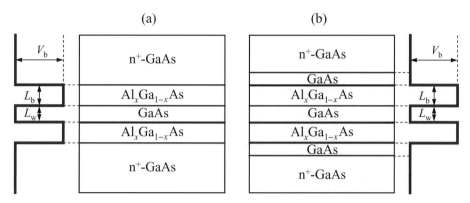

Figure 8.1 Layer designs for double-barrier resonant-tunneling structures. (a) Alternating layers: n$^+$-doped GaAs (substrate and one of the contacts), undoped AlGaAs (barrier), undoped GaAs (quantum well), undoped AlGaAs (barrier), n$^+$-doped GaAs (top contact). (b) The same as (a) except for two additional undoped spacer layers of GaAs between the contacts and the barriers. V_b is the barrier height, L_b the barrier thickness, and L_w the quantum well width.

quantum transport must be used to describe the carriers. A very important example of such nanoscale diodes is the so-called *double-barrier resonant-tunneling diode*. In this section, we consider this type of nanoscale quantum diode.

The physics underlying the resonant-tunneling effect

We start by considering a double-barrier heterostructure as an example of resonant-tunneling diodes. Figure 8.1 depicts a sequence of layers in such a structure. The top and bottom parts of the structure are doped regions, while the barriers and well layers are undoped. Figure 8.1(a) shows a specific structure in which a quantum-well layer of GaAs is embedded between two Al$_x$Ga$_{1-x}$As barrier layers. The top and bottom regions of doped GaAs serve as contacts. A slightly different design is shown in Fig. 8.1(b), where two additional spacer layers separate the doped regions and the double-barrier part of the structure. The purpose of these spacer layers is to prevent scattering of tunneling electrons by impurities in the contact regions. The thicknesses of the well, barriers, and spacers may be varied substantially. Inside the quantum well, several quantized levels can exist. In fact, these levels are quasi-bound states, because there is a small but finite probability of the electron tunneling out of the well. Quantum-mechanical tunneling is responsible for the finite lifetimes of the electrons in those levels and leads to some broadening of the quantum-well states.

Thus, a resonant-tunneling diode can be thought of as a system with two contacts, with three-dimensional electron states, and a quantum well, with a two-dimensional electron system. These three subsystems are weakly coupled through tunneling. Energy-band diagrams of the structure are presented in Fig. 8.2 for three different voltage biases. Figure 8.2(a) corresponds to the equilibrium case, when no voltage is applied. In the well under consideration, there exists at least one quasi-bound level; the case of a single level of energy, ε_1, is depicted in Fig. 8.2. Actually, ε_1 is the bottom of the lowest two-dimensional

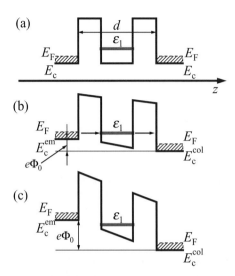

Figure 8.2 Energy band diagrams of double-barrier resonant-tunneling structures: (a) equilibrium conditions; (b) in-resonance; and (c) off-resonance. The electron potential energy and populations of the energy band in contact n$^+$-regions are sketched in the left and right parts of the figure. $E_{\mathrm{c}}^{\mathrm{em}}$ and $E_{\mathrm{c}}^{\mathrm{col}}$ denote the positions of the bottoms of the conduction bands under electric bias.

subband because there exists in-plane free electron motion. The parameters of the diode are chosen to be such that, in the non-biased state of the diode, the quasi-bound level, ε_1, lies above the Fermi energy, E_{F}, in the contacts, as in Fig. 8.2(a). By applying a voltage bias to the contacts, one can produce a downward shift of the level in the well. For electrons with arbitrary energies, the probability of tunneling through the double-barrier structure is very small. The structure is designed to prevent the thermal transfer of electrons over the barriers. Therefore, the only situation favorable for transmission of electrons through the structure is when the quasi-bound level lies below the Fermi energy, E_{F}, but above the conduction-band bottoms of the contacts. In this case, those electrons from the emitter (the contact on the left) whose kinetic energy of the in-plane (perpendicular) motion, $E_{\perp} = \hbar^2 k_z^2 / (2m^*)$, coincides with ε_1 are transmitted through the structure with finite probability. This is the so-called *resonant-tunneling process*, which has the important attribute of exhibiting negative differential resistance.

Before analyzing this effect, we recall that the electrical properties of a simple conductor with a linear current–voltage characteristic are characterized by a resistance, R, according to the formula

$$I = \Phi_0 / R,$$

where Φ_0 is the applied voltage and I is the current. If a conductor has a nonlinear current–voltage characteristic, one can introduce the so-called *differential resistance*, R_{d}:

$$R_{\mathrm{d}} = \left(\frac{\mathrm{d}I}{\mathrm{d}\Phi} \right)^{-1}.$$

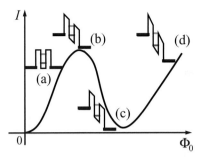

Figure 8.3 Resonant-tunneling diode current–voltage characteristic: the portions labeled by (a), (b), and (c) correspond to the physical situations illustrated in the previous figure. Portion (d) shows what happens with the current when the second quasi-bound state enters into resonance with the electrons from the emitter.

The term negative differential resistance is used to denote that $R_d < 0$; this corresponds to the unusual effect of a decrease of the current when the applied voltage increases.

   The following qualitative model of resonant tunneling explains the appearance of negative differential resistance. Before the level of the well reaches the resonance position, the current through the diode is very small because it is controlled by non-resonant tunneling and by transport over barriers, both of which have low probabilities of occurring. When the bias corresponds to the case of resonance energies, as depicted in Fig. 8.2(b), the transmission coefficient and the electric current through the diode increase sharply. Further increase in the current with increasing voltage bias continues until the resonance level passes the bottom of the emitter's conduction band. There are no electrons (the conduction band is above the Fermi level, E_F), as shown in Fig. 8.2(c), to tunnel resonantly and the current decreases in spite of the increase in voltage bias. Consequently, the current–voltage characteristic of the structure contains a segment that exhibits negative differential resistance. The general form of the current–voltage characteristic is presented in Fig. 8.3. At large bias, further increase of the current can be realized either by shifting other quasi-bound states so that they are resonant with electron energies in the emitter, or by substantial evolution of the potential profile and non-resonant transport through and over the barrier.

Quantitative characteristics of the resonant-tunneling effect

The resonant transmission of the electrons through the double-barrier diode can be provided by physically different processes. Conceptually, the simplest way is direct quantum-mechanical tunneling, which corresponds to the *coherent tunneling process*.

Coherent tunneling

In this case, the electron is characterized by a single wavefunction throughout the whole quantum structure. Traditionally, the cathode and the anode of the diode are referred to as the *emitter* and *collector*, respectively. An electron entering the structure from the

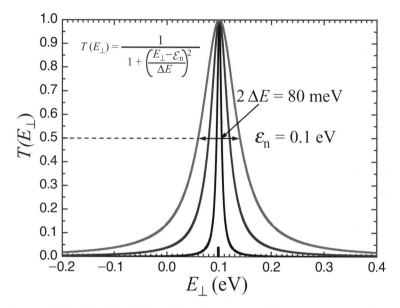

Figure 8.4 The full width, $2\,\Delta E$, of the half-maximum of the transmission peak.

emitter is described by the wavefunction $\psi^{\mathrm{em}}(\mathbf{r}, z)$, which has the form of an incident plane wave,

$$\psi^{\mathrm{em}}(\mathbf{r}, z) = A^{\mathrm{em}} \mathrm{e}^{\mathrm{i}(\mathbf{kr} + k_z^{\mathrm{em}} z)},$$

and an outgoing wave that has passed through the structure,

$$\psi^{\mathrm{em}}(\mathbf{r}, z) = B^{\mathrm{em}} \mathrm{e}^{\mathrm{i}(\mathbf{kr} + k_z^{\mathrm{col}} z)};$$

here, $\mathbf{k}$ and $\mathbf{r}$ are two-dimensional in-plane vectors, and z is directed perpendicular to the layers. The quantities k_z^{em} and k_z^{col} are the z-components of the wavevector in the emitter and collector, respectively. In an ideal layered structure, the vector $\mathbf{k}$ is conserved because there are no forces acting in plane directions; thus, $\mathbf{k}$ is the same for incident and outgoing waves. The relation of these wavevectors, the corresponding energies, $E_\perp^{\mathrm{em}}$ and $E_\perp^{\mathrm{col}}$, and the applied voltage, Φ_0, can be obtained from the energy conservation law:

$$E_\perp^{\mathrm{em}} - E_\perp^{\mathrm{col}} \equiv \frac{\hbar^2 \left(k_z^{\mathrm{em}}\right)^2}{2m^*} - \frac{\hbar^2 \left(k_z^{\mathrm{col}}\right)^2}{2m^*} = e\Phi_0.$$

The coefficients A^{em} and B^{em} can be found by solving the Schrödinger equation with the potential corresponding to the double-barrier structure under voltage bias. The ratio of the coefficients B^{em} and A^{em} defines the transmission coefficient:

$$T(E_\perp) = \frac{|B^{\mathrm{em}}(E_\perp)|^2}{|A^{\mathrm{em}}(E_\perp)|^2}.$$

Instead of solving the Schrödinger equation, here we give a simplified formula for $T(E_\perp)$, which allows us to introduce the basic parameters of the resonant-tunneling effect.

As follows from the qualitative analysis presented in the previous subsection, $T(E_\perp)$ has to have sharp peaks in the vicinity of resonance energies, ε_n. For a structure with symmetric barriers, we can approximate these peaks with

$$T(E_\perp) = \frac{1}{1 + \left(\dfrac{E_\perp - \varepsilon_n}{\Delta E}\right)^2}, \tag{8.1}$$

where $2\,\Delta E$ is the full width of the half maximum of the transmission peak. Equation (8.1) assumes that the probability of tunneling through the structure is unity when the electron energy, $E_\perp$, exactly coincides with a quasi-bound state, ε_n. In Fig. 8.4, $T(E_\perp)$ is plotted for several magnitudes of ΔE. One can see that the tunneling transmission coefficient can be a very sharp function of the electron energy.

Besides the transmission coefficient, it is convenient to introduce the probability of tunneling of the electrons out of the well per second, Γ; it is determined by ΔE:

$$\Gamma = \Delta E/\hbar.$$

According to the uncertainty relation we can estimate the lifetime, τ, of the electrons on the quasi-bound level between the barriers as

$$\tau = 1/(2\Gamma).$$

The tunneling probability per unit time can be expressed as the product of the attempt rate, $v_z/(2L_w)$, and the probability of tunneling through a single barrier with one attempt, T:

$$\Gamma = T\frac{v_z}{2L_w},$$

where L_w is the thickness of the well and v_z is the transverse velocity of the electron in the well, which can be estimated from the relationship

$$m^*v_z^2/2 = \varepsilon_n.$$

For the purpose of making illustrative numerical estimates, we choose the following set of parameters: the energy of the first level, ε_1, is 50 meV, the height and thickness of the barriers are $V_b = 300$ meV and $L_b = 40$ Å, and the well width is $L_w = 100$ Å. For this structure, we obtain $v_z = 5 \times 10^7$ cm s^{-1} and $\Gamma = 10^{11}$ s^{-1}. Hence, the lifetime of an electron in the well equals $\tau = 5 \times 10^{-12}$ s. Both τ and Γ depend critically on the height and thickness of the barriers.

The lifetime of the quasi-bound level depends on heterostructure parameters and material combinations. In Fig. 8.5, the lifetimes of the quasi-bound states are depicted for five particular double-barrier structures. The structures are assumed to be made of GaAs/AlGaAs, InGaAs/AlAs, InGaAs/GaAs, and InAs/AlSb. The widths of the quantum wells are equal to 46 Å, while the barriers have thicknesses ranging from 10 Å to 60 Å. The heights of the barriers vary from 0.3 eV to 1.2 eV and are marked on the lines. Figure 8.5 shows that the lifetime can vary over a very wide range.

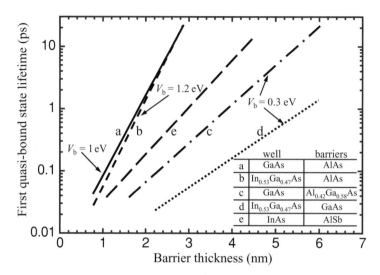

Figure 8.5 The lifetime of the first quasi-bound state in n-type double-barrier structures made from the five material systems presented in the inset. In each structure, the quantum well width is fixed at 46 Å. The energy heights of the barriers are indicated next to the curves. After T. C. L. G. Sollner *et al.*, "High-frequency applications of resonant-tunneling devices," in *Electronic Properties of Multilayers and Low-Dimensional Structures*, J. M. Chamberlain *et al.* (Eds). (New York, Plenum, 1990), pp. 283–296.

For an asymmetric double-barrier structure, we define two different transmission coefficients for the left-hand, T_l, and right-hand, T_r, barriers, respectively, so that the total transmission coefficient, T, can be approximated by

$$T(E_\perp) = \frac{4T_l T_r}{(T_l + T_r)^2} \frac{1}{1 + \left(\dfrac{E_\perp - \varepsilon_n}{\Delta E}\right)^2}, \qquad (8.2)$$

where $\Delta E \equiv \hbar\Gamma$ and $\Gamma = (T_l + T_r)v_z/(2L_w)$. For asymmetric barriers, the maximum transmission at $E_\perp = \varepsilon_n$ is less than unity. It is important to recognize that these results are obtained under the assumption that the electrons transit through the system without phase-changing scattering, so that $E_\perp$ remains constant.

Using the parameters introduced for the double-barrier resonant-tunneling structures and assuming low temperatures, it is possible to obtain the estimates for the current–voltage characteristics sketched in Fig. 8.3.

The formulas (8.1) and (8.2) describe the transmission of an electron with fixed energy $E_\perp$. In order to calculate the electric current, we should take into account all of the electrons which tunnel from the emitter and collector. We suppose that in the contact regions – heavily doped regions – the thermal equilibrium for electrons is established in a very short time. Therefore, to a reasonable degree of approximation we can assume that the Fermi distribution gives the distribution of electrons entering the double-barrier part of the structure. If a voltage bias, Φ_0, is applied, the difference between the Fermi energies of the emitter, E_F^{em}, and collector, E_F^{col}, is

$$E_F^{em} - E_F^{col} = e\Phi_0.$$

In deriving the Landauer formula in Section 6.5, we obtained the equation for the net current in a quantum device:

$$I = I^{em} - I^{col},$$

where, in terms of the present analysis, we introduced the current from the emitter to the collector, I^{em}, and the current from the collector to the emitter, I^{col}. These currents may be expressed in terms of the electron concentrations in the two electrodes, $n^{em}(E_\perp)$ and $n^{col}(E_\perp)$:

$$I^{em,col} = \frac{e}{2\pi\hbar} \int dE_\perp \ T(E_\perp) n^{em,col}(E_\perp). \tag{8.3}$$

Here, the integration runs over the energies $E_\perp$ above the bottom of the conduction band in the emitter and collector, respectively. The concentrations, n^{em} and n^{col}, are given by Eq. (7.6). Let us consider low temperatures, at which the expressions for the concentrations are simpler. Then, instead of Eq. (7.6) we obtain, for example,

$$n^{em}(E_\perp) = \frac{m^*}{\pi\hbar^2}(E_F - E_\perp). \tag{8.4}$$

The electron transverse energy is limited by $E_c^{em} < E_\perp < E_F^{em}$, otherwise the sheet density of the tunneling electrons would be equal to zero. We get the following expression for the current:

$$I = \frac{em^*}{2\pi^2\hbar^3} \int_{E_F^{em}-e\Phi_0}^{E_F^{em}} dE_\perp \left(E_F^{em} - E_\perp\right) T(E_\perp), \quad e\Phi_0 < E_F^{em}. \tag{8.5}$$

This formula takes into account the fact that, because all states in the collector with energy $E_\perp < E_F^{em} - e\Phi_0$ are occupied, tunneling is possible only for those electrons with energies satisfying the following conditions: $E_F^{em} > E_\perp > E_F^{em} - e\Phi_0$. For $e\Phi_0 > E_F^{em}$ all emitter electrons can tunnel through the barriers and we have

$$I = \frac{em^*}{2\pi^2\hbar^3} \int_{E_c^{em}}^{E_F^{em}} dE_\perp \left(E_F^{em} - E_\perp\right) T(E_\perp), \quad e\Phi_0 > E_F^{em}. \tag{8.6}$$

We suppose that the Lorentzians in Eqs. (8.1) and (8.2) are narrow relative to E_F^{em}. Then, the integration over $E_\perp$ in Eqs. (8.5) and (8.6) is straightforward and yields

$$I = \frac{em^*v_z}{2\pi\hbar^2 L_w}\left(E_F^{em} - \varepsilon_n(\Phi_0)\right)\frac{T_l T_r}{T_l + T_r}, \tag{8.7}$$

where we have used Eq. (8.2) for asymmetric barriers. This equation represents the tunneling current through the bound state with energy $\varepsilon_n(\Phi)$, which depends on the voltage bias as illustrated previously by the example of Fig. 8.3(b). A peak value of the current is reached when the bound-state energy is resonant with the bottom of the conduction band in the emitter E_c^{em}:

$$I_p = \frac{em^*v_z}{2\pi\hbar^2 L_w}E_F\frac{T_l T_r}{T_l + T_r}. \tag{8.8}$$

Table 8.1 Resonance width and collision broadening of level ε_1 of $Al_{0.3}Ga_{0.7}As$/GaAs double-barrier structures

L_w (Å)	L_b (Å)	Γ_r (meV)	Γ_r/Γ_{sc} at $T = 300$ K	at $T = 200$ K	at $T = 70$ K
50	70	1.3×10^{-2}	6×10^{-3}	1.9×10^{-2}	2.6×10^{-1}
50	50	1.5×10^{-1}	7.5×10^{-2}	8.3×10^{-1}	3.1
50	30	1.76	0.88	1.3	3.6
20	50	6.03	3.02	4.56	124

When the bound-state energy, ε_n, falls below the bottom of the conduction band in the emitter, the current drops rapidly to the value determined by off-resonance tunneling processes. This value can be estimated, if we assume that off-resonance transmission coefficients are constant:

$$I_v = \frac{em^*}{2\pi^2\hbar^3} E_F^2 T_l T_r. \tag{8.9}$$

Since for the transmission coefficients the inequality $T_l, T_r \ll 1$ is valid, the resonant current of Eq. (8.8) is much greater than the off-resonance current of Eq. (8.9). The results obtained for the coherent mechanism of tunneling through a double-barrier structure support the qualitative discussion given at the very beginning of this section.

Sequential tunneling

Another process responsible for the resonant-tunneling effect is the so-called *sequential tunneling process*. In the sequential tunneling scheme, electron transmission through the structure is regarded – somewhat artificially – as two successive transitions: first from the emitter to the quantum well and then from the well to the collector. It is important to highlight the main difference between the previously studied coherent mechanism and the sequential mechanism of resonant tunneling. The first mechanism excludes any electron collisions during the transition from the emitter to the collector. The second scenario applies even when there is electron scattering inside the quantum well. Although the coherent and sequential processes result in the same behavior of the double-barrier resonant structures, it is possible and instructive to separate and compare these processes. In the previous discussion, we defined a broadening, Γ, for the quasi-bound state due to the tunneling process. If both the processes of tunneling and scattering inside the well occur, the width of the quasi-bound state increases. Let us introduce the width of the quasi-bound state, Γ_r, as the *full width* of the half maximum of the transmission peak. Let the collision broadening of this state be Γ_{sc}. From the previous discussion, we conclude that coherent tunneling dominates if $\Gamma_r > \Gamma_{sc}$, whereas the sequential processes dominate if $\Gamma_r < \Gamma_{sc}$. Table 8.1 illustrates different regimes of tunneling for several specific $Al_{0.3}Ga_{0.7}As$/GaAs double-barrier structures at zero bias. In this table, L_w and L_b are the thicknesses of the wells and barriers, respectively. The ratio Γ_r/Γ_{sc} is calculated by using the estimate $\Gamma_{sc} = \hbar/\tau$, where τ is the scattering time. From the results of Chapter 6, this time may be either calculated from first

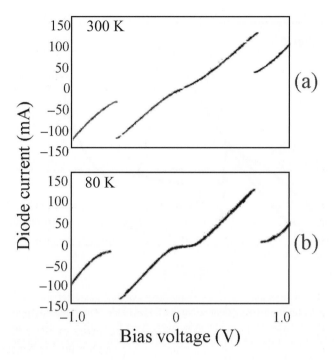

Figure 8.6 Measured current–voltage characteristics of a symmetric $Al_{0.48}In_{0.52}As/Ga_{0.47}In_{0.53}As$ double-barrier resonant-tunneling structure at 300 K (a) and 80 K (b). After Fig. 1, F. Capasso, S. Sen *et al.*, "Quantum-effect devices," in S. M. Sze (Ed.), *High-Speed Semiconductor Devices* (New York, Wiley, 1990). Reprinted with permission of John Wiley & Sons, Inc.

principles or deduced from experimental mobility measurements. For Table 8.1, the following values are assumed: when $T = 300$ K, $\mu = 7000$ cm^2 V^{-1} s^{-1}, and $\tau \simeq 3 \times 10^{-13}$ s; when $T = 200$ K, $\mu = 2 \times 10^4$ cm^2 V^{-1} s^{-1}, and $\tau = 10^{-12}$ s; and when $T = 70$ K, $\mu \geq 10^5$ cm^2 V^{-1} s^{-1}, and $\tau \geq 5 \times 10^{-12}$ s. If $\Gamma_r/\Gamma_{sc} > 1$, elastic collisions can be neglected and resonant tunneling is highly coherent. From Table 8.1 one can see that this type of tunneling is typical for the case of low temperatures and thin barriers. In the opposite limit, $\Gamma_r/\Gamma_{sc} \ll 1$, the tunneling processes are more likely to be sequential. It is important to emphasize that the scattering affects also the magnitude of the transmission coefficient. The maximum of the transmission probability decreases by the factor $\Gamma_r/(\Gamma_r + \Gamma_{sc})$ when scattering takes place. This explains why resonant tunneling is washed out at high temperatures and in structures with defects and impurities.

Negative differential resistance under resonant tunneling

As we have already seen from qualitative discussions, resonant-tunneling structures manifest strongly nonlinear current–voltage characteristics. In particular, it is very important that there is a portion of these characteristics for which the current decreases while the voltage increases. Typical current–voltage characteristics for a double-barrier structure of AlInAs/GaInAs are presented in Fig. 8.6. The results are shown for two

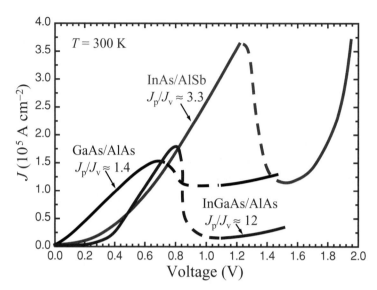

Figure 8.7 The room-temperature current density versus voltage characteristics and peak-to-valley ratios, J_p/J_v, for high-speed double-barrier RTDs made from three different material systems: GaAs/AlAs, InAs/AlSb, and InGaAs/AlAs. After E. R. Brown, "Resonant tunneling in high-speed double barrier diodes," in *Hot Carriers in Semiconductor Nanostructures: Physics and Applications*, J. Shah (Ed.) (Boston, MA, AT&T and Academic, 1992), pp. 469–498.

temperatures, $T = 80$ K and $T = 300$ K. The structure is symmetric (similar barriers, equal thicknesses of spacers, and doping of contacts); thus, there is an almost antisymmetric current–voltage characteristic. For the lower temperature, there is clearly seen a portion with almost zero current at finite voltage biases, which corresponds to the position of the resonant quasi-bound level above the Fermi level of the emitter and very small non-resonance and overbarrier currents. When the level is shifted below the bottom of the emitter's conduction band ($\Phi_0 \approx \pm 0.75$ V), the current drops to almost zero values. At room temperature, small overbarrier currents exist at any finite voltage bias and the current drop is also considerably less than in the former case. An important parameter for characterizing N-type negative differential resistance is the ratio of the maximum and minimum values of the current – the so-called *peak-to-valley ratio*. Achieving large peak-to-valley ratios greatly enhances the possibility of applications utilizing negative differential resistance. For the case presented in Fig. 8.6, the peak-to-valley ratio is 15 at 80 K. At room temperature, the ratio decreases to about 4, but still remains sufficient for applications. The peak-to-valley ratio depends not only on the physical nature of the negative differential resistance, but also on many technological and design factors. Although the development of the double-barrier system is still in progress, for the most optimized and perfect structures, a peak-to-valley ratio of about 20 or even higher may be achieved at room temperature.

Comparisons of the current–voltage characteristics for various double-barrier structures and the peak-to-valley ratios obtained for $T = 300$ K are presented in Fig. 8.7. Three types of heterostructures currently used for resonant-tunneling diodes – GaAs/AlAs,

InGaAs/AlAs, and InAs/AlSb – are characterized by current densities exceeding 10^5 A cm^{-2} and exhibit negative conductance at room temperature. The largest peak-to-valley ratio is reached for InGaAs/AlAs structures (above 10 at room temperature), but the highest current densities are, typically, for InAs/AlSb, $J_m \approx 4 \times 10^5$ A cm^{-2}. For GaAs/AlAs structures, these parameters are relatively modest: the peak-to-valley ratio is about 1.4 and $J_m \approx 1.5 \times 10^5$ A cm^{-2}.

Another important parameter of any system exhibiting negative differential resistance is the characteristic time of the processes responsible for the negative differential resistance. This time determines the physical upper frequency limit at which the negative differential resistance disappears. For the previously considered resonant-tunneling device, the frequency limit is not easy to estimate because this type of electron transport has no classical analogue. Careful analysis of the frequency properties of the tunneling can be done only by numerical self-consistent calculations involving the time-dependent Schrödinger equation, the kinetic equations describing processes in the contacts, and Poisson's equation. This complex problem has not yet been solved. However, by modulating the barriers by a small time-dependent voltage and examining the time-dependent response of the system, it has been shown that the characteristic time of the tunneling processes can be estimated by the following formula:

$$\tau_{tr} = m^* \int_0^d \frac{dz}{\sqrt{2m[V(z) - E_\perp]}} + \frac{2\hbar}{\Gamma} \equiv \frac{d}{v_g} + \frac{2\hbar}{\Gamma}, \tag{8.10}$$

where d is the total thickness of the two barriers and the well, and v_g can be interpreted as the electron group velocity. If the perpendicular energy of the incident electron, $E_\perp$, is much greater than the resonance width, Γ, it can be shown that the total transit time through the structure is given approximately by τ_{tr}. The first term represents the *semiclassical transit time* across the structure and the second term is the so-called *phase time*.

For a typical example of a symmetric resonant-tunneling structure with 17-Å-thick AlAs barriers and a 45-Å-thick well, the quasi-bound level has an energy $\varepsilon_1 \approx 0.13$ eV; hence, $2\hbar/\Gamma = 0.45 \times 10^{-12}$ s. For a drift velocity $v_d \geq 10^7$ cm s^{-1}, the first term gives only 0.8×10^{-13} s. Thus, $\tau_{tr} \approx 0.5 \times 10^{-12}$ s. Hence these quantum devices are ultrafast, with response times in the subpicosecond range.

A resonant-tunneling diode as a microwave oscillator

The application of the resonant-tunneling effect in high-frequency oscillators is based on the existence of the negative differential resistance. In order to review the principles of using negative differential resistance for obtaining electrical oscillations, we consider the simplest electric circuit containing a resistance, R_d, a capacitance, C, and an inductance, L, as represented by Fig. 8.8(a). Let us introduce the resistance R_d as the ratio

$$R_d = \frac{\Delta \Phi}{\Delta I},$$

where ΔI is the change in the current through the resistor when the voltage drop is changed by $\Delta \Phi$. Thus, in fact, R_d is the *differential resistance* that can have both positive

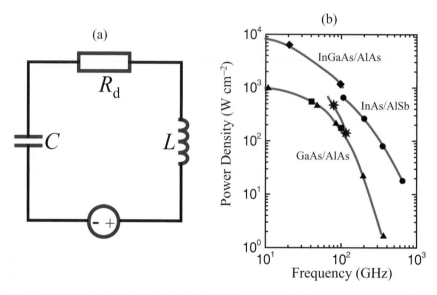

Figure 8.8 (a) The simplest electric circuit. (b) The generated microwave power per unit cross-section of double-barrier resonant-tunneling diodes as a function of the frequency for the same three devices as in Fig. 8.7. (b) After E. R. Brown, "Resonant tunneling in high-speed double barrier diodes," in *Hot Carriers in Semiconductor Nanostructures: Physics and Applications*, J. Shah (Ed.), (Boston, MA, AT&T and Academic, 1992), pp. 469–498.

and negative sign. If the alternating current, $\tilde{I}$, and voltage, $\tilde{\Phi}$, applied to the circuit are

$$\tilde{I} = I_0 e^{-i\omega t} \quad \text{and} \quad \tilde{\Phi} = \Phi_0 e^{-i\omega t},$$

one has the following relation between the magnitudes I_0 and Φ_0:

$$I_0 = \frac{1}{Z(\omega)}\Phi_0, \quad Z(\omega) = R_d - i\left(\omega L - \frac{1}{\omega C}\right), \tag{8.11}$$

where $Z(\omega)$ is the impedance of the circuit. The impedance – complex resistance – includes the active resistance, R_d.

Even if $\Phi_0 = 0$, oscillations can exist in the circuit at frequencies for which $Z(\omega_c) = 0$:

$$\omega_c = -i\frac{R_d}{2L} \pm \sqrt{\frac{1}{LC} - \left(\frac{R_d}{2L}\right)^2}. \tag{8.12}$$

For $L/C > (R_d/2)^2$ and for a positive resistance R_d, there are oscillations with the frequency $\sqrt{1/(LC) - [R_d/(2L)]^2}$ and damping characterized by

$$\gamma = -\text{Im}\{\omega_c\} = R_d/(2L).$$

Here, γ is positive and $\text{Im}\{\omega\}$ denotes the imaginary part of ω_c.

If the real part of the impedance is negative,

$$\text{Re}\{Z(\omega)\} = R_d < 0, \tag{8.13}$$

it follows that γ is negative and the steady state of the circuit is unstable with respect to the generation of voltage oscillations. Thus, if we want to realize voltage oscillations at a desired frequency, ω, the differential resistance at this frequency should be negative. The fact that double-barrier resonant-tunneling structures have extremely small electron transit times means that these devices exhibit negative differential resistance up to ultra-high frequencies. In Fig. 8.8(b), the generated microwave power per unit cross-section of double-barrier resonant-tunneling diodes is presented as a function of the frequency for the same three devices as were considered in Fig. 8.7. The generated power decreases with frequency as a result of the previously discussed finite transit time of tunneling electrons. Absolute values of the power correlate with the current densities achieved for these devices. It is seen that oscillations up to 1 THz $= 1000$ GHz (i.e., 10^{12} oscillations per second) have been reached for nanoscale quantum devices of this type.

Finally, the resonant-tunneling structures are the simplest quantum devices. They exhibit strongly nonlinear current–voltage characteristics with negative differential resistance. Since these devices have nanoscale sizes, they exhibit extremely short transit times for carrier transport through the structures. These properties allow one to exploit resonant-tunneling structures for the generation of ultra-high-frequency electromagnetic oscillations. Indeed, an oscillation frequency of about 1 THz has already been reached in resonant-tunneling diodes.

8.3   Field-effect transistors

The previously analyzed two-terminal devices – diodes – are the simplest electronic devices, for which the current is controlled by the diode bias and vice versa. A useful function can be performed mainly due to nonlinearity of current–voltage dependences. In contrast, in three-terminal devices known as *transistors* there exists the possibility of controlling the current through two electrodes by varying the voltage or the current through a third electrode. Depending on the principle of operation, transistors can be associated with one or other of two large classes: (i) *field-effect transistors* and (ii) *potential-effect transistors*.

Devices controlled by the field effect

Devices of the first group are *field-effect* or *voltage-controlled devices*. A common feature of these devices is that a voltage is applied to a controlling electrode – a *gate* – which is *capacitively coupled* to the active region of the device. A capacitive coupling means that by applying a voltage to the gate one creates a transverse electric field in the conducting channel, but no useful current flows through the gate. The gate electrode is spatially and electrically separated from carriers in the active region by an insulator or a depletion region (a region where electrons are absent). The gate electrode controls the resistance of the active region and, consequently, the current between two other terminals, which are known as the *source* and the *drain*.

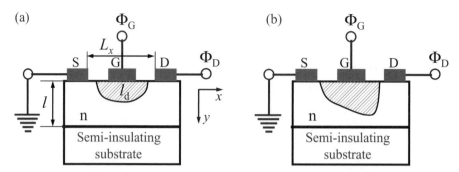

Figure 8.9 The MESFET model. The conducting channel of width l is supposed to be grown on a semi-insulating substrate. The length of the channel is L_x. The depleted region, $l_d(x)$, under the gate, G, is marked by a shadow region. In case (a) there is no voltage bias applied between the source, S, and the drain, D; in case (b) a positive voltage is applied to the drain. There are no electrons in the depletion region.

There exist several different types of field-effect transistor (FET). Before discussing the most important types of FET, including those based on nanostructures, we will explain the operation principles by using the simplest FET, the so-called *metal–semiconductor FET* (MESFET).

Figure 8.9 presents a simple model for a field-effect transistor. It is assumed that the active region of the device is made of an n channel, which can be fabricated, for example, by homogeneous doping. The source and drain are heavily doped n^+ regions, which are assumed to serve as Ohmic contacts. The gate upon the active layer forms a so-called *Schottky barrier*, which produces a depletion region, as discussed in Section 7.2 (see the subsection on the control of charge transfer). The formation of the depletion region under the gate is illustrated by Fig. 8.9. Such a design is typical for commonly used semiconducting materials such as GaAs and InP. From the bottom, the active region is restricted by a non-conducting (semi-insulating) substrate.

If no voltage is applied to the contacts, the depletion region is almost uniform along the active region, as shown in Fig. 8.9(a). The characteristics of the depletion region are determined by the built-in (Schottky) voltage. If a negative voltage is applied to the gate and there is still no voltage between the source and drain, the depletion region extends farther into the active region and decreases the width of the channel. At some voltage the channel is *completely* pinched off. Let the gate voltage be fixed and a small positive voltage be applied to the drain. A current will flow through the channel in the Ohmic linear regime. If the drain voltage is increased further, it will affect the distribution of the potential in the device: the width of the depletion region increases near the drain end of the channel, as shown in Fig. 8.9(b), as a result of the increasing potential difference between the gate and the drain. At a certain drain voltage, the channel begins to pinch off at the drain end. At this voltage the current saturates and remains nearly constant with further increase in the drain voltage. This behavior is illustrated in Fig. 8.10, where typical current–voltage characteristics of a MESFET are presented for various gate voltages. It is clear that, if the gate voltage is negative, the channel becomes narrow and the current is suppressed, whereas, for positive gate voltage, the depletion region under the gate

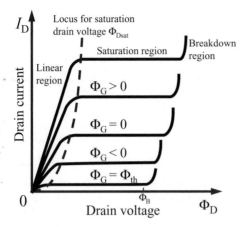

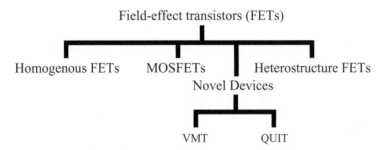

Figure 8.10 The drain-current versus drain-voltage characteristics for the MESFET at different gate voltages. The linear and saturation regions are indicated. The breakdown effect occurs at high voltages Φ_B and is presented schematically.

Field-effect transistors (FETs)

Homogenous FETs   MOSFETs   Heterostructure FETs

Novel Devices

VMT   QUIT

Figure 8.11 The family tree of FETs consists of three groups of conventional FETs that have much in common in their technologies, properties, and applications and new groups of novel devices. The abbreviations used in this chart are defined as follows: MOSFET metal–oxide–semiconductor FET; VMT, velocity-modulation transistor; and QUIT, quantum-interference transistor.

becomes narrower, the channel opens wider, and the current has to be larger. Thus, in comparison with a diode, the more complex configuration of the electrostatic potential in a FET gives rise to strongly nonlinear source–drain current–voltage characteristics. Qualitatively, two operational regimes are possible, the linear region and the saturation region, as indicated in Fig. 8.10. For both regimes, there is effective current control by the gate voltage. At very large source–drain voltages, electric breakdown can occur.

In summary, the phenomenon of controlling the resistance of the conduction channel by an external voltage (by a field) is the basic principle of any FET.

The FET-family devices

The FET family can be classified as shown in Fig. 8.11, where we give also short explanations for the abbreviations used to denote these FETs. The FET family of transistors

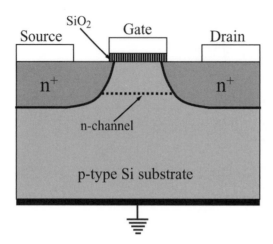

Figure 8.12 A sketch of an Si-MOSFET.

is divided into three large groups of devices used in practice, as well as a group of potentially new devices based on recent developments. The first group includes the variations of metal–semiconductor FETs discussed previously. The second group consists of metal–oxide–semiconductor FETs (MOSFETs). In contrast to the previously considered MESFETs, in MOSFETs, the gate and conducting channel are separated by a dielectric (an oxide) instead of a Schottky barrier. The MOSFETs are typically fabricated by using a metal–oxide–silicon structure, with the oxide layer being silicon dioxide. In Fig. 8.12 we present a sketch of a Si MOSFET. It should be said that Si MOSFETs are microelectronic devices of the utmost importance. A third group of FETs in Fig. 8.11 incorporates heterostructures and differs in the design of the doping and conducting channel. For example, the donor-layer heterostructure FET (HFET) subgroup has one or several doped layers, which supply electrons to the conducting channel. If the channel is formed on a single heterointerface (see Section 7.2) using a single donor layer the device is called a modulation-doped FET (MODFET). Such MODFETs have the largest electron and hole mobilities. That is why they are also called *high-electron-mobility transistors* (HEMTs). Examples of material systems used for HFETs are as follows: n^+-AlGaAs/GaAs (unstrained) heterostructures, strained layer n^+-InGaAs channels on GaAs, and n^+-InAlAs/InGaAs heterojunctions grown on InP. Although the latter two structures are more complex, they offer better transport parameters in the conducting channel.

   Figure 8.13(a) shows schematically one of the variants of a HFET – a recessed-gate n^+-AlGaAs/GaAs HFET. The device is fabricated on a semi-insulating GaAs substrate upon which an undoped buffer layer of GaAs is grown. A thin undoped AlGaAs spacer layer and an n-doped AlGaAs layer are grown on the buffer layer. Source, drain, and gate electrodes are fabricated on the top of the structure. Under the electrodes, two heavily n^+-doped regions serve as the contacts to the two-dimensional electron gas formed under the heterojunction. Such a two-dimensional electron-gas channel was studied in Section 7.2 in detail. The gate length varies from 1.0 μm to 0.1 μm, or less, depending on the speed needed for applications.

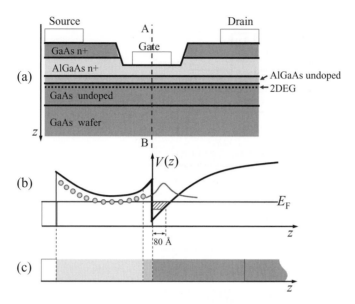

Figure 8.13 A recessed-gate n^+-AlGaAs FET. (a) The cross-section of the HFET. (b) The energy band diagram along the line A–B; the Fermi level, E_F, and electron confined states are shown. In the donor layer, there are two depletion regions for a thick barrier: one is due to the Schottky built-in voltage; the second is due to the conduction-band offset at the junction. (c) The sequence of layers along the line A–B.

The dimensions in the vertical directions can be different. This facilitates the fabrication of two types of HFET: normally turned-on and normally turned-off devices. (Both types of heterostructures were studied in Section 7.2.) Figure 8.13(b) illustrates the energy-band coordinate dependence for the cross-section of the structure under the gate. The major element affecting the physical picture is the Schottky depletion layer under the gate. The depletion layer is different for different thicknesses of the barrier layer. For a barrier layer of small thickness, the Schottky barrier depletes even the quantum well, leading to the normally turned-off state in the absence of the gate voltage. For larger barrier thicknesses, the Schottky barrier does not prevent the formation of the conducting channel, and we obtain the normally turned-on state of the HFET as illustrated in Figure 8.13(b).

Let us briefly review the heterostructures employed for FETs based on Si technology. Although many combinations of materials are being investigated with the goal of fabricating high-quality Si-based heterostructures, the Si/SiGe system is the most studied and developed and has already found various device applications. Properties of SiGe alloys and Si/SiGe heterostructures were studied in Chapter 4. These heterostructures can be used for the creation of two-dimensional electron and hole gases and for the improvement of device parameters.

In Si/SiGe systems, a low-dimensional n-type channel can be obtained if a strained Si layer is grown upon a strained SiGe layer. Figure 8.14(a) depicts the cross-section of an n-channel modulation-doped Si/SiGe FET. The device is similar to the previously discussed AlGaAs/GaAs FET with selective doping and improved channel confinement.

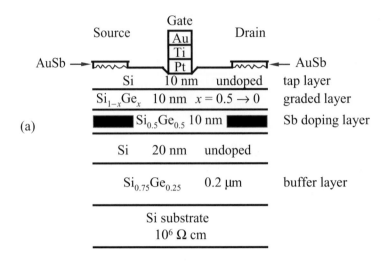

(a)

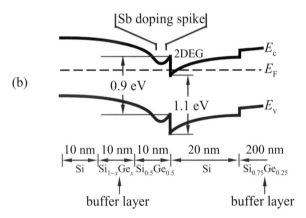

(b)

Figure 8.14 The Si/SiGe-based n-channel MODFET. (a) The cross-section of the device; the electron channel is formed in the strained Si layer. (b) Energy-band diagram of the device. After H. Daembkes, W. Goodhue *et al.*, "The n-channel SiGe/Si modulation-doped field-effect transistor," *IEEE Trans. Electron Devices*, **ED-33**, 633–638 (1986). © IEEE.

The electron channel is formed in the undoped strained Si layer situated between two SiGe layers. The lower $Si_{0.75}Ge_{0.25}$ buffer layer is grown on Si and is strained and undoped. The upper $Si_{0.5}Ge_{0.5}$ layer is thin and δ-doped by donors. Next, the $Si_{1-x}Ge_x$ layer is graded, with x varying from 0.5 to 0. The top thin Si layer is undoped. Such a structure is chosen to create the electron channel and avoid degradation of the characteristics of the strained layer. Figure 8.14(b) shows the energy-band diagram for this device. The electron channel on the $Si_{0.5}Ge_{0.5}$/Si heterointerface is marked. The $Si_{0.5}Ge_{0.5}$ donor layer is δ-doped, which leads to the formation of a spike in the potential profile.

There are many other schemes for Si-based HFETs that combine the advantages of heterostructure bandgap engineering and selective-doping methods. Motivated by the

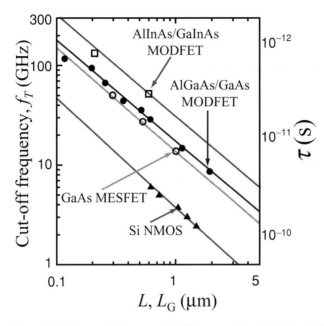

Figure 8.15 The cut-off frequency in GHz versus the device length L or the gate length L_G in µm for three groups of FETs: MOSFETs, MESFETs, and MODFETs. The latter are represented by AlGaAs/GaAs and AlInAs/GaInAs material combinations. The parameters are given at 300 K. After Fig. 25, S. J. Pearton and N. J. Shah, "Heterostructure field-effect transistors," in S. M. Sze (Ed.) *High-Speed Semiconductor Devices* (New York, Wiley, 1990). Reprinted with permission of John Wiley & Sons, Inc.

current wide spread use of Si-based devices, worldwide developments in these areas are now in progress.

One of the key parameters of contemporary devices is the maximum frequency of efficient operation – the cut-off frequency. At fixed device dimensions, the cut-off frequency depends on material characteristics and device design. As a basis for making comparisons, Fig. 8.15 illustrates representative parameters of all three of these groups of FETs. One can see that the HFET class exhibits superb high-frequency performance. Currently, the highest speed is reached for AlInAs/GaInAs HFETs with short gates. This type of HFET with a nanoscale gate of the length $L_G = 45$ nm has a high cut-off frequency, above 450 GHz.

Nanowire FETs

The previously–analyzed HFETs use heterostructures with two-dimensional electrons. The next step in reducing the number of dimensions of the device is to exploit nanowires. Semiconductor nanowires and carbon nanotubes are attractive components for nanoscale FETs.

For example, nanowire FETs can be configured by depositing the nanomaterial onto an insulating substrate surface, and making source and drain contacts on the ends of the

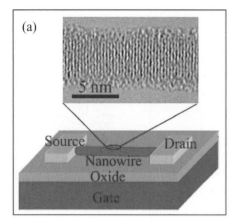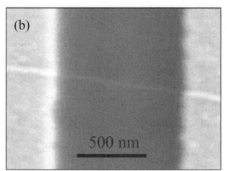

Figure 8.16 (a) A schematic diagram of a Si FET with the nanowire, the metal source, and drain electrodes on the surface of a SiO_2/Si substrate. (b) A scanning electron micrograph of a Si nanowire FET; the scale bar is 500 nm. The 5-nm diameter Si wire image was obtained by using a high-resolution transmission electron microscope. Reprinted with permission, from Y. Cui, Z. Zhong *et al.*, "High performance silicon nanowire field effect transistors," *Nano Lett.*, **3**, 149–152 (2003). © American Chemical Society.

nanowire. Figure 8.16(a) illustrates this approach. There, we show a schematic diagram of a Si-nanowire FET with the nanowire, the metal source, and drain electrodes on the surface of the SiO_2/Si substrate. The image of the 5-nm-diameter Si wire was obtained by using a high-resolution transmission electron microscope. In Fig. 8.16(b), a scanning electron micrograph of a Si-nanowire FET is shown; the scale bar represents 500 nm. This approach may serve as the basis for hybrid electronic systems consisting of nanoscale building blocks integrated with more complex planar silicon circuitry.

An extremely small FET may be built on the basis of carbon nanotubes. Indeed, despite their nanosize diameters, the carbon nanotubes are sufficiently robust and may be long enough to allow electrical connections to lithographically defined metallic electrodes, therefore making it possible to probe directly the electrical properties of these nanometer-size structures. Electrical connections to nanotubes can be achieved either by depositing a metal electrode over the top of the tubes (*end-contacted* samples), or by placing the tubes on the top of prepared metal leads (*bulk-contacted* samples).

An advanced nanotube-based FET is shown in Fig. 8.17. The depicted nanotube FET is fabricated on SiO_2/Si substrates. The silicon is doped heavily to serve as a back gate. The dielectric SiO_2 of thickness 10 nm covers most of the area of the substrate and electrically isolates the Si substrate and the wire. Two metal Pd electrodes were deposited to fabricate the source and drain. This technology provides the nanotube FET with a back gate shown in Fig. 8.17(a). Additionally, by using atomic layer deposition, an 8-nm-thick HfO_2 film can be deposited to cover the device from above. Then, 20 nm of Al is deposited onto the dielectric (HfO_2) film to create the top gate, as shown in Fig. 8.17(b). Thus, we have presented nanowire FETs with single and double gates. For both devices, the total tube length between metal electrodes was 2 μm, and the top-gated section length was 0.5 μm.

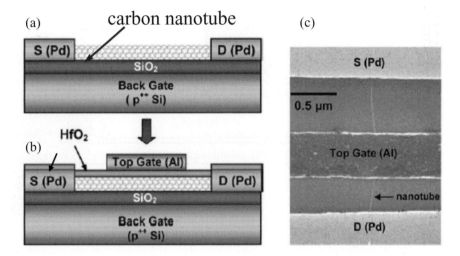

Figure 8.17 Schematic diagrams of a carbon-nanotube FET: (a) two metal (Pd) electrodes are deposited onto a carbon nanotube placed on a 10-nm-thick layer of SiO_2 that insulates it from the back gate; (b) 8-nm-thick HfO_2 serves as an insulator for the top gate; and (c) a scanning electron micrograph of the carbon nanotube FET. Reprinted with permission, from A. Jarvey, J. Guo *et al.*, "Carbon nanotube field-effect transistors with integrated ohmic contacts and high-*k* gate dielectrics," *Nano Lett.*, **4**, 447–450 (2004). © American Chemical Society.

In Fig. 8.17(c), an electron-scanning micrograph of a carbon-nanotube FET is shown. This nanotransistor works and has good performance.

In conclusion, the nanowire devices discussed here have great potential for applications in nanoelectronics.

Velocity-modulation transistors

For all of the previously analyzed FETs, the gate voltage controls the number of conducting electrons, and the speed of operation is determined by the transport time of electrons along the conducting channel. An alternative principle of device operation consists of modulating electron mobility, or, possibly, electron drift velocity, without significant changes in the number of current carriers. To achieve such a modulation of the drift velocity one can use spatially nonuniform (selective) carrier scattering across the active channel. Then, a redistribution of the carriers across the channel in response to a gate voltage would lead to control of the velocity. The time of such a redistribution across a narrow channel can be considerably smaller than that of the longitudinal transport. A three-terminal device with modulation of the drift velocity by an external voltage has been referred to as the *velocity-modulation transistor* (VMT).

The characteristic time of charge transfer in a channel of thickness about 100 Å can be estimated to be as short as 10^{-13} s. Thus, for ultra-high-frequency operation, the velocity-modulation effect should be the dominant factor controlling electric signals.

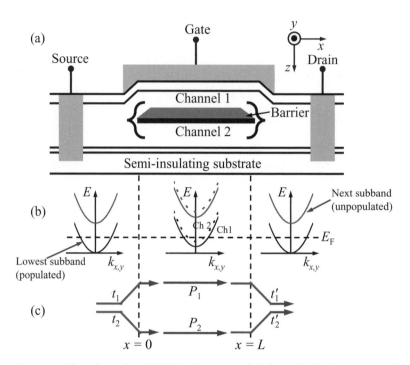

Figure 8.18 The schematics of QUIT with two parallel channels: (a) the structure of the device; (b) electron subbands in different regions of the device and the Fermi energy (under equilibrium); (c) an illustration of the electron paths in the device; t_1, t_2 and t_1', t_2' indicate the amplitudes of split and interfering waves, respectively. P_1 and P_2 denote factors of different phase shifts in the channels.

Quantum-interference transistors

The remaining class of novel nanoscale FETs presented in Fig. 8.11 is the so-called *quantum-interference transistor* (QUIT). It is based on quantum ballistic electron transport. According to the classification given in Section 6.2, for such a transport regime, one has to make the coherence length, l_ϕ, greater than the characteristic device scale, L:

$$l_\phi > L; \qquad (8.14)$$

that is, this device is a *mesoscopic* device. The principle of operation of a QUIT is the *control of the interference pattern of conducting electrons by an external voltage*.

Remember that interference is one of the most important effects in the physics of waves. We studied the interference effect in Sections 2.4. Examples of interference patterns are given in Figs. 2.3 and 2.7. In a QUIT, these patterns can be arranged as a result of interference of the waves traveling through two or more channels (arms).

A schematic diagram of a two-channel quantum-interference device is shown in Fig. 8.18(a). Basically, it is an ordinary FET with a short channel and a barrier parallel to the current. The barrier is embedded in the middle region of the device. The barrier splits the main channel into two channels: 1 and 2. There are also two contacts to the main channel – the source and the drain. The gate is placed on the top of the device. Owing to the gate, the symmetry of channels 1 and 2 is broken if a gate voltage is applied. The distance between contacts, L, satisfies the inequality of Eq. (8.14). The widths of the main channel and the split channels are small, so that there is a quantization of transverse electron states in the z direction. Two-dimensional subbands are illustrated in Fig. 8.18(b); the subbands, $E(k_x, k_y)$, and the Fermi energy, E_F, are plotted for three major device regions. For simplicity, one can assume that only the lowest subband is populated in each of the device regions. If a voltage is applied to the gate, in the middle region, the bottoms of the subbands, ε_1 and ε_2, for channels 1 and 2 are generally different. The electron energy spectra are of the form

$$E = \varepsilon_1 + \frac{\hbar^2}{2m^*}\left(k_{x,1}^2 + k_y^2\right) = \varepsilon_2 + \frac{\hbar^2}{2m^*}\left(k_{x,2}^2 + k_y^2\right), \tag{8.15}$$

where E is the energy of an incident electron, and k_x and k_y are the components of its wavevector in the plane of the device. Obviously, the energy E does not change along the channel. The k_y-component is also conserved because it is assumed that there are no forces along the y direction. Thus, from Eq. (8.15), it follows that only the k_x-components in the channels can be different.

Let an electron be injected from the source into the left region of the main channel. Its wavefunction is

$$\psi_L = \chi_L(z)e^{i(k_x x + k_y y)},$$

where $\chi_L(z)$ is the wavefunction of the lowest subband in the left region. We can introduce the amplitudes of the waves transmitted into channels 1 and 2 as t_1 and t_2. These amplitudes determine the electron wavefunctions inside the middle device region:

$$\psi_M = \begin{cases} t_1 \chi_{M,1} e^{i(k_{x,1}(x-x_L)+k_y y)}, & \text{upper channel,} \\ t_2 \chi_{M,2} e^{i(k_{x,2}(x-x_L)+k_y y)}, & \text{lower channel,} \end{cases}$$

where $\chi_{M,1}(z)$ and $\chi_{M,2}(z)$ are the wavefunctions of the lowest subbands in channels 1 and 2, respectively, and x_L is the coordinate of the left end of the channels. Since, in general, the potential profiles in the two channels are different, the wavevectors are different: $k_{x,1} \neq k_{x,2}$. Then, let t_1' and t_2' characterize the waves transmitted from the channels into the right device region; see Fig. 8.18(c). The wavefunction inside the right region can be written as

$$\psi_R = \left(t_1 P_1 t_1' + t_2 P_2 t_2'\right)\chi_R(z)e^{i(k_x(x_L-x_R)+k_y y)}, \tag{8.16}$$

where $P_1 \equiv \exp(ik_{x,1}L)$, $P_2 \equiv \exp(ik_{x,2}L)$, and $\chi_R(z)$ is the wavefunction of the transverse motion at the right end of the channels. For the sake of simplicity, we neglect multiple reflections from the ends of the barriers.

The wavefunctions χ_L and χ_R are normalized to unity; thus, the total transmission coefficient for the device is

$$T(E) = |t_1' P_1 t_1 + t_2' P_2 t_2|^2. \tag{8.17}$$

We can assume that both channels, 1 and 2, are symmetric with respect to $z = 0$ in the absence of a gate voltage. Thus, for the lowest occupied subbands we find

$$t_2 = t_1 \qquad \text{and} \qquad t_2' = t_1'.$$

Then, Eq. (8.17) gives us

$$|T(E)|^2 = 2|t_1 t_1'|^2 (1 + \cos\theta), \tag{8.18}$$

where

$$\theta = (k_{x,2} - k_{x,1})L \tag{8.19}$$

is the relative phase shift of the two channels. If we introduce the average electron velocity,

$$v_x = \frac{\hbar(k_{x,1} + k_{x,2})}{2m^*},$$

we can represent the difference between the wavevectors as follows:

$$k_{x,2} - k_{x,1} = \frac{\varepsilon_1 - \varepsilon_2}{\hbar v_x}. \tag{8.20}$$

Now, the phase shift takes the form

$$\theta = \frac{L}{v_x}\frac{\varepsilon_1 - \varepsilon_2}{\hbar}.$$

The origin of the phase shift is obvious: if $\varepsilon_1 \neq \varepsilon_2$, a difference between the kinetic energies in the two channels gives rise to different phases of the waves coming into the right device region; of course, these different phases lead to quantum interference.

For channels that are symmetric with respect to $z = 0$, in the absence of a gate voltage, we find $\varepsilon_1 = \varepsilon_2 = \varepsilon_0$, and the phase shift equals zero. If we apply a gate voltage, the potential energy as a function of the transverse coordinate is modified:

$$V(z) = V_0(z) - e\Phi(z).$$

Here, $\Phi(z)$ is the potential induced by the applied voltage. This leads to subband energies

$$\varepsilon_1 = \varepsilon_0 - e\langle \chi_{M,1}|\Phi|\chi_{M,1}\rangle,$$
$$\varepsilon_2 = \varepsilon_0 - e\langle \chi_{M,2}|\Phi|\chi_{M,2}\rangle.$$

Using Eqs. (8.20) and (8.19), we find the phase shift,

$$\theta = \frac{L}{v_x}\frac{e\Phi_{12}}{\hbar}, \tag{8.21}$$

where $\Phi_{12} \equiv \langle \chi_{M,2}|\Phi(z)|\chi_{M,2}\rangle - \langle \chi_{M,1}|\Phi(z)|\chi_{M,1}\rangle$. The value Φ_{12} represents the difference between the average potentials in the channels. This difference determines the transmission through the device.

To calculate the electric current, one can use the results of Section 6.5, where quantum ballistic transport was analyzed. Equation (6.120) defines the device conductance, G. At low temperatures, these results lead to the Landauer formula of Eq. (6.121), which, in the case under consideration, has the form

$$G = \frac{2e^2}{h} T(E) = \frac{4e^2}{h} |t_1 t_2|^2 (1 + \cos\theta).$$ (8.22)

The second term in the brackets is due to electron interference. One can see that the interference controls the device conductance. If the phase shift of split waves $\theta = 0$ ($\Phi_{12} = 0$), the conductance reaches the maximum, $G_{\max} = 8e^2 |t_1 t_2|^2 / h$. If $\theta = \pi$, i.e.,

$$e\Phi_{12} = \frac{\hbar \pi v_x}{L},$$ (8.23)

the interference is destructive and the conductance vanishes.

Let us introduce the characteristic transit time through the channels, $t_{\mathrm{tr}} = L/v_x$. If we set $L = 2000$ Å and $v_x = 2 \times 10^7$ cm s^{-1}, we get $t_{\mathrm{tr}} = 1$ ps, and for the destructive potential difference we get $\Phi_{12} \approx 2$ mV. The shorter channels require a larger gate voltage. Obviously, the transit time, t_{tr}, determines the cut-off frequency of the device, $\omega_{\mathrm{cf}} = 2\pi/t_{\mathrm{tr}}$. Since the device channels are not doped and can be made quite short, these devices portend operation up to the terahertz frequency region.

Our simple model of the quantum-interference transistor allows us to compare this device with a conventional FET. If the latter is in a normally turned-on state with the Fermi energy of electrons in the conducting channel equal to E_F, we can estimate the threshold voltage needed to deplete the FET channel as

$$e\Phi_{\mathrm{th,FET}} \approx E_F = \frac{\hbar^2 k_F^2}{2m^*}.$$ (8.24)

For quantum devices described by Eq. (8.23), we find

$$e\Phi_{\mathrm{th,QUIT}} \approx e\Phi_{12} = \frac{\pi\hbar v_x}{L} = \frac{\pi\hbar^2 k_F}{m^* L} = \frac{\hbar^2 k_F^2}{2m^*} \frac{2\pi}{k_F L} = e\Phi_{\mathrm{th,FET}} \frac{\lambda_F}{L},$$ (8.25)

where λ_F is the de Broglie wavelength of electrons in the Fermi level. Hence $\lambda_F \ll L$; one can see that the quantum device can operate with a significantly smaller controlling gate voltage.

Another design of a structure for a quantum-interference transistor is sketched in Fig. 8.19. The structure is T-shaped and consists of a channel connecting the source (grounded), drain, and a transverse arm (stub). The transverse arm has finite dimensions and has a gate on the end of the arm; this end is referred to commonly as the "top" of the device. If the length, L_2, and width, L_3, are less than the coherence length, l_ϕ, the reflection of the electron wave from the arm produces de Broglie wave-interference patterns. If the width of the main channel, L_1, also is small in comparison with l_ϕ, the pattern extends across the channel and determines the transmission coefficient of electrons through the device, i.e., the source–drain conductance. A voltage applied to the gate changes the penetration length, L^*, of the electron wave into the arm and, as a consequence, the interference pattern. Estimates show that the gate voltage can effectively control the conductance of such a T-shaped device when the dimensions L_1, L_2, and L_3 are approximately several hundred Å or less.

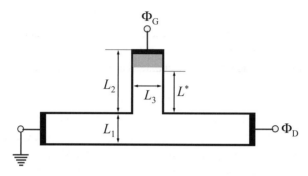

Figure 8.19 The schematics of a T-shaped QUIT. The source (grounded), drain, and gate electrodes are indicated.

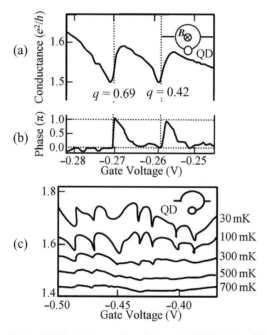

Figure 8.20 Experimental results on the QUIT effect: (a) and (b) for the two-channel QUIT $T = 0.03$ K, (c) for the "T-shaped" structure, both fabricated from a high-mobility AlGaAs/GaAs heterostructure. (a) Modulation of the device conductance by the gate voltage. (b) The phase shift as a function of the gate voltage. (c) The same as in (a) for the "T-shaped" device at five different temperatures. Reprinted with permission from K. Kobayashi, H. Aikawa *et al.* "Fano resonance in a quantum wire with a side-coupled quantum dot," *Phys. Rev.* B, **70**, 035319 (2004). © 2004 by the American Physical Society.

As a practical realization of the QUIT effect, we present the results obtained for an AlGaAs/GaAs structure. The channels were made from high-mobility GaAs. In one of the channels a GaAs quantum dot was embedded under the gate. This device corresponds to the scheme in Fig. 8.18(a). If one of the channels were blocked, one would obtain the T-shaped structure, as in Fig. 8.19. Both cases also are illustrated schematically in the insets of Fig. 8.20. The low-temperature mobility of the two-dimensional electron gas was as high as 9×10^5 cm^2 V^{-1}s^{-1}, and the measured electron concentration was 2×10^{11} cm^{-2}. These parameters correspond to a mean free path of about 8 μm. Low temperatures (0.03–0.7 K) ensure that the dephasing length l_ϕ is even greater. In the case under consideration, the lengths of the channels are about 1 μm, and the quantum-dot (two-dimensional) size is 0.2 μm × 0.2 μm. In Fig. 8.20(a), the QUIT conductance is presented as a function of the gate voltage (in units of the quantum conductance, e^2/h). One can see that the conductance is periodically modulated by the gate voltage, in accordance with the previously developed model of the QUIT. Special experiments with applied magnetic field, B, make it possible to deduce the phase shift of the electrons in the QUIT. The results are presented in Fig. 8.20(b). The modulation of the conductance for a T-shaped device is shown in Fig. 8.20(c). Measurements at various temperatures illustrate the smearing out of the interference patterns due to the temperature dispersion of the energy of the electrons entering the device.

Summarizing, a pronounced QUIT effect is observed for devices with submicrometer scale at low temperatures. The scaling down of device dimensions will allow significantly increased working temperatures.

In conclusion, the quantum-interference effect provides a new principle of operation for three-terminal devices. These devices are now in an early stage of development. However, these approaches portend effective control by a small voltage as well as the development of very-high-speed transistors.

8.4   Single-electron-transfer devices

Before analyzing examples of the devices based on single-electron transport studied in Section 6.5, we shall overview briefly a relatively simple technique for the formation of structures with low-dimensional electron gas.

The split-gate technique

Having a two-dimensional electron gas on an interface, or in a quantum well, one can further lower the electron-gas dimensionality by various methods. One such method is the so-called *split-gate technique*. The principles of this technique can be explained as follows. Typically, two-dimensional electrons are separated from the surface of the sample by a wide-bandgap dielectric layer. It can be a SiO$_2$ layer on Si, an AlGaAs barrier layer on GaAs, etc. Figure 8.21(a) illustrates such a structure. A modulation-doped barrier layer decreases electron scattering by the donors and results in high electron mobility. A thin GaAs layer grown on the top of this structure is used as additional electrical isolation from the metal gates. Let a metal strip – a gate – be deposited onto the top of this structure. The distribution of the potential energy for the case of a negative applied gate voltage is presented in Fig. 8.21(b). According to this energy scheme, two-dimensional electrons are repelled from the region beneath the metal strip; their Fermi energy, E_F, is below the lowest subband energy. As a result, the region under the gate becomes completely depleted, as depicted in Fig. 8.21. Now it is clear that, by using several gates, possibly

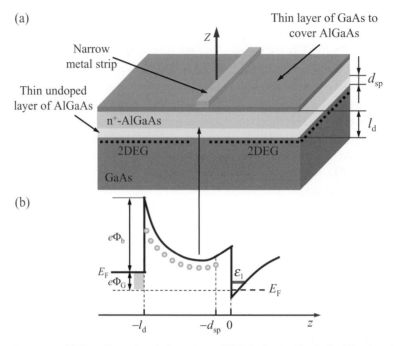

Figure 8.21 (a) Two-dimensional electron gas (2DEG, denoted by dashed lines) at the interface of a modulation-doped GaAs/AlGaAs heterostructure with a depletion region under the negatively biased narrow metal strip (gate). (b) The band diagram of the AlGaAs/GaAs heterostructure in the gate region. The lowest subband is shown by the ε_1 line. Φ_b is the built-in Schottky voltage and Φ_G is the gate voltage.

of different forms, one can create various configurations of regions occupied by the electrons. It is possible to form wires, dots, rings, cavities, etc. for the two-dimensional electrons. For example, if two closely placed parallel metal strips are fabricated on the top of the heterostructure, then, by applying negative voltage to these two gates, we can form two side barriers for the electrons and confine them into a channel. If the channel is narrow enough, the two-dimensional electrons can be quantized in the second direction and we obtain a quantum wire; such a wire is shown in Fig. 8.22. The confinement of the electrons to the dots, wires, rings, etc. can be accomplished by a heterojunction on one side and electrostatics on all other sides. The split-gate technique has successfully been exploited for measurements of transport regimes in various quantum structures, for example, in quantum point contacts and electron waveguides.

A considerably more sophisticated design of patterning of a two-dimensional gas to a shape desirable for single-electron applications is presented in Fig. 8.23. The main features of the design shown in Fig. 8.23(a) are the following: (i) the Ohmic contacts (OC) to the two-dimensional electron gas (contacts to two electron reservoirs (R)) and (ii) a system of gates, which create electrostatic tunnel barriers (TB) and confine electrons into a quantum dot (QD). The tunnel barriers are formed when the voltages applied to the gates are negative with respect to the voltages applied to the contacts. The barriers should be high enough to decouple the quantum dot and the reservoirs. In Fig. 8.23(b), the

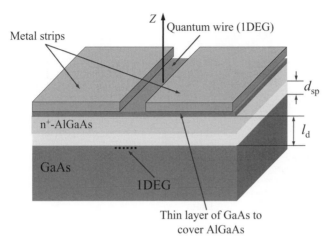

Figure 8.22 A quantum wire at the interface of a GaAs/AlGaAs heterostructure with two depletion regions under the negatively biased metal strips.

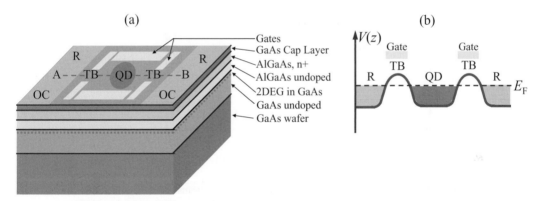

Figure 8.23 (a) Design patterning of two-dimensional gas to a shape desirable for single-electron applications: OC, ohmic contacts; R, reservoir of electrons; QD, quantum dot; and TB, tunnel barrier. (b) The potential profile, $V(z)$, along the line A–B. Ohmic contacts, OC, are not shown since they are outside of the depicted region.

resulting potential profile is depicted schematically. Split-gate techniques and resulting structures are used to observe single-electron effects. Indeed, if the quantum-dot-like structure is small enough, it can be considered to be a single-electron box. A voltage applied to the contacts (the source and the drain) induces transfer of electrons through the structure one by one, as studied in Chapter 6. In general, the flexible split-gate technique is a powerful method to realize single-electron-transport devices.

Single-electron transistors

Single-electron devices like the one sketched in Fig. 8.23 are two-terminal devices, i.e., they are diode-type devices. It is possible to introduce an additional gate and create

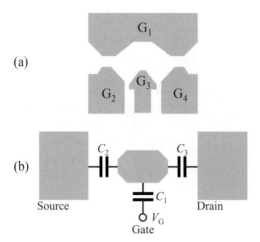

(a)

(b)

Source　　　　　　　Drain

V_G
Gate

Figure 8.24 (a) The design of a quantum-dot single-electron transistor; (b) an electric circuit with a quantum-dot single-electron transistor.

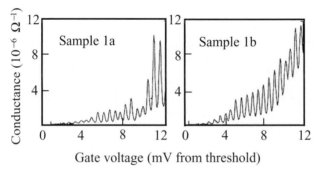

Gate voltage (mV from threshold)

Figure 8.25 Modulation of conductance in single-electron transistors. Reprinted with permission, from M. A. Kastner, "The single electron transistor," *Rev. Mod. Phys.*, **64**, 849 (1992).

a three-terminal device, a single-electron transistor. In Fig. 8.24(a), the design of the metallic electrodes on the top of a heterostructure with a two-dimensional electron gas is shown. The gates G_1, G_2, and G_4 form a quantum dot as above; the gate G_3 addition-ally controls the size and shape of the dot, changing its properties. The single-electron transistor works as follows.

The electron transfer is determined by two factors: the Coulomb charging of the dot and the quantized energy levels in the dot. If the drain is biased with respect to the source, an electric current occurs in the regime of single-electron transfer. By applying the voltage to the gate and changing the quantum-dot parameters, one can change the conditions of electron tunneling and affect the source–drain current. Examples of modulation of the conductance in single-electron transistors by the gate voltage are presented in Fig. 8.25. The devices have almost the same geometry. Their dimensions are large enough to have a number of quantized levels. In Fig. 8.25 each peak in the conductance corresponds to transfer of one electron, when an energy level enters into resonance with the electron

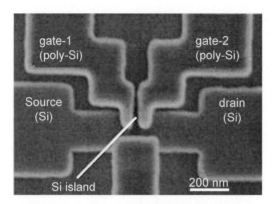

Figure 8.26 A scanning electron microscope image of a single-electron transistor. From Y. Ono, A. Fujiwara *et al.*, "Manipulation and detection of single electrons for future information processing," *J. Appl. Phys.*, **97**, 031101 (2005). Reused with permission from Yukinori Ono, Akira Fujiwara, Katsuhiko Nishiguchi, Hiroshi Inokawa, and Yasuo Takahashi, *Journal of Applied Physics*, **97**, 031101 (2005). © 2005 American Institute of Physics.

states in the contacts. Though the conductance versus gate-voltage dependences are different, i.e., not reproducible, the peak spacing is the same for both devices. It is determined by the change in the gate voltage required to change the charging energy of the quantum dots by one electron. The figures show clearly that the electric current is modulated significantly by the gate voltage. Thus, for transistors with single-electron transport, strong control of very small electric current may be possible. The problem of fabrication of *reproducible* devices requires further improvements in technology.

A single-electron pump and turnstile

In the single-electron transistor considered previously, the barriers which separate the electrodes and the dot practically do not change when a gate voltage is applied (instead, quantum-dot parameters are modified). However, variation of the barriers is possible and it adds a new function to the single-electron devices. As will be shown below, a device with tunable barrier can work both as a *pump* and as a *turnstile*. Consider such a case for the example of silicon-based devices. A scanning electron microscope image of such a device is shown in Fig. 8.26; this single-electron transistor is made in a very thin Si layer. The conducting channel, source, and drain of the device are clearly seen in the figure. The fine gates are made of polycrystalline Si, which is a good conducting material. Three gates form a dot. Two of them, gates 1 and 2, tune the barriers between the dot and the electrodes. Figure 8.27 illustrates the sequence for the pump operation. We start with a Coulomb-blockade state with an electron in the dot (state I). Then, by applying a negative bias to gate 1 and making the left barrier higher, we may close the left channel (state II). During this action, we keep the dot potential nearly constant by applying a positive control bias to gate 2. Then, we raise the island potential so that an electron in the dot is ejected to the right channel, ending up with a new Coulomb-blockade state with $n - 1$ electrons (state III). Next, we open the left channel and close the right channel, keeping

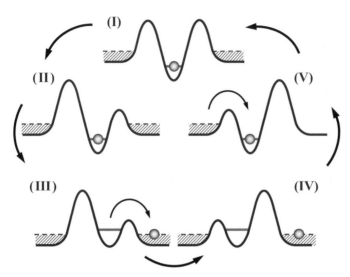

Figure 8.27 The sequence for pump operation.

the island potential nearly constant (state IV), and then lower the island potential so that an electron can enter from the left channel (state V). Thus, during one period only one electron can be controllably transferred through the device. This system can be viewed as a *single-electron pump*, since it transfers the electrons one by one without applying a bias between the source and drain.

The electron pump can have a number of applications. One of them is a current etalon. Indeed, the average current flowing through the device is

$$I = -ef,$$

where f is the frequency of variation of the voltage on the gates. Since the frequency can be measured with very high accuracy, the device can serve as the *electric-current standard*. The accuracy of such a standard is much higher than the analogous characteristic of any other current standards.

The *turnstile operation* is realized by making a small change in the previously described regime of operation. The detailed procedure for the turnstile is illustrated schematically in Fig. 8.28. In the turnstile mode, with a non-zero source–drain voltage, we first close gate 2 so that an electron enters only from the left side (stage I). We next close both gates (stage II), and then open gate 2 so that an electron is emitted to the right (stage III). Finally, we again close both gates (stage IV), and open gate 1, reaching the initial state. This procedure can be accomplished by applying an AC bias to each gate with the phase shift of π. Noteworthy is that, in this turnstile procedure, at least one of the channels is always closed, which is not the case in the pump operation mode. This is advantageous because high resistance effectively prevents the tunneling of two electrons – one of the major error sources for single-electron-transfer regimes. Figure 8.29 shows the drain-current versus drain-voltage characteristics in the aforementioned turnstile mode at several frequencies, $f = 0.001, 0.5$, and 1.0 MHz. Staircases quantized in units of ef are observed both for

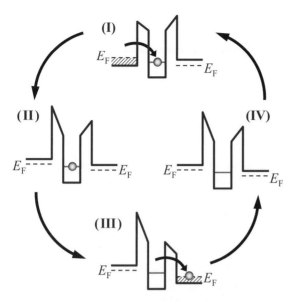

Figure 8.28 Schematics of a turnstile operation. From Y. Ono, A. Fujiwara *et al.*, "Manipulation and detection of single electrons for future information processing," *J. Appl. Phys.*, **97**, 031101 (2005). Reused with permission from Yukinori Ono, Akira Fujiwara, Katsuhiko Nishiguchi, Hiroshi Inokawa, and Yasuo Takahashi, *Journal of Applied Physics*, **97**, 031101 (2005). © 2005 American Institute of Physics.

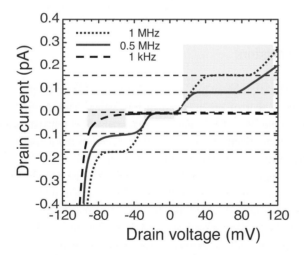

Figure 8.29 Drain current versus drain voltage characteristics of a turnstile device. 1 pA = 10^{-12} A, where pA denotes a picoampere. From Y. Ono, A. Fujiwara *et al.*, "Manipulation and detection of single electrons for future information processing," *J. Appl. Phys.*, **97**, 031101 (2005). Reused with permission from Yukinori Ono, Akira Fujiwara, Katsuhiko Nishiguchi, Hiroshi Inokawa, and Yasuo Takahashi, *Journal of Applied Physics*, **97**, 031101 (2005). © 2005 American Institute of Physics.

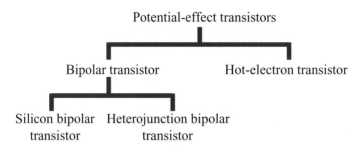

Figure 8.30 The family tree of potential-effect transistors. The devices are set apart into two large groups: bipolar and hot-electron transistors.

positive and for negative drain voltages. This is the evidence of single-electron transfer by the turnstile operation. The levels of the current plateaus for the positive drain voltages are exactly equal to ef to within the accuracy of the measurement system of about 10^{-2}. In these experiments, the working currents were extremely small.

In conclusion, the main principle of the FETs – current control by means of a voltage applied through capacitively coupled electrode(s) – can be successfully applied to single-electron devices. By combining the field effect and the single-electron phenomena, one can manipulate the device states using extremely small electric currents.

To date, single-electron experiments have been conducted at low temperatures (typically below 20 K). However, with further scaling down of semiconductor technology to nanosizes, we may expect that single-electron effects and devices will be exploited at higher temperatures, including under room-temperature conditions.

8.5   Potential-effect transistors

According to the classification given in Section 8.3, a second large family of electronic three-terminal devices is that of potential-effect transistors. This family is shown in Fig. 8.30. In contrast to the case of FET devices, potential-effect transistors are current-controlled. The controlling electrode is *resistively coupled* to the active region of the device and the carriers are separated by an energy barrier.

The most important representative of this class is the bipolar transistor, which was invented in 1947, and has undergone considerable and persistent transformation. Currently, the bipolar transistor provides high speed of operation in most circuit applications.

Detailed analysis of the physics of bipolar transistors is certainly beyond the scope of this book. Instead, at the end of this chapter we provide references to books on bipolar transistors. However, we shall consider the working principles of this transistor, at least very briefly.

p–n Junctions

The key element of the transistor is the so-called *p–n junction*. In the simplest case it is a junction of two regions in a semiconductor: one side is doped by donors and contains

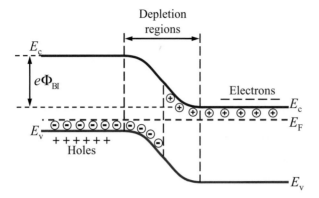

Figure 8.31 A p–n junction with abrupt coordinate dependence of doping concentration.

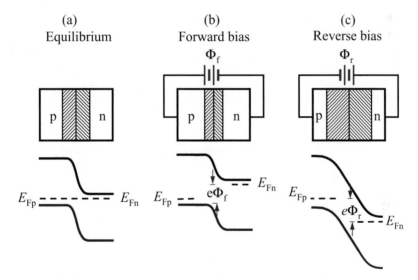

Figure 8.32 The scheme of a p–n junction (a) at equilibrium; (b) under forward bias Φ_f; and (c) under reverse bias Φ_r. For non-equilibrium cases, (b) and (c), instead of one Fermi level, there are two quasi-Fermi levels, one for holes, E_{Fp}, and one for electrons, E_{Fn}, as shown.

electrons; the other side is doped by acceptors and contains the holes. In Fig. 8.31, the scheme of such a p–n junction is presented. An abrupt transition between the p- and n-doped sides of a semiconductor is assumed. The charge transfer across the junction lines up the various Fermi levels and results in a Fermi level, E_F. It depletes two narrow regions in the p- and n-doped sides and creates a built-in electrostatic potential, Φ_{BI}. The potential barriers of height $e\Phi_{BI}$ prevent the penetration of electrons into the p-type side of the structure and of holes into the n-type side of the structure. No electric current passes between p and n regions under such an equilibrium condition; see Fig. 8.32(a). As is clear from this discussion, the potential profile which appears at the p–n junction and the possibility of controlling of it are characteristic for potential-effect devices.

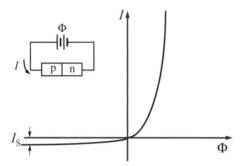

Figure 8.33 The current–voltage characteristic of the p–n junction.

Now, let a voltage be applied to the p–n junction. The corresponding electric circuits are shown in the upper part of Fig. 8.32. There are two quite different cases. The *forward bias* of the p–n junction corresponds to a negative potential, $\Phi = -\Phi_f (\Phi_f > 0)$, applied to the n-type side. Potential barriers both for the electrons and for the holes decrease and are equal to $e(\Phi_{BI} - \Phi_f)$. This case is illustrated by Fig. 8.32(b). A positive potential (the *reverse bias* $\Phi_r > 0$ of the junction) leads to an increase of the potential barrier between the n and p parts of the structure, as illustrated by Fig. 8.32(c). Now the barrier height equals $e(\Phi_{BI} + \Phi_r)$. The applied voltage is dropped primarily across the depletion regions because their resistances are much greater than those of neutral n and p regions of the junction. The voltage breaks down the equilibrium; in particular, there is no longer a common Fermi level. Instead of the true Fermi level, *quasi-Fermi* levels E_{Fn} and E_{Fp} can be introduced for better understanding of changes in the potential relief, as shown in Fig. 8.32. As soon as $E_{Fn} \neq E_{Fp}$, an electric current starts to flow between the n and p parts of the structure. For a forward applied bias, the current increases exponentially as the potential barrier decreases, and the junction becomes strongly conducting. For an applied reverse bias, the current is small and saturates at large biases. The current–voltage characteristic of the p–n junction is presented in Fig. 8.33. The main features of this characteristic are a rectifying behavior and a strong nonlinearity. This nonlinearity is used in numerous applications of the p–n diode.

Bipolar transistors

A homostructure bipolar transistor consists of two p–n junctions. Usually one of them – the emitter junction – is forward-biased, while the other – the collector junction – is reverse-biased. The energy-band diagram for a bipolar transistor is shown schematically in Fig. 8.34. This diagram clearly illustrates the formation of a potential for electron and hole motion. The control of this potential is the key element of the bipolar transistor. The device depicted is of the n–p–n type, though almost all results are applicable also to p–n–p devices. This diagram can be easily understood in terms of the charge-transfer effect between the emitter (left n region), the base (middle p region), and the collector (right n region). The voltage bias is supposed to be such that an electron travels from the left n region across the p region to the right n region. An applied forward

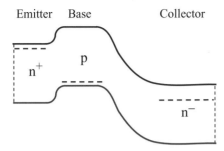

Figure 8.34 An energy band diagram of a homostructure n–p–n bipolar transistor along the direction of electron travel. Quasineutral emitter, base, and collector regions are indicated. The emitter junction is forward-biased, and the collector junction is slightly reverse-biased.

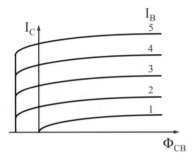

Figure 8.35 The collector current I_C in a bipolar transistor for five different base currents, I_B.

bias to the emitter junction lowers the energy barrier for an electron flowing from the emitter to the base. Simultaneously, a similar flow of holes from the base to the emitter appears. Those electrons that overcome the barrier travel across the base. Recombination of the minority carriers (electrons) occurs in the base, but, if the thickness of the base region is less than the diffusion length of the minority carriers, these electron losses are negligible. When the electrons reach the collector junction, they are swept away into the collector due to the high electric field within the depletion region of the latter junction and form the collector current. In the bipolar transistor, each of the conducting regions (the emitter, the base, and the collector) is provided by an electrical contact. Thus, this is a three-terminal device. The operation of a bipolar transistor is based on the principle of controlling the current by injecting minority carriers. For example, in an n–p–n device, the injection of minority carriers from a forward-biased n–p junction (emitter–base) into the base provides the current through the base contact and the controlling function of the collector current. In Fig. 8.35, the collector current versus the collector–base voltage, Φ_{CB}, is shown for various currents, I_B, through the base contact (compare this with the current–voltage characteristics of the FETs presented in Fig. 8.10).

The carriers travel from the emitter to the collector perpendicular to junctions; thus, the carrier transit time through the base determines the speed of operation and the cut-off frequency of the device. Device scaling in order to reduce the transit time in a

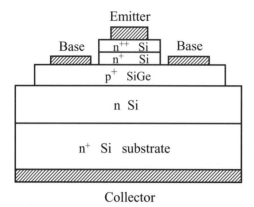

Figure 8.36 The schematic cross-section of an n–p–n Si/SiGe heterojunction bipolar transistor. n^{++}, n^{+}, p^{+}, and n denote high, intermediate (n^{+} and p^{+}), and relatively low doping, respectively. The p^{+} -SiGe layer is strained.

homojunction bipolar transistor can not be achieved while simultaneously realizing all the requirements for high performance; such requirements include high doping levels both in the emitter and in the base, and a small base region. The use of heterostructures helps to solve these problems and to improve device performance. Thus, one can divide this transistor class into two large and important groups: homojunction and heterojunction bipolar transistors, as shown in Fig. 8.30. As an example of the latter subclass of bipolar transistors, consider the Si/SiGe device. In Fig. 8.36, the cross-section of a Si/SiGe heterostructure n–p–n bipolar transistor is shown. The device is fabricated on an n^{+}-Si substrate (subcollector with doping of about 2×10^{19} cm^{-3}) contacted by an electrode. The collector layer is doped to approximately 2×10^{16} cm^{-3} and has a width of about 0.3 μm. The p^{+} base is fabricated of a $Si_{1-x}Ge_{x}$ alloy with x ranging from 0.2 to 0.3, and has a width of typically 30–50 nm. The base doping is in the range from 2×10^{18} to 5×10^{19} cm^{-3}. Two-sided metallic electrodes provide the direct contact with the base region. The emitter is made of n^{+}-Si with doping of about 5×10^{17} cm^{-3}. A heavily doped layer, marked as n^{++}-Si, is placed at the top of the emitter to provide a good contact to the metallic electrode. Owing to the very short base and other peculiarities of the heterojunctions, for such a device an extremely high cut-off frequency of 75 GHz at room temperature is measured, whereas at liquid-nitrogen temperature, a cut-off frequency as high as 94 GHz is reported for this type of bipolar transistor. The very recent record results for this type of bipolar transistor were reported by IBM: the silicon–germanium transistor hits 500 GHz. More detailed analysis of heterostructure bipolar transistors can be found in references presented at the end of this chapter.

Hot-electron transistors

Another group of potential-effect devices is represented by *hot-electron transistors*, as illustrated in Fig. 8.30. These devices employ an emitter–barrier–base–barrier–collector structure. The hot electrons are injected over, or through, an emitter barrier into a

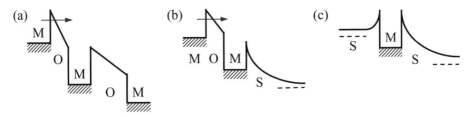

Figure 8.37 Energy-band diagrams for metal-base ballistic-injection transistors. Metal, oxide, and semiconductor layers are marked by M, O, and S, respectively. Semiconductor layers are doped. The diagrams are presented under operational bias conditions.

narrow base region. Injected carriers have a large velocity and transit through the base almost ballistically; that is, they transit the base without significant scattering. The current through the device is controlled by changing the height of a collector barrier. This principle is used in a number of different configurations and combinations of materials.

Consider a ballistic-injection device, which is primarily a *unipolar device*; that is, only one type of carrier, say electrons, is used. A ballistic-injection device also consists of an emitter, base, and collector. The role of the emitter is to inject electrons with high velocities into the base; the second electrode should collect these electrons. The input base voltage controls the electron injection and, therefore, the output emitter–collector current. If small changes in the input produce larger changes in the output, the device exhibits a current gain. Generally, in ballistic devices, the electrons are injected into the base with a high energy exceeding 0.1 eV. This should lead to a decrease in the time of flight through the base region. Another advantage of ballistic devices is related to their unipolar character; this means that it is possible to choose the fastest type of majority carriers (electrons) and avoid the participation of slower minority holes.

In order to realize a transistor faster than a bipolar or field-effect transistor, several schemes of ballistic-injection devices have been proposed. They differ in terms of the physical mechanisms of electron injection as well as in the materials used in the devices. The first and, perhaps, the simplest device is a metal–oxide–metal–oxide–metal heterostructure. In Fig. 8.37(a), this device is shown under a bias. Other similar device structures are the metal–oxide–metal–semiconductor structure and the semiconductor–metal–semiconductor (SMS) structure presented in Figs. 8.37(b) and 8.37(c), respectively. The principle of operation is the same for these three structures. Consider, for example, the case of the SMS structure. One can see a close analogy with the bipolar transistor: a forward-biased semiconductor–metal junction serves as the emitter, a second metal–semiconductor junction serves as the collector, and a metal layer is the base. Both junctions are, in fact, Schottky diodes, one is forward-biased and the other is reverse-biased. Under such bias conditions, electrons are injected over the Schottky barrier with energies substantially exceeding the thermal energy in the base. If the base is narrow, the electrons "fly" across the base region without losing their energy. Their subsequent destiny is defined by the base–collector bias: a lowering of the collector barrier increases the fraction of electrons coming into the collector electrode and, consequently, the collector current.

Various materials have been used for the structures shown in Fig. 8.37. In particular, the Si/CoSi$_2$/Si structure was investigated for the SMS scheme. An α factor as high as 0.6 was reported for this case. A common disadvantage of metal-base transistors is the high electron reflection from the metal. This reflection is, mainly, a quantum-mechanical effect in nature and, therefore, can not be avoided. It occurs even for ideal semiconductor–metal interfaces.

Semiconductor heterostructures may be used in other ways to realize ballistic-injection devices. Let us consider n-type devices. In semiconductor-based ballistic-injection devices, the emitter, base, and collector are doped regions separated by two barriers. The barriers can be formed by growing layers of materials with a positive conduction-band offset as shown in Fig. 8.38(a). This structure actually uses four heterojunctions. Another kind of barrier can be produced by planar acceptor doping in a homostructure as illustrated in Fig. 8.38(b).

Consider the typical parameters of structures for ballistic-injection devices. If the height of the emitter barrier is V_b, the velocity of electrons injected into the base can be estimated as $v_B \approx \sqrt{2V_b/m^*} = 5.9 \times 10^7 \sqrt{\tilde{V_b}/(m^*/m_0)}\,\mathrm{cm\,s^{-1}}$, where $\tilde{V_b}$ is the barrier energy in eV. For GaAs we can assume $\tilde{V_b} \approx 0.3$ eV and $m^* = 0.067m_0$. For the velocity we get $v_B = 1.3 \times 10^8\,\mathrm{cm\,s^{-1}}$. This value is appreciably larger than the characteristic electron velocity in devices such as FETs and bipolar transistors. Another important feature is that the injected electrons exhibit velocity spreading in a very narrow velocity cone. Actually, the average value of the lateral component of the electron momentum in the emitter is $p_{\parallel,E} \sim \sqrt{2m^* k_B T}$, where T is the device temperature. Owing to the lateral translation symmetry this component does not change under electron injection through the barrier. Thus, the characteristic angle for the velocity spreading can be evaluated as follows:

$$\theta \sim \frac{v_{\parallel,B}}{v_B} = \frac{p_{\parallel,B}}{m^* v_B} = \frac{p_{\parallel,E}}{m^* v_B} = \sqrt{\frac{k_B T}{V_b}}.$$

For example, in the case of liquid-helium temperatures and $V_b \approx 0.3$ eV, this equation gives $\theta \approx 6°$.

Collisions in the base reduce the number of ballistic electrons. If the electron mean free path in the base is l_e and the base width is L_B, the fraction of ballistic electrons collected by the collector is estimated to be $\alpha \approx \exp(-L_B/l_e)$. Thus, the base region should be quite narrow. In this case, the base region has to be heavily doped to reduce the resistance for the base current.

A limitation of the base-doping technique comes about as a result of the fact that a high doping gives rise to additional electron scattering and quenches the ballistic regime of electron motion. Specifically, for III–V compounds, if impurity concentrations exceed 10^{18} cm^{-3} electron scattering becomes very strong.

As a result, for AlGaAs/GaAs devices, base regions with doping of about 10^{18} cm^{-3} and with widths of 30–80 nm are used. Optimization of the structure parameters facilitates

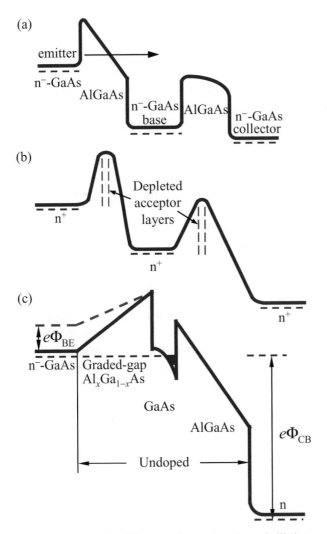

Figure 8.38 Energy-band diagrams for semiconductor ballistic transistors. (a) The AlGaAs/GaAs structure with four heterojunctions and a doped base. Electrons are injected into the base due to tunneling. This kind of device is referred to as a tunneling hot-electron-transfer amplifier. (b) A homostructure electron device. Barriers are formed by planar p-doping. The base is n-doped. (c) A device with an undoped base. After Fig. 12, S. Luryi, "Hot-electron transistors," in S. M. Sze (Ed.), *High-Speed Semiconductor Devices* (New York, Wiley, 1990). Reprinted with permission of John Wiley & Sons, Inc.

the realization of a ballistic device with a transfer ratio of $\alpha = 0.9$ at liquid-helium temperature.

Figure 8.38(c) presents another design for a ballistic device. The emitter barrier is made with a graded composition and the base region is *undoped*. The base is induced by the electric field of the collector, which leads to the formation of a two-dimensional electron gas at the undoped interface. This type of ballistic device is referred to as the *induced-base transistor*. The advantages of the induced-base transistor are the following.

The two-dimensional electron gas in the base is characterized by a high mobility and large two-dimensional electron concentrations of up to 2×10^{12} cm^{-2}. Both effects cause a low base resistance; thus, the base can be chosen to be very narrow, typically 100 Å. This results in a high fraction of ballistic electrons coming into the collector. For example, a well-designed AlGaAs/GaAs induced-base transistor results in $\alpha \approx 0.96$ even at room temperature. Similar induced-base transistors have been fabricated using InGaAs/AlGaAs and p-doped Ge/SiGe heterostructures.

In conclusion, hot-electron transistors with nanoscale base regions are characterized by a high operation speed. An electron transit time of the order of 0.1–0.5 ps corresponds to a cut-off frequency in the range 2–10 THz.

8.6   Light-emitting diodes and lasers

So far we have studied *electronic* nanoscale devices, i.e., a class of devices that exploits electrical properties of nanostructures and operates with electric input and output signals. Another class is composed of *optoelectronic devices*, which are based on both electrical and optical properties of materials and work with optical and electric signals.

In this section, we will analyze two very important classes of optoelectronic devices: light-emitting diodes and laser diodes. As their titles imply, the devices were invented to produce light with certain properties. In particular, the energy of the electric current flowing through these diodes is transformed into light energy. These optoelectronic devices have a huge number of applications and deserve consideration in detail.

In Chapters 2 and 3, among various different wave fields, we introduced and studied electromagnetic fields, of which light waves are a particular example. In addition, we studied electrons in atoms (Chapters 2 and 3) and solids (Chapter 4). However, it was supposed that these two physical entities do not interact with each other. In fact, light–matter interaction gives rise to fundamental physical phenomena. Historically, light–matter interactions provided some of the first evidence for the quantum nature of matter. Remember that an electromagnetic field consists of an infinite number of *modes* (waves), each of which is characterized by a wavevector and a specific polarization. According to quantum physics (see Chapter 3), each mode may be described in terms of a harmonic oscillator of frequency ω. Correspondingly, the energy separation between levels of this quantum-mechanical oscillator is $\hbar\omega$; see Eq. (2.41). This oscillator may be in the non-excited state, which manifests the ground-state or *zero-point vibrations* of the electromagnetic field. The oscillator may be excited to some higher energy level. If the Nth level of the oscillator is excited, there are N quanta (photons) in the mode under consideration. The classical description of an electromagnetic wave is valid at large numbers of photons: $N \gg 1$.

Besides quantization of the energy, the quantum nature of electromagnetic fields is revealed in the equilibrium statistics of photons. Indeed, photons obey the so-called

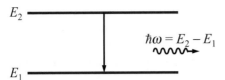

Figure 8.39 A two-level system interacting with light.

boson statistics, or Bose–Einstein statistics, which gives the average number of photons of some chosen mode under equilibrium in the form

$$N = \frac{1}{\exp[\hbar\omega/(k_B T)] - 1},\tag{8.26}$$

with T being the ambient temperature. Equation (8.26) is also known as *Planck's formula*. Clearly, the Bose–Einstein statistics of photons differs drastically from Fermi statistics, which we used for electrons (see Chapter 6). In particular, it permits the accumulation of an arbitrary number of photons in any mode. For example, from Eq. (8.26) it follows that at $k_B T/(\hbar\omega) \gg 1$ the phonon number $N \approx k_B T/(\hbar\omega) \gg 1$, whereas the number of electrons occupying any state can not exceed 1.

After these introductory remarks on quantum properties of electromagnetic fields, we review briefly the three major processes involving a quantum material system and light: absorption, spontaneous emission, and stimulated emission.

Photon absorption and emission

In order to visualize these processes, we consider a simple quantized system with two energy levels E_1 and E_2 as depicted in Fig. 8.39. The occupancies of the energy levels of this system correspond to particular states of a system of the electrons. The charged electrons interact with the electromagnetic field. This interaction results in transitions between quantum states of the system. These transitions are frequently referred to as *phototransitions*. According to the quantum theory, the system can change its energy as a result of interaction with electromagnetic waves exclusively of the frequency

$$\omega = (E_2 - E_1)/\hbar.\tag{8.27}$$

If the lowest energy level E_1 is occupied, the wave can excite the system into the upper level E_2 and the electromagnetic energy must decrease. One can describe this process as the *absorption of one photon* because the energy of the electromagnetic field decreases by $E_2 - E_1$. If the system occupies the upper level E_2, it can make a transition to level E_1 as a result of interaction with the electromagnetic field. Then, the electromagnetic energy increases by $E_2 - E_1$. This process represents the *emission of a photon* with energy $\hbar\omega$. When activated by an external electromagnetic wave, the latter process is called *stimulated emission*. It is important that, for stimulated emission, each emitted photon has the energy, direction, polarization, and even phase coinciding *precisely* with those of the stimulating wave.

Both processes, absorption and stimulated emission, can be described in terms of an interaction with a classical electromagnetic wave. The rates of these processes are proportional to the intensity of the wave, or, according to Table 2.1, proportional to the number of photons $N_{\mathbf{q},\mathbf{b}}$ of a given wavevector $\mathbf{q}$ and polarization $\mathbf{b}$. These rates can be written as

$$R_{\text{abs}} = B_{12} N_{\mathbf{q},\mathbf{b}} n_1, \tag{8.28}$$

$$R_{\text{st.em}} = B_{21} N_{\mathbf{q},\mathbf{b}} n_2, \tag{8.29}$$

where n_1 and n_2 are the numbers of particles in the system occupying the levels 1 and 2, respectively; B_{12} and B_{21} are kinetic coefficients describing these processes. The physical meanings of these coefficients will be addressed in the subsequent discussion.

The two processes of absorption and emission are insufficient to describe the whole picture of the interaction between radiation and matter. For example, let us apply these two processes only for thermal equilibrium conditions, for which the ratio of the populations of the two levels is

$$\frac{n_2}{n_1} = e^{-(E_2 - E_1)/(k_{\text{B}} T)} = e^{-\hbar\omega/(k_{\text{B}} T)} \tag{8.30}$$

(see Section 6.3 and particularly Eq. (6.14)). Using Eqs. (8.28)–(8.30) one can see that $R_{\text{abs}} \neq R_{\text{st.em}}$ at any temperature T. This result is in contradiction with the expected equilibrium between the system and the field. According to the Einstein theory, there is an additional quantum radiative transition in the system with the *spontaneous emission of a photon* of the same mode. The rate of this process is

$$R_{\text{sp.em}} = A_{21} n_2, \tag{8.31}$$

where A_{21} is the coefficient or rate of spontaneous emission. The spontaneous process does not depend on the intensity of the electromagnetic wave and takes place even in the absence of this wave. According to quantum electrodynamics, the excited material system spontaneously emits a photon as a result of the interaction between the electrons and the zero-point vibrations of electromagnetic fields. The zero-point vibrations of electromagnetic fields were introduced briefly in Section 3.3.

In contrast to the case of stimulated emission, a photon produced by the spontaneous process has an arbitrary phase. Moreover, this process produces photons with different directions of $\mathbf{q}$ and polarizations, but the energy is fixed, i.e., it produces photons of different modes of the same frequency.

Now we can apply the results of Eqs. (8.28)–(8.31) to thermal equilibrium. Under equilibrium conditions, the total rate of photon emission has to be equal to the rate of photon absorption; thus we will have

$$R_{\text{abs}} = R_{\text{sp.em}} + R_{\text{st.em}}. \tag{8.32}$$

Using the Planck formula of Eq. (8.26) and the ratio of n_2/n_1 of Eq. (8.30), and substituting the expressions for R_{abs}, $R_{\text{sp.em}}$, and $R_{\text{st.em}}$ into Eq. (8.32), one can find the relation

$$B_{21} - A_{21} = (B_{12} - A_{21}) e^{\hbar\omega/(k_{\text{B}} T)}. \tag{8.33}$$

Because this relation has to be satisfied at any arbitrary temperature T, one can obtain two equalities:

$$A_{21} = B_{21} = B_{12}. \tag{8.34}$$

Thus, we have established the existence of three basic processes for the resonant interaction of radiation and matter: absorption, stimulated emission, and spontaneous emission. Moreover, we have found the relations between the coefficients determining the rates of these processes. It is worth emphasizing that all three processes are related to interactions with photons of the same mode.

The sum of the stimulated and spontaneous emission rates, as determined by Eqs. (8.29) and (8.31), gives the total emission rate of photons for a fixed mode:

$$R_{em} = A_{21}(1 + N_{q,b})n_2. \tag{8.35}$$

From this equation, one can see that stimulated emission dominates over spontaneous emission for a fixed mode, $\{q, b\}$, if the number of photons, $N_{q,b}$, is sufficiently larger than 1. However, there is spontaneous emission of a great number of other modes with the same frequency but with different directions of q and different polarizations b. This total spontaneous emission can be the dominant radiative process even if stimulated emission is the most important process for a particular mode.

Now, let us compare absorption and stimulated emission by calculating the rate of increase of the number of photons in some fixed modes:

$$\left(\frac{dN_{q,b}}{dt}\right)_{st} \equiv R \equiv R_{st.em} - R_{abs} = B_{21}N_{q,b}(n_2 - n_1). \tag{8.36}$$

This result shows that, if

$$n_2 - n_1 > 0, \tag{8.37}$$

stimulated emission dominates over absorption. Evidently, under equilibrium conditions, the opposite is true.

The inequality of Eq. (8.37) is the criterion for *population inversion*. If a population inversion is achieved, electromagnetic waves with the resonance frequency can be amplified when passing through the material medium. This process of amplification of the radiation due to population inversion is the key mechanism underlying the operation of a laser (*light amplification by stimulated emission of radiation*). The medium where the population inversion occurs is often called the *active medium*.

Actually, Eq. (8.36) describes an increase/decrease in the photon number with time. Amplification/absorption of light waves can be naturally described as an increase/decrease of their intensity with spatial coordinate along the direction of their propagation. The relation between the photon numbers and the intensities of the light waves was introduced in Section 2.4 (see also Table 2.1). The corresponding equation for the intensity can be derived easily from Eq. (8.36), if we consider a light pulse propagating through the medium. Let the maximum of the photon number for this pulse be characterized by the coordinate $x(t)$. Then, in Eq. (8.36) the derivative with respect to time can be calculated as

$$\frac{\mathrm{d}}{\mathrm{d}t} = \frac{\mathrm{d}x}{\mathrm{d}t}\frac{\mathrm{d}}{\mathrm{d}x} = c_\mathrm{m}\frac{\mathrm{d}}{\mathrm{d}x},$$

where c_m is the velocity of light in the medium. Instead of the photon number $N_\mathrm{q,b}$, we introduce the light intensity $\mathcal{I} = c_\mathrm{m}\hbar\omega N_\mathrm{q,b}/V$, where V is the volume of the medium. Now we can rewrite Eq. (8.36) in the form

$$\frac{\mathrm{d}\mathcal{I}}{\mathrm{d}x} = \alpha\mathcal{I}. \tag{8.38}$$

Here,

$$\alpha = \frac{1}{c_\mathrm{m}}B_{21}N_\mathrm{q,b}(n_2 - n_1)$$

represents the gain coefficient. The solution of Eq. (8.38) is

$$\mathcal{I} = \mathcal{I}_0\mathrm{e}^{\alpha x},$$

with $\mathcal{I}_0$ being the intensity at a point $x = 0$. Thus, under population inversion the gain coefficient, α, determines the exponential increase of the light intensity. If $\alpha < 0$, then, instead of amplification, absorption of light occurs.

A medium with population inversion can be used in two ways. First, an active medium can *amplify* an external light beam in accordance with the previously discussed dependence of $\mathcal{I}(x)$. Second, an active medium can *generate* a light beam itself, if the proper optical feedback is provided. Here, the optical-feedback phenomenon implies that some portion of the energy of the light which is amplified in the active medium is returned into the medium for further amplification. In an active medium with optical feedback, spontaneous emission gives birth to the initial photons. These "seed" photons are amplified due to the stimulated emission, and then they partially return to be amplified again. As a whole, this process leads to the generation of coherent laser light.

Typically, optical feedback is realized by placing the active medium in an optical resonator. In the simplest case, the resonator consists of plane or curved mirrors, which provide repeated reflections, and some kind of "trap" of the light – a *cavity* – in the region between the mirrors. The optical waves which can be trapped in this cavity compose the *resonator modes*. A universal characteristic of a resonator mode is the *quality factor*, Q, which can be defined as

$$Q = \omega \times \frac{\text{field energy stored in the cavity}}{\text{power dissipated in the resonator}}. \tag{8.39}$$

Dissipation of electromagnetic energy is caused by many factors: absorption by the mirrors or by matter inside the cavity, transmission of light through the mirrors, light scattering, radiation out of the resonator as a result of the diffraction of light, etc. The power dissipated and the quality factor may be different for different modes.

In principle, in a resonator there is a large number of modes with different frequencies and polarizations, and all possible propagation directions. Excitation of many modes would lead to extremely incoherent emitted light. To avoid such an effect, one can employ so-called *open optical resonators*. The simplest open resonator consists of two

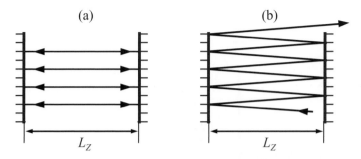

Figure 8.40 Longitudinal and transverse modes in the planar Fabry–Pérot resonator.
(a) Longitudinal modes. Strictly perpendicular light rays do not escape from the resonator.
(b) Transverse modes. Slightly inclined light rays eventually escape from the resonator and have poor quality factors.

plane mirrors parallel to each other, which are finite in their transverse dimensions; an example is given by the so-called Fabry–Pérot etalon shown in Fig. 8.40. In this resonator, most of the modes propagating through the cavity are lost in a single traversal of the cavity since the mirrors are inclined with respect to the mode-propagation directions. This implies that most of the modes, so-called *transverse modes*, have a very low quality factor. Only the waves propagating perpendicular to the mirrors can be reflected back and travel from one mirror to another without escaping from the resonator. These waves correspond to the so-called *longitudinal modes* of the resonator. Thus, for finite dimensions of mirrors only longitudinal modes can have a high quality factor. Their loss and diffraction are caused by absorption by the mirrors, by transmission through the mirror, and by wave diffraction on the sides of the mirrors. The wave-diffraction losses can be made much smaller than can those arising from other loss mechanisms. Thus, the open resonator provides for strong discrimination between modes. Relatively few of these modes have a high quality factor, Q. According to the definition of Eq. (8.39), they are capable of accumulating the light energy. The photons corresponding to these high-quality modes can be generated effectively by the active medium.

At this point, we end our discussion of the simple two-level model of the optical medium and start to consider more realistic systems.

Interband emission and absorption in semiconductors

Semiconductors constitute a material system that may be used in practice to realize controllable light emission. Thus, we shall consider the mechanisms of absorption and emission of photons in semiconductors. Among these mechanisms, the most important is *interband (band-to-band) phototransitions*. Absorption of a photon can result in the creation of an electron in the conduction band and a hole in the valence band, i.e., an electron–hole pair. The inverse process is radiative electron–hole recombination, which results in the emission of a photon (see Fig. 8.41).

Figure 8.42 illustrates the dependence of the absorption coefficient on photon energy and wavelength for various semiconductors. From Fig. 8.42, one can see that the

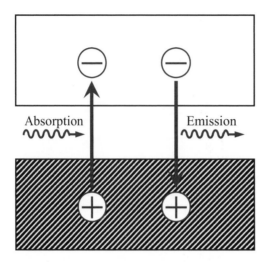

Figure 8.41 Absorption and emission of photons in a semiconductor: band-to-band transitions.

absorption increases sharply in the short-wavelength region. Let E_g be the bandgap of the semiconductor. Then the material is relatively transparent for $\hbar\omega < E_g$. For $\hbar\omega > E_g$, the semiconductor exhibits strong absorption; accordingly, $\hbar\omega_g = E_g$ corresponds to the *absorption edge*. The shape of the absorption edge depends significantly on the structure of the electron bands. Remember that in Section 4.4 we defined direct- and indirect-bandgap semiconductors. Direct-bandgap semiconductors such as GaAs have a more abrupt absorption edge and a larger absorption value than do indirect-bandgap materials, of which Si provides an example. We can introduce the so-called *bandgap wavelength*, or *cut-off wavelength*, $\lambda_g = 2\pi c\hbar/E_g$. If E_g is given in eV, the bandgap wavelength in micrometers is

$$\lambda_g = \frac{1.24}{E_g}. \tag{8.40}$$

The values of E_g and λ_g for various III–V semiconductor materials are apparent from the curves plotted in Fig. 8.42. One can see that interband transitions in III–V compounds cover a wide range from infrared to visible spectra. Optical activity in this spectral region is crucial for optoelectronic applications of these materials.

A photon absorbed during an interband transition excites an electron from the valence band to the conduction band, i.e., it creates an electron–hole pair as depicted in Fig. 8.43(a). The inverse process – the phototransition of an electron from the conduction band to the valence band – is referred to as the *radiative recombination* (annihilation) of an electron and a hole; this is depicted in Figs. 8.43(b) and 8.43(c). According to the general properties of phototransitions studied at the begining of Section 8.6, there exist two such processes: spontaneous and stimulated emission, as illustrated by cases (b) and (c), respectively, in Fig. 8.43.

From the above analysis it is clear that, to obtain intense interband emission, one should provide nonequilibrium electrons and holes. Moreover, large concentrations of

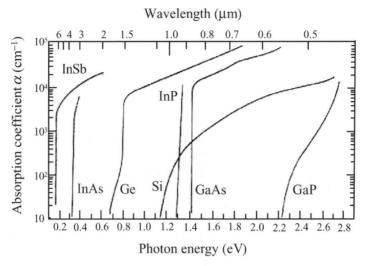

Figure 8.42 Absorption coefficient versus photon energy and wavelength for interband phototransitions in various semiconductors. After G. E. Stillman, V. Robbins *et al.*, "III–V compound semiconductor devices: optical detectors," *IEEE Trans. Electron Devices*, **ED-31**, 1643–1655 (1984). © IEEE.

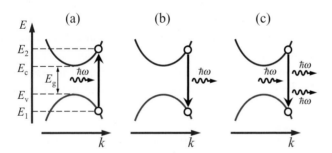

Figure 8.43 (a) The absorption of a photon results in excitation of the electron–hole pair. (b) Spontaneous emission, which can be interpreted as electron–hole recombination. (c) Stimulated emission.

both electrons and holes in the same spatial region correspond to the population inversion necessary for stimulated emission. Indeed, according to Section 4.4, the presence of the holes means "empty" electron states in the valence band. This explains the appearance of the population inversion between the conduction and valence bands under conditions of large electron and hole concentrations.

However, nonequilibrium carriers quickly relax due to various interband relaxation mechanisms. When the electron and hole concentrations are highly nonequilibrium, the characteristic lifetime of these excess carriers is small. To describe the nonequilibrium electrons and holes, we introduce the density of the pumping (excitation) rate, $\mathcal{R}_{\text{pump}}$, which represents the number of electron–hole pairs excited in a unit volume per unit time. For the case when the radiation mechanism is the decay of electron–hole pairs, one can write

$$R_{\text{pump}} = Bn^2 = \frac{n}{\tau_R(n)}, \tag{8.41}$$

where n is the concentration of electron–hole pairs, B is a parameter, and $\tau_R = 1/(Bn)$ is the radiation lifetime. For the concentrations of about 10^{18} cm^{-3} which are encountered in practice and are necessary to realize the needed significant population inversion and to achieve lasing conditions, this time is less than 10 ns. The pumping rate, R_{pump}, necessary to induce such a concentration is estimated to be

$$R_{\text{pump}} \geq 10^{26} \text{ cm}^{-3} \text{ s}^{-1}. \tag{8.42}$$

To obtain an idea of how intense this pumping is, we can perform the following estimates. Suppose that the bandgap, E_g, is about 1 eV, then the pumping rate of Eq. (8.42) corresponding to the density of *the pumping power* given in the previous example is $E_g R_{\text{pump}} \geq 16$ MW cm^{-3}. This is a huge pumping power!

At this point, an important spectral property of light amplification in semiconductor materials will be discussed. For the previously analyzed two-level model, phototransitions and light amplification/absorption were possible only at a fixed photon energy $\hbar\omega = E_2 - E_1$. This implies that in such a two-level system the amplification and absorption occur in a very narrow spectral range near the value $(E_2 - E_1)/\hbar$. On the contrary, in semiconductors, the spectral range within which amplification and absorption are possible is restricted only from below, $\omega > E_g/\hbar$. This results in a wide spectral band of optical activity. In Fig. 8.44, we show the spectral dependences of the gain/loss coefficient for GaAs for various electron–hole concentrations at room temperature. The left boundary of these dependences coincides approximately with the bandgap value 1.42 eV. The spectral range for which the gain coefficient, α, is positive becomes wider and its maximum moves to larger photon energies as the electron/hole concentration increases because the number of inversely populated states increases with the concentration. Amplification in a wide spectral range allows one to build a laser with considerable tuning of the frequency.

The following two methods can be applied to generate large nonequilibrium carrier concentrations. The first method is *optical excitation* or *pumping*, whereby external light, often incoherent, with a photon energy larger than the bandgap is absorbed and creates nonequilibrium electron–hole pairs. Then, because of the short time associated with the intraband relaxation, the electrons and holes relax to the bottom of the conduction band and the top of the valence band, respectively, where they are accumulated. If the rate of optical pumping is sufficiently large, the necessary level of interband emission can be reached. In particular, inversion between the bands can be induced and light amplification and laser generation become possible. The optical pumping corresponds to a conversion of one kind of radiation, not necessarily coherent, into coherent radiation with a lower photon frequency. One employs optical pumping in cases when electric-current pumping either is not possible or is ineffective. This pumping method is often used to test prototype laser structures before the design of the current pumping system.

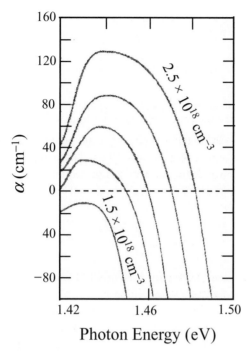

Figure 8.44 The amplification coefficient versus the photon energy in a GaAs light amplifier for five different carrier concentrations. The temperature is 300 K. The results are given for five values of the concentration with equal steps of 0.25×10^{18} cm^{-3}. After J. Singh, *Semiconductor Devices: An Introduction* (New York, McGraw-Hill, 1994), Fig. 11.16.

Laser diodes

A much more convenient method that can be applied to achieve the interband emission is electron and hole injection in devices with p–n junctions. The physics of a p–n junction was studied in the previous section. There, we found that, under forward bias of a p–n junction, it is possible to inject the electrons from the n part of the junction and the holes from the p part of the junction into the same spatial region. Such a double injection can give rise to large nonequilibrium electron and hole concentrations. Often, the n and p parts are separated by a narrow intrinsic (undoped) i region. The corresponding structure is called a p–i–n device.

In Fig. 8.45, we show the case of forward biasing of a p–i–n structure – the so-called *flat-band* condition, when there are no potential barriers for the electrons and holes, and the maximum possible carrier injection into the i region is realized. In the case of direct-bandgap semiconductors, double injection provides for the intense emission of light.

It is easy to estimate the electric currents which are necessary to obtain a given concentration of carriers under double injection. We define the *active region*, where both nonequilibrium electrons and holes are present. In the case of a p–i–n structure, the active region coincides with the i region. Let the length of this region be l. We introduce

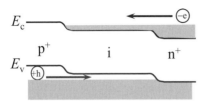

Figure 8.45 The double injection of the electrons and holes, which leads to the formation of interband population inversion in an active region.

a cross-sectional area A, through which the current I is injected into the p–n region of a diode. Then, at steady state, the rate of injection of electrons and holes into a unit volume per unit time can be expressed as

$$\mathcal{R} = \frac{I}{eAl} = \frac{J}{el}, \tag{8.43}$$

where $J = I/A$ is the injection current density. Injection leads to the accumulation of nonequilibrium electron–hole pairs with concentration

$$n = \tau\mathcal{R} = \frac{\tau}{el}J, \tag{8.44}$$

where τ is the total lifetime of the nonequilibrium pairs in the active region. In general, due to the small electron and hole lifetimes, the active region is quite narrow. For GaAs, the length of the active region is estimated to be from 1 μm to 3 μm, depending on temperature, crystal quality, etc. Now, we can use Eq. (8.43) to estimate the current density necessary to achieve the pumping rate of Eq. (8.42): $J \approx (1.6\text{–}4.8) \times 10^3$ A cm^{-2}. The current densities obtained are very large. These estimates indicate that, in order to operate with acceptable current levels, the cross-sections of real emitting diodes have to be very small.

From Eqs. (8.43) and (8.44), it follows that another possible way of decreasing the pumping electric current is to reduce the thickness of the active region l. This thickness is one of the critical parameters for injection pumping of light-emitting devices. Indeed, as understood from the previous discussion and Fig. 8.45, an excess carrier concentration in the active region always leads to carrier diffusion out of this region: the electrons diffuse through the active region to the p part of device, while the holes diffuse to the n part. In general, "diffusive leakage" of the carriers is limited by their finite lifetime $\tau(n)$. In Section 6.2 we found that, if the diffusion coefficient D is given, the diffusion length during a time τ can be estimated as $L_D = \sqrt{D\tau}$. In the case under consideration, τ is the excess carrier lifetime and the length L_D corresponds to the average distance of electron (hole) transfer before recombination. Obviously, the minority carriers (the electrons in the p part and the holes in the n part) recombine immediately with the majority carriers (the holes in the p part and the electrons in the n part). Thus, the width of the region with the excess carrier concentration can not be less than the diffusion length L_D, i.e., $d > L_D$. Since the diffusion length of electrons and holes in direct-bandgap materials is of the order of a few micrometers, it is not possible to make the active region shorter than a

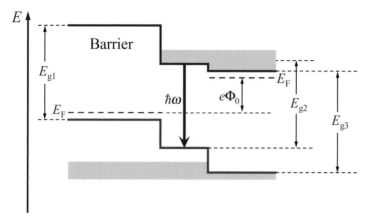

Figure 8.46 The energy-band diagram of a double heterostructure for light amplification/
generation. The applied voltage Φ_0 induces the flat band conditions; E_{g1}, E_{g2}, and E_{g3} are the
bandgaps in different regions of the structure. The energy states filled by the electrons are
denoted by the shaded areas. Thus, in some spatial region there are electrons in the conduction
band and depopulated states (holes) in the valence band.

few micrometers. This conclusion is valid for homostructure (bulk-like) semiconductors.
Heterostructure technologies open the way to different solutions to this problem.

In order to localize nonequilibrium electrons and holes in a smaller active region, one
can employ two heterojunctions. The basic idea of using a double heterostructure is to
design potential barriers on both sides of the p–n junction – this prevents electrons and
holes from diffusing. The potential profile of a double heterostructure is sketched in
Fig. 8.46. The heterostructure consists of three materials with bandgaps E_{g1}, E_{g2}, and
E_{g3}. The band offsets are chosen appropriately to design a structure with a barrier for
electrons in the left part of the structure before the p region and a barrier for holes in the
right part before the n region. The middle i region with the bandgap E_{g2} is accessible
by both types of carriers and it serves as the active region. Figure 8.46 corresponds
to the case of the flat-band condition of a p–n junction with a double heterostructure
embedded in the depletion region. In this case, it is not the diffusion length but the
distance between the barriers that determines the size of the active region. As a result,
the size can be as small as 0.1 μm and the critical electric current is smaller by one
order of magnitude or more compared with that of a conventional homostructure p–n
junction.

If the thickness of the double heterostructure is decreased further, the influence of quantum effects on carrier motion becomes important. For the structure shown in Fig. 8.46,
quantum effects do not lead to any advantages because the advantages can be obtained if
heterostructures are designed so that quantum confinement applies both to electrons and
to holes. The simplest case of quantum confinement can be achieved if a quantum-well
layer is embedded in an active region of a type-I heterostructure. Three possible designs
of active regions exhibiting quantum confinement are sketched in Fig. 8.47. Cases (a)
and (b) correspond to the resultant confinement in a single quantum well, while case
(c) corresponds to confinement in multiple quantum wells. For these designs, electrons

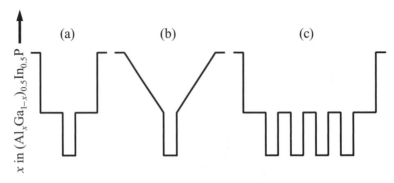

Figure 8.47 Composition profiles of $(Al_xGa_{1-x})_{0.5}In_{0.5}P$ heterostructures providing simultaneously quantum confinement of the carriers and optical confinement: (a) a single quantum well and a step-like refractive-index heterostructure, (b) a single quantum well and a graded-index optical confinement structure, and (c) a multiple quantum well and a step-like index structure. After P. S. Zory Jr., *Quantum Well Lasers* (Boston, MA, Academic, 1993).

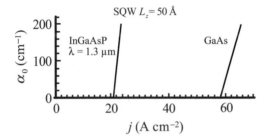

Figure 8.48 The amplification coefficient α_0 as a function of the pumping current density for single InGaAsP and GaAs quantum wells of width 50 Å. After P. S. Zory Jr., *Quantum Well Lasers* (Boston, MA, Academic, 1993).

and holes that are either generated by external light or injected from p and n regions move in barrier layers and then are captured in the active region and quantum wells. The characteristic time of this capture is less than 1 ps. Escape processes require additional energy and have low relative probabilities of occurrence. Carriers in the quantum wells relax to the lowest energy states available. This results in the accumulation of both types of carriers in an extremely narrow active region, which is typically 100 Å wide or even narrower. A similar situation can be realized if quantum wires or dots are embedded in the active region.

A positive effect of electron confinement in quantum wells is illustrated in Fig. 8.48. In this figure, the results for the gain factor α_0 as a function of the current density J are presented for two material systems: InGaAsP/InP and AlGaAs/GaAs heterostructures. It can be seen that the gain coefficient always increases with the current. The current for which α_0 becomes positive, i.e., a population inversion is established, can be defined as the threshold current. The gain coefficient is shown for heterostructures with 50-Å quantum wells. This system has low values of threshold currents for both heterostructures: 20 A cm^{-2} and 60 A cm^{-2} for InGaAs/InP and AlGaAs/GaAs,

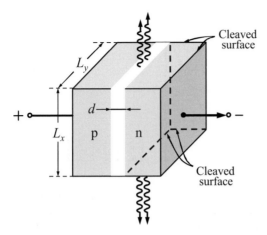

Figure 8.49 The scheme of an injection laser with two cleaved facets that act as reflectors. After B. E. Saleh and M. C. Teich, *Fundamentals of Photonics* (New York, Wiley, 1991). Reprinted with permission of John Wiley & Sons, Inc.

respectively. For both material systems, the $\alpha_0(J)$ curves increase sharply with the current density above the threshold. As discussed previously, lasing can be achieved if a light amplifier is supplied with a path for optical feedback. For an injection laser, the feedback is usually obtained by cleaving the crystal planes normal to the plane of the p–n junction. Figure 8.49 depicts a device with two cleaved surfaces forming an optical resonator. For the light reflected from the crystal boundaries, we define the reflection coefficient by

$$r = \frac{\mathcal{I}_r}{\mathcal{I}_{in}},$$

where $\mathcal{I}_{in}$ and $\mathcal{I}_r$ are the intensities of incident and reflected light, respectively. The reflection coefficient for an air–semiconductor boundary is

$$r = \left(\frac{n_{ri} - 1}{n_{ri} + 1}\right)^2,$$

where n_{ri} is the refractive index of the semiconductor material. Since semiconductors usually have large refractive indexes the coefficients r are large enough. The intensity of the light transmitted through this mirror is

$$\mathcal{I}_{out} = (1 - r)\mathcal{I}_{in}.$$

Let two cleaved surfaces be characterized by two coefficients, r_1 and r_2. After two passes through the device, the light intensity is attenuated by the factor $r_1 \times r_2$. We can define an effective overall distributed coefficient of optical losses:

$$\alpha_r \equiv \frac{1}{2L_x} \ln\left(\frac{1}{r_1 r_2}\right).$$

Here, L_x is the distance between the cleaved surfaces. In principle, there can be other sources of optical losses in the resonator. Let them be characterized by the absorption

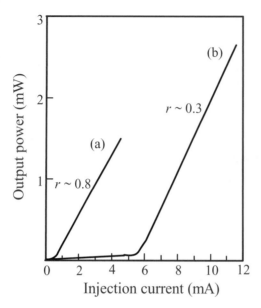

Figure 8.50 The laser output versus the pumping current for a GaAs single-quantum-well laser. After B. E. Saleh and M. C. Teich, *Fundamentals of Photonics* (New York, Wiley, 1991). Reprinted with permission of John Wiley & Sons, Inc.

coefficient, α_s. Then, the total loss coefficient is

$$\alpha_{\text{threshold}} = \alpha_r + \alpha_s.$$

If α is the gain coefficient of some light mode in this resonator, we can write the *criterion for laser oscillations* as

$$\alpha \geq \alpha_{\text{threshold}} = \alpha_r + \alpha_s. \tag{8.45}$$

For injection lasers, the criterion of Eq. (8.45) is a condition imposed on the magnitude of the injection current density, J.

In order to have an idea of the order of magnitude of optical losses in laser diodes, we consider an example. For light of visible and near-infrared ranges, the typical refractive index can be estimated as $n_{\text{ri}} \approx 3.2$–$3.5$. Thus, the reflection coefficient is $r \approx 0.3$. Let a diode have the geometry presented in Fig. 8.49 with dimensions such that $d \times L_x \times L_y = 1\,\mu\text{m} \times 200\,\mu\text{m} \times 200\,\mu\text{m}$. Then, we can estimate the radiation losses: $\alpha_r \approx 60\,\text{cm}^{-1}$. The total current through the laser diode is 1–2 A. The threshold currents for lasing for the quantum-well-based structures are considerably smaller and are typically of the order of tens of mA. In Fig. 8.50, the output light power is presented as a function of the injection current for an AlGaAs laser with a single 100-Å quantum well embedded in the active region. This particular laser design has a threshold current of about 8 mA for the optical feedback due to the cleaved (uncoated) end facets (the reflection coefficient is $r \approx 0.3$). The light reflection can be improved through the use of special reflecting coatings on the end facets. In the latter case, the threshold current decreases below 1 mA, as shown in Fig. 8.50 for the reflection coefficient $r \approx 0.8$. The same figure illustrates

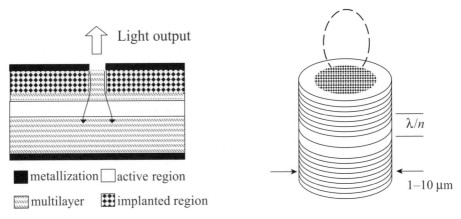

metallization ☐ active region

▧ multilayer ⊞ implanted region

Figure 8.51 Surface-emitting lasers.

also a typical output light power in the injection lasers: from 1 mW to tens of milliwatts depending on the particular diode and resonator design.

Surface-emitting lasers

Thus far, in considering quantum-well lasers we have assumed that the laser is designed for light propagation along the quantum-well layers. Another possible design uses light propagating perpendicular to the layers in a so-called *vertical* geometry. The amplification of light passing through a quantum-well layer can be defined as

$$\frac{\mathcal{I}_{\text{out}} - \mathcal{I}_{\text{in}}}{\mathcal{I}_{\text{in}}} \equiv \beta.$$

Here, $\mathcal{I}_{\text{in}}$ and $\mathcal{I}_{\text{out}}$ are the input and output light intensities, respectively. The quantity β is estimated through the gain coefficient as $\beta(\omega) \approx \alpha(\omega)L$, where L is the width of the quantum well. It is easy to see that β is typically very small. For example, if $\alpha_0 = 100 \text{ cm}^{-1}$ and $L = 100 \text{ Å}$, we get $\beta = 10^{-4}$. In order to obtain laser oscillations in a vertical-geometry structure, one should employ a multiple-quantum-well structure and provide near-perfect mirrors with extremely high reflectivities. Figure 8.51(a) depicts schematically such a *surface-emitting laser*. The laser design includes an active region providing for high light gain, dielectric multilayers, metallic contacts, and implanted regions, which form the light output. Layered dielectric mirrors give very high reflection while the active region contains a multiple-quantum-well structure. The lateral sizes of this laser can be reduced to the 1–10-μm range. A decrease in the surface area of the diode leads to a considerable decrease in the magnitude of the threshold current. In the case of the quantum-well structures just considered, we can assume a characteristic current density of about 100 A cm^{-2}; accordingly, if the lateral sizes are each 5 μm, the pumping surface area is 2.5×10^{-7} cm^2 and the threshold current equals 25 μA.

Surface-emitting quantum-well lasers offer a new advantage of high-packing density. Nowadays, technology allows one to fabricate an array of about 10^6 surface-emitting

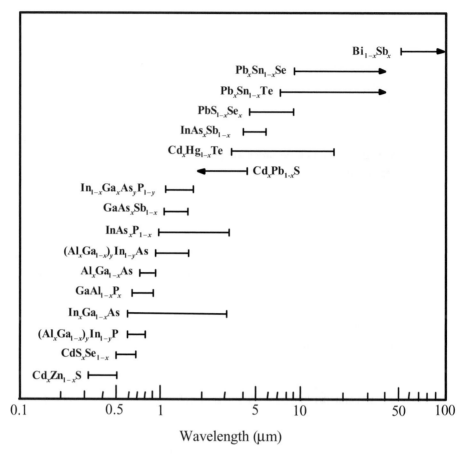

Figure 8.52 Spectral regions of lasers with various semiconductor media. After P. L. Derry, A. Yariv *et al.*, "Ultralow-threshold graded-index separate-confinement single quantum well buried heterostructure (Al, Ga)As lasers with high reflectivity coatings," *Appl. Phys. Lett.*, **50**, 1773–1775 (1987). Reused with permission from Pamela L. Derry, Amnon Yariv, Kam Y. Lau, Nadav Bar-Chaim, Kevin Lee, and Jan Rosenberg, *Applied Physics Letters*, **50**, 1773 (1987). © 1987 American Institute of Physics.

electrically pumped *microlasers*. Microlasers can operate at room temperature and threshold currents are below 0.1 mA.

Blue and ultraviolet quantum-well lasers

As we emphasized previously, the wavelength and the energy quanta of light generated by a semiconductor laser basically depend on the energy bandgap of the semiconductor material selected. In Fig. 8.52, the spectral characteristics of the lasers from various semiconductors are shown. For example, one can see that the most developed lasers based on III–V compounds cover the wavelength range from approximately 0.6 μm to 3 μm.

Operation in the very interesting range of shorter wavelengths requires the use of semiconductors with larger energy bandgaps. Well-known examples of such *wide-bandgap semiconductors* are representative II–VI compounds: CdS, CdSe, ZnCdSe, etc. By using different quantum-well designs, the blue injection lasers were realized on the basis of these materials. However, the major problems associated with this class of quantum-well lasers are the device degradation and short lifetimes of the devices. The electrical power dissipated in II–VI diodes is too high and causes rapid degradation as a result of the generation of intrinsic defects. Much improvement must be realized in this field in order for the group II–VI-based devices to have wide practical application.

Recently, another class of heterostructure materials has been studied in detail – the nitrides of group III. It is difficult to obtain these materials in the usual single-crystal form. However, thin nanometer-scale layers can be grown successfully with good quality by applying the growth methods presented in Chapter 5. The nitrides of group III include InN, GaN, AlN, and all possible ternary and quaternary alloys. For InN and GaN, the energy bandgaps are about 0.9 eV and 3.5 eV, respectively, while that for AlN is 6 eV. Thus, by changing the In and Al content in the alloys $In_yAl_xGa_{1-y-x}N$, one can increase the bandgap from about 1 eV to 6 eV and realize lasers with wavelengths spanning the range from 1.3 μm (near-infrared light) to 0.2 μm (deep-ultraviolet light). The alloys $In_xGa_{1-x}N$ produce green, blue, and violet light.

Light-emitting diodes

Although stimulated emission from the injection laser diode is very important, practically, sub-threshold operation of the diode – when only spontaneous light is emitted – is in many cases advantageous and has a number of applications. This mode of operation does not require feedback to control the power output, which facilitates operation over a wide range of temperatures, and is reliable and inexpensive. Diodes operating with spontaneous light emission are called *light-emitting diodes.*

The important characteristic of the light-emitting diode is the spectral distribution of emission. The spectrum of emission is determined, primarily, by the electron/hole distributions over energy, which can be approximately described by corresponding Fermi functions, as studied in Chapter 6. Thus, the ambient temperature, T, defines both the spectral maximum ("the peak") and the spectral width of emission. The peak value of the spectral distribution can be estimated as

$$\hbar\omega_m = E_g + \tfrac{1}{2}k_B T.$$

The full width at half maximum of the distribution is $\Delta\omega \approx 2k_B T/\hbar$ and is independent of ω. In terms of the wavelength, λ, we obtain

$$\Delta\lambda = \left[\lambda_m^2/(2\pi c)\right]\Delta\omega,$$

or

$$\Delta\lambda = 1.45\lambda_m^2 k_B T, \tag{8.46}$$

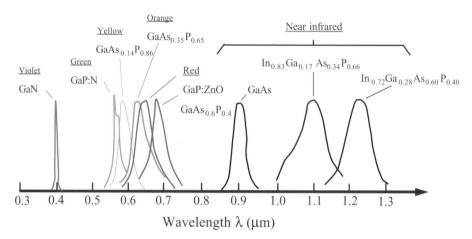

Figure 8.53 The spectra of light-emitting semiconductor diodes with different bandgaps. After Fig. 13 on page 696 of S. M. Sze, *Physics of Semiconductor Devices* (New York, Wiley, 1981). Reprinted with permission of John Wiley & Sons, Inc.

where λ_m corresponds to the maximum of the spectral distribution, $\Delta\lambda$ and λ_m are expressed in micrometers, and $k_B T$ is expressed in eV. Figure 8.53 shows the spectral density as a function of the wavelength for light-emitting diodes based on various materials. The spectral density is normalized so that its maximum equals 1 for all samples. For these different materials, the spectral linewidths increase in proportion to λ^2, in accordance with Eq. (8.46). From Fig. 8.53, one can see that light-emitting diodes cover a wide spectral region from the infrared – about 8 μm for InGaAsP alloys – to the near ultraviolet – 0.4 μm for GaN. They are, indeed, very universal light sources.

Similarly to the case of lasers, the parameters of light-emitting diodes may be considerably enhanced by using heterostructures, particularly quantum wells. Light-emitting diodes may be also designed either in a *surface-emitting configuration*, or in an *edge-emitting configuration*. These configurations are illustrated in Figs. 8.54 and 8.55, respectively. Surface-emitting diodes radiate from the face parallel to the p–n junction plane. The light emitted in the opposite direction is either absorbed by a substrate, or reflected by metallic contacts. The edge-emitting diodes radiate from the edge of the junction region. Usually, surface-emitting diodes are more efficient.

Since the diodes under consideration radiate through spontaneous emission, the spatial patterns of the emitted light depend only on the geometries of the devices. Different lenses can improve the emission pattern. Usually, the edge-emitting diodes have narrower emission patterns.

Light-emitting diodes find many applications, ranging from common lighting systems to signal processing and light communications.

Unipolar intersubband quantum-cascade lasers

So far in this chapter, we have considered light emission and laser action based on *interband* phototransitions involving both electrons and holes. Another type of photo

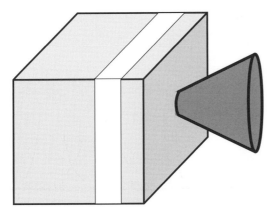

Figure 8.54 A surface-emitting diode.

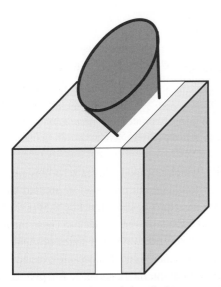

Figure 8.55 An edge-emitting diode.

transition – *intraband* absorption – is allowed in an ideal crystal system with heterojunctions. Since the latter phototransitions are drastically different from interband transitions, laser action associated with these intraband transitions should differ in a fundamental way from that studied for the laser schemes considered above. First of all, an intraband-transition laser should employ only one type of carrier, i.e., it is a *unipolar device*. Second, it should be based on electron transitions between confined states arising from the quantization in semiconductor heterostructures. In order to create a population inversion between two confined states, one needs to provide for (i) electron injection into a higher lasing state and (ii) depletion of a lower lasing state. For this purpose, a vertical scheme of electron transport has been proposed. This scheme is illustrated in Fig. 8.56(a). The proposed heterostructure is a superlattice with a complex design for each period. Each of the periods consists of four AlInAs barriers, forming three GaInAs quantum

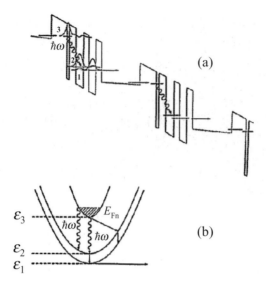

Figure 8.56 (a) Two periods of the 25-stage staircase coupled-well region of a quantum-cascade laser under operational conditions. The laser phototransitions are indicated by arrows. They occur between levels (subbands) 3 and 2 with the photon energy of 295 meV. Level 2 depopulates through level 1 and subsequent tunneling. The energy separation between levels 2 and 1 is 30 meV. (b) Energy dispersion for subbands 1, 2, and 3, phototransitions, and intersubband-scattering processes (straight lines). Reprinted with permission from J. Faist, F. Capasso *et al.*, "Quantum cascade laser," *Science*, **264**, 553–556 (1994). © 1994 AAAS.

wells, and a graded AlInGaAs region, which is doped. Under zero-bias conditions, the overall band diagram resembles a sawtooth structure. Under an applied electric field, the band diagram takes on a staircase structure as shown in Fig. 8.56(a). The barriers form *three coupled quantum wells* with three quasi-bound levels. These three levels are labeled in Fig. 8.56(a) by 1, 2, and 3. Each of the confined states originates from one of the wells. The structure is chosen so that there is a considerable overlap between the wavefunctions of the upper state, 3, and the intermediate state, 2. The same holds for wavefunctions of states 2 and 1. Under a voltage bias, the potential in the doped regions is almost flat, as shown in Fig. 8.56(a). The electrons are injected from the doped regions through the barrier in the confined state 3 of the first quantum well. From this state, they relax primarily to state 2. There are two processes of relaxation: phonon emission and photon emission. In Fig. 8.56(b), these processes are shown for electrons with various values of the in-plane wavevector, $\mathbf{k}$. The three indicated subbands, $\varepsilon_{1,2,3}(\mathbf{k})$, correspond to the three confined states. The straight arrows represent intersubband phonon relaxation. The third confined state, 3, is selected to provide depletion of state 2 as fast as possible. Thus, in this manner we have a three-level scheme whereby the upper level is pumped by the direct injection of electrons from the doped region. The second level is depleted due to strong coupling with the lowest level 1. From level 1, electrons escape to the next doped region. Then the processes are repeated in each subsequent period of the superlattice. One can say that the carriers make transitions down through such a *cascade structure*.

To describe the processes in the cascade structure we need to define the numbers of electrons in three states: n_3, n_2, and n_1. The criterion for population inversion between levels 2 and 3 should be

$$n_3 > n_2.$$

We can write simple balance equations for n_3 and n_2:

$$\frac{dn_3}{dt} = -\frac{1}{e}J - \frac{n_3}{\tau_{32}}, \tag{8.47}$$

$$\frac{dn_2}{dt} = \frac{n_3}{\tau_{32}} - \frac{n_2}{\tau_{21}}, \tag{8.48}$$

where J is the density of the injection current, and τ_{32} and τ_{21} are relaxation times between the states 3 and 2, and 2 and 1, respectively. In Eq. (8.48), we neglect the inverse $1 \rightarrow 2$ process since state 1 can be regarded as almost empty as a result of fast electron escape to the doped region. For the steady-state conditions, we obtain the concentrations

$$n_3 = -\frac{1}{e}J\tau_{32} \quad \text{and} \quad n_2 = n_3\frac{\tau_{21}}{\tau_{32}},$$

and population inversion

$$\Delta n \equiv n_3 - n_2 = -\frac{1}{e}J\tau_{32}\left(1 - \frac{\tau_{21}}{\tau_{32}}\right). \tag{8.49}$$

Thus, to create a population inversion, one should design the laser so that

$$\tau_{21} < \tau_{32}. \tag{8.50}$$

In order to fabricate such unipolar laser structures with vertical electron transport, very precise sophisticated semiconductor technology is necessary.

For a particular device structure (Fig. 8.57) with an optical path of about 700 μm and the mirror reflectivity $r_1 = r_2 = 0.27$, the laser output-current characteristics at various temperatures are shown in Fig. 8.58. The insets of Fig. 8.58 show the current–voltage characteristics and temperature dependence of the laser threshold current. The laser threshold current can be approximated by $I_{th} = C \exp(T/112)$, where the constant C is about 900 mA and T is measured in degrees Kelvin. From Fig. 8.58, it follows that the output power reaches tens of milliwatts.

The emission energy is in the range 275–310 meV. Spectra of the laser output for various currents at $T = 80$ K are presented in the right-hand inset of Fig. 8.58. For this case, the threshold current is about 1.06 A. This inset clearly demonstrates a sharp narrowing of the emission spectra above the laser threshold: the spectra reduce to a sharp peak at $I = 1.1$ A $> I_{th}$.

Thus, the unipolar cascade laser is drastically different from the lasers based on intersubband phototransitions. The properties of intraband phototransitions and, consequently, the properties of the unipolar laser are determined to a large degree by quantum confinement; accordingly, this novel laser can be tailored for operation in

GaInAs (Sn-doped)	$n = 2.0 \times 10^{20}\,cm^{-3}$	20.0 nm	Contact layer
GaInAs	1.0×10^{18}	670.0	
AlGaInAs (graded)	1.0×10^{18}	30.0	
AlInAs	5.0×10^{17}	1500.0	Waveguide cladding
AlInAs	1.5×10^{17}	1000.0	
AlGaInAs (digitally graded)	1.5×10^{17}	18.6	
Active region	undoped	21.1	
GaInAs	1.0×10^{17}	300.0	
AlGaInAs (digitally graded)	1.5×10^{17}	14.6	
AlGaInAs (digitally graded)	1.5×10^{17}	18.6	Waveguide core
Active region	undoped	21.1	×25
GaInAs	1.0×10^{17}	300.0	
AlGaInAs (digitally graded)	1.5×10^{17}	33.2	
AlInAs	5.0×10^{17}	500.0	Waveguide cladding
Doped n⁺-InP substrate			

Figure 8.57 A schematic cross-section of the cascade laser structure. The whole structure consists of 500 layers. Reprinted with permission from F. Capasso, "Quantum cascade laser," *Science*, **264**, 553–556 (1994). © 1994 AAAS.

the spectral region from the middle infrared to submillimeter waves. Quantum cascade lasers are very sensitive to the ambient temperature and typically work at reduced temperatures.

In conclusion, we have shown that nanostructures play a key role not only for downscaled electrical devices, but also for optical devices. They facilitate improvements in bipolar injection lasers and make it possible to realize cascade-laser structures, which work on the basis of innovative concepts.

8.7   Nanoelectromechanical system devices

So far, we have concentrated on the electron properties of nanostructures and have shown that electronic effects on the nanoscale can be exploited for electrical devices. Mechanical properties of nanostructures are very different from those of bulk samples. One can use both electronic and mechanical properties on the nanoscale to develop a new class of devices – nanoelectromechanical devices. Fabrication of nano-electromechanical

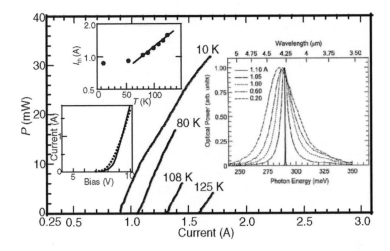

Figure 8.58 The measured optical power P from a single facet of the quantum-cascade laser with the structure presented in Fig. 8.57 and an optical cavity length of 1.2 mm. The results are given for four different temperatures. The insets on the left show the dependence of the laser threshold current as a function of temperature and the current–voltage characteristics of the device. The inset on the right of the figure shows the output spectrum for a 1.2-mm-long laser below and above threshold ($I_{th} = 1.06$ A) at 80 K heat-sink temperature. Above 0.4 A the luminescence peak was found to grow superlinearly with current due to optical gain causing the line to narrow. After F. Capasso, "Quantum cascade lasers: a unipolar intersubband semiconductor laser," in *Proceedings of the International Conference on the Physics of Semiconductors* (Singapore, World Scientific, 1995), pp. 1636–1640.

systems (NEMSs) was described in Section 5.9. Here, we will discuss a few particular NEMS devices.

Resonators. Parametric amplification

Parametric resonator NEMSs serve as mechanical amplifiers in a narrow frequency range. The basic idea of a mechanical parametric resonator can be illustrated by the following simple example. Consider a simple parallel-plate capacitor, in which one plate is a part of the mechanical resonator, while the second plate is fixed, as shown in Fig. 8.59. A displacement, z, of the resonator plate changes the spacing in the capacitor and thus the device capacitance, $C = \epsilon_0 S/(d + z)$, where S and d are the area of the plates and the equilibrium distance between them, respectively. Since the displacements are small, we can use the approximation

$$C \approx \frac{\epsilon_0 S}{d}\left(1 - \frac{z}{d} + \frac{z^2}{d^2}\cdots\right).$$

If a time-dependent voltage $V(t)$ is applied to the capacitor, its electrical energy becomes

$$E_{el} = \frac{1}{2}CV^2(t).$$

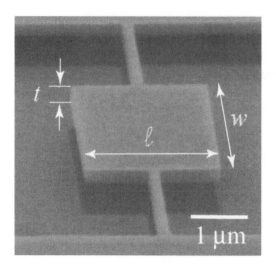

Figure 8.59 A mechanical parametric resonator fabricated by the SIMOX technique. Characteristic dimensions of the resonator are indicated. From A. N. Cleland, *Foundations of Nanomechanics* (Berlin, Springer-Verlag, 2003).

Let M and K be the mass and the spring constant of our resonator, respectively. Then, Newton's second law for the resonator plate takes the form

$$M\frac{d^2z}{dt^2} + M\gamma\,\frac{dz}{dt} + Kz = f = -\frac{dE_{el}}{dz}. \tag{8.51}$$

Here, we introduce the term $M\gamma\,dz/dt$ describing the damping of the resonator. The force acting on the resonator is

$$f = -\frac{\epsilon_0 S}{d^2}V^2(t)\left(1 + \frac{2z}{d}\right); \tag{8.52}$$

i.e., it is proportional to the square of the driving voltage, $V(t)$. Importantly, the force contains a term proportional to the displacement z. We can characterize the resonator by the frequency of vibration, $\omega_0 = \sqrt{K/M}$, and the quality factor, $Q = \omega_0/\gamma$. Introducing the notation

$$\Delta K(t) = \frac{2\epsilon_0 S V^2(t)}{d^3} \tag{8.53}$$

and

$$F_D(t) = -\frac{\epsilon_0 S V^2(t)}{d^2}, \tag{8.54}$$

we can rewrite the force equation as

$$\frac{d^2z}{dt^2} + \frac{\omega_0}{Q}\frac{dz}{dt} + \left(\omega_0^2 + \frac{\Delta K(t)}{M}\right)z = F_D(t). \tag{8.55}$$

The physical significance of each of the quantities $\Delta K(t)$ and $F_D(t)$ is obvious: $\Delta K(t)$ describes a parametric modulation of the spring constant by an applied voltage, while $F_D(t)$ is the displacement-independent driving force. The latter equation is the simplest basic equation for the effect of parametric resonance: the parameters ΔK and F_D have the same time dependence; i.e., they are both proportional to $V^2(t)$. If the voltage depends on time harmonically, $V \propto \sin(\omega t)$, we find that both the driving force and the spring constant oscillate with the driving frequency, which varies with time as $\cos(2\omega t)$. In general, a NEMS can be fabricated with two or more capacitor plates; then, different voltages can be applied to these capacitors. Thus, to generalize the analysis we may assume that the spring-constant modulation, ΔK, and the driving force, F_D, have independent modulation frequencies, ω_P and ω_D:

$$\Delta K = \Delta K_0 \sin(\omega_P t);$$

$$F_D = F_0 \cos(\omega_D t + \phi).$$

Here, ϕ denotes a phase shift between these two dependences. More specifically, the basic equation now becomes:

$$\frac{d^2 z}{dt^2} + \frac{\omega_0}{Q} \frac{dz}{dt} + \left(\omega_0^2 + \frac{\Delta K_0}{M} \sin(\omega_P t) \right) z = F_0 \cos(\omega_D t + \phi). \qquad (8.56)$$

This equation is the so-called *Mathieu equation with damping*. Analysis of this equation yields the following results. First, let the parametric modulation be absent ($\Delta K_0 = 0$), then the equation describes vibrations of a resonator with frequency ω_0. For small damping ($Q \gg 1$), vibrations will manifest a strong resonance at the driving frequency, $\omega_D \approx \omega_0$:

$$z(t) = A \cos(\omega_0 t) + B \sin(\omega_0 t), \qquad (8.57)$$

$$A = \frac{F_0 Q}{K} \sin\phi, \qquad B = \frac{F_0 Q}{K} \cos\phi. \qquad (8.58)$$

This result shows that, for an oscillator with a quality factor $Q \to \infty$, the vibration magnitude diverges.

If a parametric modulation is tuned by a small driving force, additional strong resonances arise at $\omega_P = 2\omega_0/n$, with n being an integer. That is, the parametric modulation resonates for all submultiples of the frequency $2\omega_0$. Consider, for example, the case of $n = 1$, i.e., $\omega_P = 2\omega_0$ and $\omega_D = \omega_0$. In the limit of high Q, it is possible to find a solution in the form of Eq. (8.58) with

$$A_p = \frac{F_0 Q}{K} \frac{\sin\phi}{1 - Q\,\Delta K_0/(2K)}, \qquad B_p = \frac{F_0 Q}{K} \frac{\cos\phi}{1 + Q\,\Delta K_0/(2K)}. \qquad (8.59)$$

Thus, the oscillations depend essentially on the parametric coupling parameter ΔK_0, and the amplitudes of the mechanical vibrations can be controlled by these parameters. By comparing these results with and without parametric coupling, i.e., Eqs. (8.59) and (8.58), we can define the parametric gain,

$$G = \frac{\sqrt{A_p^2 + B_p^2}}{\sqrt{A^2 + B^2}}.$$

Using the above formulas for the coefficients A, B, A_p, and B_p, we find

$$G = \left(\frac{\cos^2\phi}{[1 + Q\,\Delta K_0/(2K)]^2} + \frac{\sin^2\phi}{[1 - Q\,\Delta K_0/(2K)]^2} \right)^{1/2}.$$

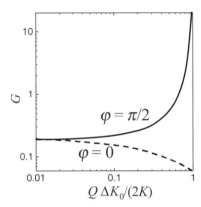

Figure 8.60 The gain/loss coefficient G as a function of the quality factor Q for different phase shifts between spring-constant modulation and the driving force. From A. N. Cleland, *Foundations of Nanomechanics* (Berlin, Springer-Verlag, 2003).

For $\Delta K \neq 0$, the gain depends on the relative phase ϕ. If $\phi = 0$, π, 2π, ..., the amplitudes of the vibrations are suppressed and $G < 1$. If $\phi = \pi/2$, $3\pi/2$, $5\pi/2$, ..., the vibrations are amplified and $G > 1$. Formally, if $\Delta K_0 \rightarrow 2K/Q$, the amplitude diverges with $G \rightarrow \infty$. This phenomenon is called *parametric resonance*. Both de-amplification and amplification regimes are illustrated by Fig. 8.60. We can conclude that, if the quality factor, Q, is large, an amplification of the amplitude of vibrations can be obtained even for small modulations of the driving force.

The resonator of a micromechanical parametric oscillator, shown in Fig. 8.59, is fabricated by the SIMOX technique. The dimensions of the suspended resonator plate are 4 μm × 4 μm × 0.2 μm. The substrate is grounded. A metal is deposited onto the resonator plate. The resonator is driven by a voltage applied between the plate and the substrate. The fundamental frequency of the resonator is $\omega_0/(2\pi) = 485$ kHz. Thus, the amplified signal is at this frequency, while the parametric drive is at $\omega_P = 2\omega_0$. These parameters correspond to the lowest resonance in the Mathieu equation (8.56). To detect and measure displacements, one can use reflection of a laser beam from the substrate and the resonator. Interference of these two reflected signals yields good displacement sensitivity. In Fig. 8.61, the measured square of the oscillation amplitude is presented as a function of the pump amplitude for such a parametric oscillator. The maximum amplification, G, achieved is about 10.

The resonance properties of NEMSs will undoubtedly be employed in a broad range of applications. Obviously, one of the principal areas will be signal processing in the very-high-frequency (VHF), ultra-high-frequency (UHF), and microwave-frequency bands.

Mechanically detected magnetic resonance imaging

Another promising application of NEMSs is mechanically detected magnetic resonance imaging (MRI). It is well known that the phenomenon of nuclear magnetic resonance is widely used for diagnostic purposes in medicine. The conventional inductive detection

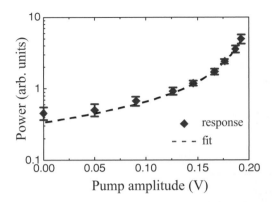

Figure 8.61 The square of the oscillation amplitude ("the power of oscillations") versus the pumping amplitude for the parametric resonance oscillator shown in Fig. 8.59. From A. N. Cleland, *Foundations of Nanomechanics* (Berlin, Springer-Verlag, 2003).

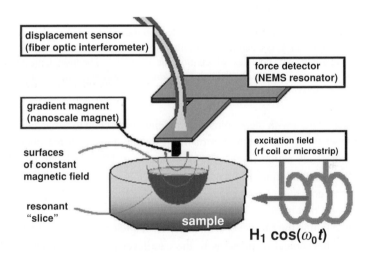

Figure 8.62 The principal scheme of mechanically detected magnetic resonance imaging. After M. L. Roukes, "Nanoelectromechanical systems," in *Technical Digest of the 2000 Solid-State and Actuator Workshop*, Hilton Head Island, SC, pp. 1–10.

techniques still take about 10^{14}–10^{16} nuclei to generate a measurable signal of the magnetic resonance. This means that state-of-the-art MRI in research laboratories attains, at best, a maximal resolution (minimum voxel size) of the order of 1 μm. For standard clinical MRI, the resolution yielded by commercial instrumentation is much poorer, with a voxel size of about 1 mm.

Mechanically detected MRI, now commonly called *magnetic resonance force microscopy* (MRFM), is significantly more sensitive than conventional MRI. There are several principal components of a MRFM instrument, which are presented in Fig. 8.62. An antenna structure in the form of a coil or a microstripline generates a radio-frequency field of frequency ω_0. The static magnetic field created by a miniature magnet splits the spin states of impurities within a sample and provides a resonance interaction of the

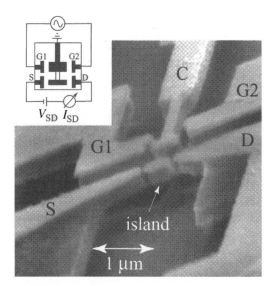

Figure 8.63 The electron shuttle – a device with electron transport provided through mechanical motion of the cantilever. C is a nanosized cantilever, S and D are the source and drain contacts, and G1 and G2 are the gate contacts, which move the cantilever. The inset depicts the electric circuit of the structure. From A. N. Cleland, *Foundations of Nanomechanics* (Berlin, Springer-Verlag, 2003).

radio-frequency field with these spins. The magnet is attached to a mechanical resonator of a cantilever type. The interaction of the resonant spins with the magnet results in a time-varying back-action force upon this cantilever. This force can be detected by a sensor with high resolution of displacements, such as an optical interferometer. All of these elements constitute a resonant force sensor. Such a sensor can detect the extremely weak forces exerted by the resonant spins upon the mechanical system. Microscopy is realized by scanning the cantilever with attached magnet over the sample. By correlating the resonant mechanical response with the cantilever position, one can obtain spatial imaging of spin density.

The resonance properties of NEMSs provide key advances for MRFM. Indeed, utilization of a nanometer-scale ultra-high-frequency mechanical resonator makes it feasible to couple directly the spin precession and mechanical vibrations and to improve drastically the resolution of MRI.

The field of MRFM is still very much in its infancy. Sustained efforts are required in order to take it from a scientific demonstration to a useful technique for high-resolution MRI. With its potential for atomic resolution, such efforts are of great potential importance, especially for biochemical applications.

Coupling of electron transport and mechanical motion. The electron shuttle

The coupling of electron transport and mechanical motion of NEMSs gives rise to new effects, which can be useful for a number of applications. Consider an example of such a coupling, which can be called the *electron shuttle*. The structure contains a metallized cantilever suspended between two metallic electrodes, as shown in Fig. 8.63, where an

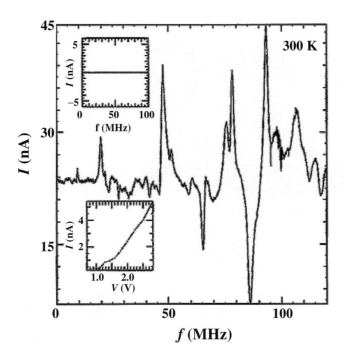

Figure 8.64 Electric current in the electron shuttle versus frequency in the gate circuit. Upper inset: the cantilever is blocked and there is no current. Lower inset: the shuttle current as a function of the source–drain voltage at a given frequency in the gate circuit. From A. N. Cleland, *Foundations of Nanomechanics* (Berlin, Springer-Verlag, 2003).

electron micrograph of the electron shuttle is presented. Two gates are fabricated to drive the cantilever electrostatically. The cantilever can be driven by the voltage, $V_{G_1 G_2}$, so that the cantilever can nearly contact each of the spatially separated electrodes. A voltage V_{SD} is applied to the electrodes, which can be considered as a source and a drain. If the frequency of the voltage applied to the gates, $V_{G_1 G_2}$, does not coincide with the resonance frequency of the cantilever, the amplitude of cantilever vibrations is small. Accordingly, the cantilever does not mechanically contact the source and drain electrodes, and electron transport is suppressed. As soon as the frequency $V_{G_1 G_2}$ matches the mechanical resonance, the cantilever contacts the source and drain during each cycle of the mechanical motion. When the cantilever contacts the electrodes successively, the metallized end of the cantilever charges and discharges, and transfers electrons between the source and the drain. In Fig. 8.64, results of measurements of the source–drain current are presented; several peaks in the current correspond to different resonant vibration modes of the cantilever. Only for these resonances are the amplitudes of the vibrations large enough to support electron transfer through the nanostructure. If the motion of the cantilever is blocked, no current is observed, as seen from the upper inset to Fig. 8.64. The lower inset shows the magnitude of one of the current peaks as a function of V_{SD} at a given resonance frequency. Thus, this NEMS actually presents an electron shuttle. The number of electrons transferred depends on the dimensions of the metallized island

Figure 8.65 A set of SiC NEMSs. Submicrometer double-clamped SiC beams exhibit fundamental resonance frequencies from 2 to 134 MHz. From M. L. Roukes, "Nanoelectromechanical systems," in *Technical Digest of the 2000 Solid-State and Actuator Workshop*, Hilton Head Island, SC, pp. 1–10.

on the cantilever. By decreasing the size of the island, it becomes feasible to transfer a single electron at a time.

Following these examples of NEMSs with different functions, we consider the basic parameters that determine the performance of any NEMS. Actually, these parameters are practically the same as for an electronic device: the response time or characteristic frequency, the quality factor (energy loss), the operating power, the signal-to-noise ratio (sensitivity), etc.

Frequency

We start by considering a NEMS as a mechanical vibrator at a natural angular frequency, ω_0. The frequency can be approximated by $\omega_0 = \sqrt{K/M}$, where K is an effective spring constant and M is the mass of the mechanical vibrator. Underlying these simplified "effective" terms is a complex set of elasticity equations that govern the mechanical response of these objects. If we reduce the size of the mechanical device while preserving its overall shape, then the fundamental frequency, ω_0, increases as the vibrator's linear dimension decreases. This is obvious because the mass is proportional to the volume of the mechanical vibrator, while the effective spring constant for flexural deformations increases with the decrease of its length. This increasing frequency effect is important because a high frequency response translates directly to a fast temporal response to applied forces. It also implies that a fast response can be achieved without the expense of making stiff structures. Moreover, a small spring constant provides very high mechanical responsivity.

The shape of the vibrations and resulting frequencies depend on the way the beams are clamped. Three variants can be realized in a particular device: (i) both ends clamped or free, (ii) both ends pinned, and (iii) a suspended beam clamped at one end (the cantilever). In Fig. 8.65, images of double-clamped SiC beams are shown. The beams are of

Table 8.2 Estimated fundamental frequency versus geometry for SiC, Si, and GaAs mechanical resonators

Boundary conditions	Resonator dimensions $L_x \times L_y \times L_z$ in (µm)		
	$10 \times 0.2 \times 0.1$	$1 \times 0.05 \times 0.05$	$0.1 \times 0.01 \times 0.01$
Material	SiC, Si, GaAs	SiC, Si, GaAs	SiC, Si, GaAs
Both ends clamped or free	12, 7.7, 4.2 MHz	590, 380, 205 MHz	12, 7.7, 4.2, GHz
Both ends pinned	5.3, 3.4, 1.8 MHz	260, 170, 92 MHz	5.3, 3.4, 1.8 GHz
Cantilever	1.9, 1.2, 0.65 MHz	93, 60, 32 MHz	1.9, 1.2, 0.65 GHz

After M. L. Roukes "Nanoelectromechanical systems", in *Technical Digest of the 2000 Solid-State Sensor and Actuator Workshop*, Hilton Head Island, SC, pp. 1–10.

different dimensions and exhibit fundamental frequencies of flexural vibrations varying from 2 MHz to 134 MHz. Table 8.2 displays frequencies for the fundamental *flexural* modes of thin beams for various materials (SiC, Si, and GaAs) and dimensions spanning the domain from micromechanical systems to nanosystems well within the nanoregime. The last column in Table 8.2 represents the dimensions currently attainable with advanced electron-beam lithography. The materials SiC, Si, and GaAs are of particular interest, because they are available with extremely high purity as monocrystalline layers in epitaxially grown heterostructures. The numbers in Table 8.2 are rough averages for the various commonly used crystallographic orientations. It is particularly notable that, for structures of the same dimensions, Si yields frequencies of a factor of two, and SiC of a factor of three, higher than those obtained with GaAs devices. This increase reflects the increased phase velocity of sound as well as the higher stiffness of the materials.

Thus, mechanical resonators with fundamental frequencies above 10 GHz (10^{10} Hz) can now be built using surface nanomachining processes involving state-of-the-art nanolithography on the 10-nm scale.

Quality factor

Another important parameter, which characterizes the rate of energy dissipation in a NEMS, is the so-called *quality factor*, Q. If we define a decay time of the flexural vibrations of the beam in a NEMS, τ_d, then, $Q \approx \omega_0 \tau_d$. The quality factor achieved for a NEMS in a moderate vacuum is in the range from 10^3 to 10^5. This greatly exceeds the quality factor typically realized for electrical (microwave) resonators. The small degree of internal energy dissipation in a NEMS directly relates to the high attainable force sensitivity and low operating power levels.

Characteristic operating power level

Applications of NEMS resonators involve the use of a specific vibrational mode, typically one of the lowest *flexural* modes. A rough understanding of the minimum power

Table 8.3 Representative operating power levels for NEMSs
(1 attowatt $= 1$ aW $\equiv 10^{-18}$ W)

Frequency, ω_0	Quality factor, Q	Minimum power, P_{min}
100 MHz	10 000	40 aW
100 MHz	100 000	4 aW
1 GHz	10 000	400 aW
1 GHz	100 000	40 aW

After M. L. Roukes (2000) "Nanoelectromechanical systems," in
*Technical Digest of the 2000 Solid-State Sensor and Actuator Work-
shop*, Hilton Head Island, SC, pp. 1–10.

necessary to operate with a NEMS using the flexural mode can be obtained as follows.
At equilibrium the average energy of such a mode is equal to the thermal energy $k_B T$.
To perform an operation, the energy of the external input signal should be larger than the
thermal energy. The characteristic time scale for energy exchange between the mode, at
frequency ω_0, and its surroundings is just the decay time, $\tau_d = Q/\omega_0$. Thus, the minimum power of the signal which has to be applied to the system to drive it to an amplitude
above the thermal fluctuations is

$$P > P_{min} = k_B T, \qquad \tau_d = \frac{k_B T \omega_0}{Q}.$$

In Table 8.3, we present values of P_{min} for various frequencies and quality factors at room
temperature. As displayed in Table 8.3, this minimum power can be extremely small for
a NEMS. For device dimensions accessible today via electron-beam lithography, the
characteristic level is of the order of tens of attowatts (10^{-17} W!). This is many orders
of magnitude smaller than the power dissipation in contemporary systems of similar
complexity based on digital devices that work solely with electric signals.

Dynamic range of a NEMS

From Table 8.3 it is clear that NEMSs have the potential to provide new types of ultra-
low-power electromechanical signal processing. However, realization of these potential
advantages is not a simple task. To utilize the full potential of a NEMS, displacement
transduction schemes should be capable of providing resolution of the beam displacement at the level of the thermomechanical fluctuations. Indeed, at a finite temperature
there always occur mechanical fluctuations and, thus, there exist small random (chaotic)
displacements, Δz_T, of the beam. These displacements can be readily estimated by
equating the potential energy of our vibrator, $\frac{1}{2}K(\Delta z_T)^2$, to the thermal energy $k_B T$,
which gives the amplitude of such thermally induced vibrations as $\Delta z_T = \sqrt{2k_B T/K}$.
As we indicated above, the spring constant K decreases as the dimensions of the beam
diminish. Thus, thermally induced displacements scale with length, L, as $\Delta z_T \propto 1/\sqrt{L}$.

Simultaneously, the signal amplitude of vibrations for a mechanical device (in a linear operation regime) scales downward in direct proportion to its size. It is important to estimate the acceptable level of operational displacements of the suspended structure, i.e., the so-called *dynamic range* for a linear NEMS. To estimate the characteristic dynamic range, we need to define the displacement amplitude corresponding to the onset of nonlinearity. This nonlinearity implies that, in a power-series expansion for elastic potential energy, the largest term beyond the quadratic term (i.e., beyond the Hookes'-law term) becomes important. For a double-clamped beam, this condition translates into the relation $\Delta z_N \sim 0.5 L_z$. This criterion depends only upon the beam thickness, L_z, in the direction of vibrations. The linear dynamic range for a NEMS can be defined as the ratio of the nonlinearity onset, Δz_N, to the thermal displacement, Δz_T: DR $= 10 \ln(\Delta z_N / \Delta z_T)$. The following example highlights the characteristic displacements and the dynamic range in a relatively small NEMS. Consider a suspended Si structure of dimensions $0.1~\mu m \times 0.01~\mu m \times 0.01~\mu m$ with a quality factor $Q = 10^4$ at $T = 300$ K. Then, we find $\Delta z_T \approx 0.09$ nm and $\Delta z_N \approx 5$ nm. For these parameters, a high value of the dynamic range, DR, of approximately 40 is obtained.

These considerations and estimates indicate two very crucial areas for NEMS engineering necessary to provide femtowatt to picowatt regimes: (i) development of ultra-sensitive transducers that are capable of enhanced displacement resolution with increasingly higher frequencies as device sizes are progressively scaled downward and (ii) development of techniques tailored to operate over the entire dynamic range of the NEMS. As we studied previously, nowadays, probe microscopy techniques are capable of probing and measuring quantitatively estimated ultimate displacements.

In conclusion, a nanoelectromechanical system consists of a nanometer-to-submicrometer-scale mechanical resonator that is coupled to an electronic device of comparable dimensions. The mechanical resonator may have a simple geometry, such as a cantilever (a suspended beam clamped at one end) or a bridge (a suspended beam clamped at both ends) and is fabricated from materials such as silicon using lithographic and other techniques similar to those employed for fabricating integrated circuits. Because of their submicrometer and nanoscale size, the mechanical resonators can vibrate at frequencies ranging from a few megahertz to around a gigahertz. The quality factors of these resonators greatly exceed those of typical microwave resonators. A NEMS operates with low power dissipation in a wide dynamic range. These properties of NEMSs open the way to a number of applications, ranging from signal processing to novel detectors.

8.8   Quantum-dot cellular automata

As we discussed previously, the general tendency is that improvement in technology leads to progressive scaling down of electronic devices and widening of their functionality. However, as more and more devices are placed into the same area, the heat generated during a switching cycle can no longer be removed and this may result in

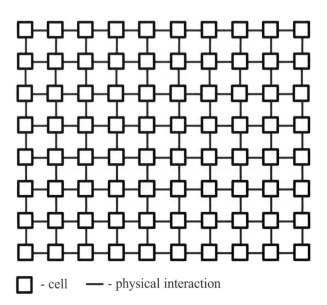

☐ - cell   ——— - physical interaction

Figure 8.66 A schematic picture of a cellular array with interconnections via physical interaction.

considerable limitation of the speed of operation. In addition, interconnections between devices can not be scaled in accordance with the device scaling because of the effects of wire resistance and capacitance. The latter phenomenon can lead to a wiring bottleneck. In such a situation, alternative approaches are needed in order to solve these problems.

Contemporary nanotechnology can support alternative electronic devices and system architectures. One such approach is based on quantum dots arranged in locally interconnected cellular-automata-like arrays. The fundamental idea of quantum-dot cellular-automata operation is to encode information using the charge configuration of a set of quantum dots. Importantly, in the quantum-dot cellular-automata approach, the information is contained in the arrangement of charges of the dots, rather than in the flow of the charges (i.e., current). It can be said that the devices interact by direct Coulomb coupling rather than via the current through the wires. Figure 8.66 illustrates a locally interconnected array of cells. Obviously, a *dense arrangement* of nanometer-scale quantum dots can provide the necessary physical interactions inside the array. These physical interactions between the elements, together with the topology of the system, determine the overall functionality of the array.

As discussed in Section 5.5, technologies for the growth and processing of quantum-dot arrays are already available. They facilitate the fabrication of quantum dots with the necessary properties, arrangement, etc.

In quantum-dot cellular automata, the building block is called a cell. Figure 8.67(a) shows such a cell. A cell is composed of (at least) four quantum dots positioned at the corners of a square. A cell contains two excess electrons, which are allowed to tunnel between neighboring quantum dots in the cell. Tunneling out of a cell is assumed to be completely suppressed by the potential barriers between the cells.

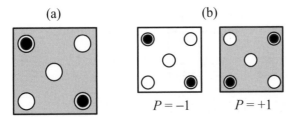

Figure 8.67 A quantum-dot cell consisting of five dots with two electrons in the cell: (a) the Coulomb repulsion causes the electron to occupy the antipodal sites; and (b) two bistable states result in different polarizations. From S. Craig, P. Lent *et al.*, "Bistable saturation in coupled quantum dots for quantum cellular automata," *Appl. Phys. Lett.*, **62**, 714 (1993). Reused with permission from S. Craig, P. Lent, D. Tougaw, and W. Porod, *Applied Physics Letters*, **62**, 714 (1993). © 1993 American Institute of Physics.

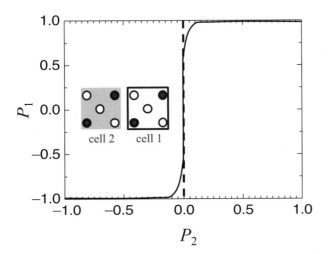

Figure 8.68 The cell–cell response. A polarized cell 1 implies the polarization of cell 2 due to the Coulomb repulsion of the electrons populating the cells. The strong nonlinearity of the cell–cell response plays the same role as the gain in a conventional digital device. From S. Craig, P. Lent *et al.*, "Bistable saturation in coupled quantum dots for quantum cellular automata," *Appl. Phys. Lett.*, **62**, 714 (1993). Reused with permission from S. Craig, P. Lent, D. Tougaw, and W. Porod, *Applied Physics Letters*, **62**, 714 (1993). © 1993 American Institute of Physics.

The Coulomb repulsion between the electrons in a cell tends to place them at antipodal sites in the square. For an isolated cell, there are two energetically equivalent arrangements of the extra electrons, which are denoted as "cell polarizations," $P = +1$ and $P = -1$. The polarization is used to encode binary information. For example, if $P = +1$ represents a binary 1, then $P = -1$ can represent a binary 0. The two polarization states of a cell will not be energetically equivalent if another cell is nearby. Figure 8.68 shows how one cell is influenced by the state of the neighboring cell. The inset illustrates two cells where the polarization P_1 is determined by the polarization of the neighbor, P_2,

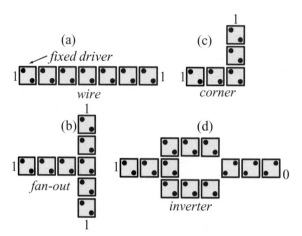

Figure 8.69 Examples of the simplest quantum-dot cellular-automata arrays: wire, corner, fan-out, and inverter. From Y. Ono, A. Fujiwara *et al.*, "Manipulation and detection of single electrons for future information processing," *J. Appl. Phys.*, **97**, 031101 (2005). Reused with permission from Yukinori Ono, Akira Fujiwara, Katsuhiko Nishiguchi, Hiroshi Inokawa, and Yasuo Takahashi, *Journal of Applied Physics*, **97**, 031101 (2005). © 2005 American Institute of Physics.

through the interaction of the electrons in the cells. Let us assume that the polarization P_2 has a given value corresponding to a certain arrangement of the charge in the cell 2. Owing to the charge repulsion, the response of cell 1 is given by the strongly nonlinear dependence presented in Fig. 8.68: a small asymmetry of charge in cell 2 is sufficient to break the degeneracy between the two possible states of cell 1 and leads to the same configuration of cell 1. Thus, in some cell arrangements, fixing the polarization of a cell at an edge by an external bias determines the polarization of the cells at other edges. The polarization is dependent on the cell configuration.

In the quantum-dot cellular-automata approach, the circuit is built by forming a tree of cells. Figure 8.69 shows some elements of the quantum-dot cellular automata. Figure 8.69(a) is a binary wire. The polarization of the leftmost cell is fixed, which represents the input. Then, all the other cells, including the output cell, align with the same polarity because it is the most energetically favorable. Flipping the polarity of the input cell results in the flipping of all the other cells. During this procedure, no direct current flows in the circuit. Cells positioned diagonally from each other tend to anti-align. This feature is employed to construct other logical elements. A fan-out, a corner, and an inverter are shown in Figs. 8.69(b), 8.69(c), and 8.69(d), respectively.

In conclusion, quantum-dot cellular-automata systems exploit the interactions between quantum dots on the nanoscale. They are able to perform all processes necessary for signal processing. No current is used in the circuits built on the basis of quantum automata. The quantum-dot cellular-automata approach presents a nanoscale alternative to conventional microelectronics and nanoelectronics.

8.9   Closing remarks

In this chapter we have considered several different nanostructure devices. It is convenient to classify these devices into two categories. The first are the devices which are based on usual classical principles of operation but considerably scaled down. This scaling down facilitates the improvement of device performance and solves some fundamental difficulties, which can not be overcome in microelectronics. These devices include field-effect transistors and bipolar transistors.

Another category is nanostructure devices based on new physical principles. Among these are resonant-tunneling diodes – the simplest quantum devices. Since these devices have nanoscale dimensional features, they have extremely short transit times for carrier transport through the structures, which results in the possibility of generating ultra-high-frequency electromagnetic oscillations up to the terahertz frequency range. As examples of new principles for three-terminal electronic devices (transistors), we analyzed the quantum-interference and hot-electron transistors, which have great potential for ultra-high-speed operation.

To illustrate the contribution of nanostructures to optical devices, we have focussed on the examples of quantum-well and quantum-wire lasers with bipolar injection, and the multilayered quantum-cascade laser. The latter is a monopolar device, i.e., it uses only one type of carrier (electrons). Exploitation of nanostructures results in a dramatic decrease of the threshold current necessary for laser generation and in a widening of emission spectra.

We have analyzed some nanoelectromechanical systems with mechanical resonators that can vibrate at frequencies ranging from a few megahertz up to 1 GHz and have quality factors greatly exceeding those of typical microwave resonators. This implies that nanoelectromechanical systems operate with low power dissipation. These properties open the way for a number of applications, ranging from signal processing to novel sensors/detectors.

Finally, we have described innovative systems for signal processing based on quantum-dot cellular automata. These automata exploit interactions between quantum dots on the nanoscale. This approach presents a nanoscale alternative to conventional microelectronics and nanoelectronics.

The description of nanostructured devices which we presented here is not complete. More information on traditional scaling down to nanosize devices, double-barrier resonant-tunneling diodes, and lasers can be found in the following books:

M. Shur, *Physics of Semiconductor Devices* (Englewood Cliffs, NJ, Prentice-Hall, 1990).

S. M. Sze, *High-Speed Semiconductor Devices* (New York, John Wiley & Sons, Inc., 1990).

V. V. Mitin, V. A. Kochelap, and M. A. Stroscio, *Quantum Heterostructures* (New York, Cambridge University Press, 1999).

Single-electron transport is described in the following references:

K. K. Likharev, "Correlated discrete transfer of single electrons in ultrasmall tunnel junctions," *IBM J. Res. Develop.*, **12**, 144 (1988).

M. A. Kastner, "The single electron transistor," *Rev. Mod. Phys.*, **64**, 849 (1992).

Y. Ono, A. Fujiwara *et al.*, "Manipulation and detection of single electrons for future information processing," *J. Appl. Phys.*, **97**, 031101 (2005).

A very detailed review of nanoelectromechanical systems is given in the text

A. N. Cleland, *Foundations of Nanomechanics* (Berlin, Springer-Verlag, 2003).

Appendix: tables of units

Table 1 SI base units

Quantity	Unit	
	Name	Symbol
Length	meter	m
Mass	kilogram	kg
Time	second	s
Electric current	ampere	A
Temperature	kelvin	K
Amount of substance	mole	mol

Table 2 SI derived units

Quantity	Unit		
	Name	Symbol	Equivalent
Plane angle	radian	rad	$m/m = 1$
Solid angle	steradian	sr	$m^2/m^2 = 1$
Speed, velocity			$m\,s^{-1}$
Acceleration			$m\,s^{-2}$
Angular velocity			$rad\,s^{-1}$
Angular acceleration			$rad\,s^{-2}$
Frequency	hertz	Hz	s^{-1}
Force	newton	N	$kg\,m\,s^{-2}$
Pressure, stress	pascal	Pa	$N\,m^{-2}$
Work, energy, heat	joule	J	$N\,m, kg\,m^2\,s^{-2}$
Impulse, momentum			$N\,s, kg\,m\,s^{-1}$
Power	watt	W	$J\,s^{-1}$
Electric charge	coulomb	C	$A\,s$
Electric potential, emf	volt	V	$J\,C^{-1}, W\,A^{-1}$
Resistance	ohm	Ω	$V\,A^{-1}$
Conductance	siemens	S	$A\,V^{-1}, \Omega^{-1}$
Magnetic flux	weber	Wb	$V\,s$
Inductance	henry	H	$Wb\,A^{-1}$
Capacitance	farad	F	$C\,V^{-1}$
Electric field strength			$V\,m^{-1}, N\,C^{-1}$
Magnetic flux density	tesla	T	$Wb\,m^{-2}, N\,A^{-1}\,m^{-1}$
Electric displacement			$C\,m^{-2}$
Magnetic field strength			$A\,m^{-1}$
Celsius temperature	degree Celsius	°C	K
Luminous flux	lumen	lm	cd sr
Illuminance	lux	lx	$lm\,m^{-2}$
Radioactivity	becquerel	Bq	s^{-1}
Catalytic activity	katal	kat	$mol\,s^{-1}$

Table 3 Physical Constants

Constant	Symbol	Value	Units
Speed of light in vacuum	c	2.9979×10^8 $\approx 3 \times 10^8$	$\mathrm{m\ s^{-1}}$
Elementary charge	e	1.602×10^{-19}	C
Electron mass	m_0	9.11×10^{-31}	kg
Electron charge to mass ratio	e/m_0	1.76×10^{11}	$\mathrm{C\ kg^{-1}}$
Proton mass	m_p	1.67×10^{-27}	kg
Boltzmann constant	k_B	1.38×10^{-23}	$\mathrm{J\ K^{-1}}$
Gravitation constant	G	6.67×10^{-11}	$\mathrm{m^3\ kg^{-1}\ s^{-2}}$
Standard acceleration of gravity	g	9.807	$\mathrm{m\ s^{-2}}$
Permittivity of free space	ϵ_0	8.854×10^{-12} $\approx 10^{-19}/(36\pi)$	$\mathrm{F\ m^{-1}}$
Permeability of free space	μ_0	$4\pi \times 10^{-7}$	$\mathrm{H\ m^{-1}}$
Planck's constant	h	6.6256×10^{-34}	J s
Impedance of free space	$\eta_0 = \sqrt{\mu_0/\epsilon_0}$	$376.73 \approx 120\pi$	Ω
Avogadro constant	N_A	6.022×10^{23}	$\mathrm{mol^{-1}}$

Table 4 Standard prefixes used with SI units

Prefix	Abbreviation	Meaning	Prefix	Abbreviation	Meaning
atto-	a-	10^{-18}	deka-	da-	10^{1}
femto-	f-	10^{-15}	hecto-	h-	10^{2}
pico-	p-	10^{-12}	kilo-	k-	10^{3}
nano-	n-	10^{-9}	mega-	M-	10^{6}
micro-	μ-	10^{-6}	giga-	G-	10^{9}
milli-	m-	10^{-3}	tera-	T-	10^{12}
centi-	c-	10^{-2}	peta-	P-	10^{15}
deci-	d-	10^{-1}	exa-	E-	10^{18}

Table 5 Conversion of SI units to Guassian units

Quantity	Si unit	Gaussian units
Length	1 m	10^2 cm
Mass	1 kg	10^3 g
Force	1 N	10^5 dyne $= 10^5$ g cm s^{-2}
Energy	1 J	10^7 erg $= 10^7$ g cm^2 s^{-2}

$$1\ \mathrm{eV} = 1.602 \times 10^{-19}\ \mathrm{J} = 1.602 \times 10^{-12}\ \mathrm{erg}$$

Index